Space and Society

Series Editor-in-Chief

Douglas A. Vakoch, METI International, San Francisco, CA, USA

Series Editors

Setsuko Aoki, Keio University, Tokyo, Japan
Anthony Milligan, King's College London, London, UK
Beth O'Leary, Department of Anthropology, New Mexico State University, Las Cruces, NM, USA

The Space and Society series explores a broad range of topics in astronomy and the space sciences from the perspectives of the social sciences, humanities, and the arts. As humankind gains an increasingly sophisticated understanding of the structure and evolution of the universe, critical issues arise about the societal implications of this new knowledge. Similarly, as we conduct ever more ambitious missions into space, questions arise about the meaning and significance of our exploration of the solar system and beyond. These and related issues are addressed in books published in this series. Our authors and contributors include scholars from disciplines including but not limited to anthropology, architecture, art, environmental studies, ethics, history, law, literature, philosophy, psychology, religious studies, and sociology. To foster a constructive dialogue between these researchers and the scientists and engineers who seek to understand and explore humankind's cosmic context, the Space and Society series publishes work that is relevant to those engaged in astronomy and the space sciences, while also being of interest to scholars from the author's primary discipline. For example, a book on the anthropology of space exploration in this series benefits individuals and organizations responsible for space missions, while also providing insights of interest to anthropologists. The monographs and edited volumes in the series are academic works that target interdisciplinary professional or scholarly audiences. Space enthusiasts with basic background knowledge will also find works accessible to them.

More information about this series at http://www.springer.com/series/11929

Cameron M. Smith

Principles of Space Anthropology

Establishing a Science of Human Space Settlement

Springer

Cameron M. Smith
Department of Anthropology
Portland State University
Portland, OR, USA

ISSN 2199-3882 ISSN 2199-3890 (electronic)
Space and Society
ISBN 978-3-030-25019-5 ISBN 978-3-030-25021-8 (eBook)
https://doi.org/10.1007/978-3-030-25021-8

Cover design: Paul Duffield

This Springer imprint is published by the registered company Springer Nature Switzerland AG
The registered company address is: Gewerbestrasse 11, 6330 Cham, Switzerland

Foreword

It is not often that one is fortunate enough to see the beginning of new science (let alone be among the fortunate few to know one of the primary progenitors of it), yet that is exactly where I am and where you, the reader, are as you pick up this volume and begin to read and understand the new science of Space Anthropology. We are in a new age of space exploration in which the first real migration of humans from Earth to space may finally begin. In this 50th anniversary year of the first Apollo moon landing, it is more than a little disappointing to realize that the space settlement and diaspora envisioned in those early, heady days have not already occurred. But this lull in spaceward expansion may have been what was needed for the technology (aerospace and biological) to mature to the point where it might actually be a realizable goal. Ultrawealthy space entrepreneurs at companies like Blue Origin and SpaceX are upending the traditional aerospace industry with the goal of establishing a permanent human presence beyond Earth—a goal very different from their more established corporate competitors. Assuming they are successful, what then? How will the first space settlements be structured to maximize their chances of success?

Once the first groups of humans emigrate beyond Earth, they will begin an inevitable biological, social, and cultural evolution that can be studied and (somewhat) predicted by this new science. Toward this goal, Dr. Cameron Smith has written this comprehensive space anthropological guide that should be required reading for every current and aspiring astronaut and those planning the coming diaspora.

Dr. Smith does not limit his examination of Space Anthropology to the Space Age, as many would assume to be the natural starting point. He comprehensively examines ancient cultures to understand their adaptive traits and technologies, and the subsequent "lessons learned" that might be applied to future human space settlements. He draws from evolutionary biology (on Earth) to consider how humans might adapt and evolve beyond the terrestrial biosphere—carefully connecting human cultural and social evolution to the biological, noting the complex interplay between them.

What might be most interesting to the field of anthropology in general (at some future date) would be a study of the divergence from Earthbound cultural norms that space settlers exhibit. Aboard a multi-generational starship, what will become of families? Will the crew have religion and religious rites? If they do not begin with a religion, might they invent, or reinvent, one or more of them over time? What of the critical skills required to keep the ship functioning—how will those born during the journey or in the colony decide what roles to assume? Assuming the crew speaks a common language at the start, how might it change over centuries of isolation? Will the crew develop new forms of communication that are uniquely relevant to the environment in which they live? What will be the physical and cultural adaptation to life in space that might have no natural sunlight, low or nonexistent gravity, long-term risk from galactic cosmic radiation, and the ever-present risk of disaster just beyond the relatively thin skin of the ark or colony wall? It is likely that each space settlement, separated from kin by perhaps light-years of distance, and each in their own unique environment, might independently evolve similarly or take separate and divergent paths. Time will tell.

For those interested in a serious examination of what it would take to implement a fully independent space settlement, whether it be in Earth orbit, on the Moon, or in an ark on a millennium-long journey to another star, understanding the possible, even likely, anthropological changes the colonists may experience is a must. There is no better resource available to inform such studies than Principles of Space Anthropology. I fully expect it to be the most-referenced resource as this exciting new field of study blossoms.

Ad Astra!

Huntsville, USA

Les Johnson
Scientist, Author, and Space Technologist

Preface

In 2012, I co-authored *Emigrating Beyond Earth* with E.T. Davies. That book presents a comprehensive anthropological and evolutionary context for human space settlement. It argues that human space settlement will be a continuation of ancient evolutionary patterns of Earth life dispersal and adaptation to new environments. It also delves into the argument of whether humanity should attempt space settlement. In just the ten years since I started writing that book (2009) and the completion of the present volume (January 2019), concrete steps towards human space settlement have been taken, largely in the proliferation and diversification of the private space industry. In this book, I do not revisit the issue of whether space settlement is warranted or morally good except in a brief Appendix; for a deeper exploration, please see Chap. 4 of *Emigrating Beyond Earth.*

In this book, I wanted to get very specific about how humanity would settle space—not just by nebulously stating 'well, we'll just adapt', but by identifying just how humanity adapts and what that means for space settlement and space settlement planning. Humanity adapts both culturally and biologically, on various timescales. In this book, I attempt to specify what we can learn about how to successfully settle space when informed about our cultural and biological tools of adaptation.

In the sections on cultural adaptation, I mention aspects of life that will change as people adjust their ways of living to new conditions beyond Earth. For example, our musical instruments, our cuisine, our literature and even religions, all will be adjusted because they will be used in situations different from the planet on which they emerged. When I mention this in casual conversation, I often find that people feel these would be trivial changes. Individually, they may be, but taken together they constitute cultural change. New species arise by the accumulation of biological novelties, and new cultures emerge by the accumulation of cultural novelties. Over time, cultures beyond Earth will diverge from those of Earth. And this is not just some by-product of any 'larger picture' because there is no larger picture; it is our artistic expression, philosophies, religions and so that we wish to preserve by the method of space settlement. We would not take up this project to preserve, say, the rocket engines that get us to space or the guidance computers that will take us to Mars and Ceres; those, magnificent as they are, are simply tools that support humanity.

From my perspective, space settlement is about cultures, and life and living things, not the tools meant to support life. For these reasons, the extensive investigation into what aspects of our cultures are likely to change when we adjust to conditions beyond Earth should not be read as somehow secondary material, they are central to the whole project of permanent space settlement.

Portland, USA Cameron M. Smith

Acknowledgements

I would like to thank the people noted below for their encouragement, which has been helpful while exploring the sometimes speculative and often unorthodox mindscapes of multigenerational futures beyond our home planet.

I thank Andreas Hein, Ph.D. (Initiative for Interstellar Studies and Ecole Centrale Supelec School of Industrial Engineering), for first reaching out to me in 2011 to participate in the 2012 Hundred Year Starship Conference in Houston, Texas. At this meeting, I delivered an exploratory paper outlining an anthropological approach to space settlement, and Icarus Interstellar's Dr. Richard Obousy asked me to estimate a viable population for interstellar voyaging. This resulted in a paper written in 2012–2013 and published in *Acta Astronautica* in 2014. Soon thereafter I formally began writing the present book with the encouragement of Mr. Les Johnson (NASA Science and Technology Office, Marshall Spaceflight Center), who kindly invited me to speak at several of his Tennessee Valley Interstellar Workshops. More recently, I have had the pleasure to collaborate with Kai Staats, M.Sc. (Arizona State University's Interplanetary Initiative), and Dr. Frederic Marin (L'Observatoire Astronomique de Strasbourg), whose work is mentioned in this book.

I also thank the late Dr. Ben Finney (University of Hawaii), whose 1985 book *Interstellar Voyaging and the Human Experience* (University of California-Berkeley Press) paved the way for my own research in this field: in a 2012 email, he welcomed me 'to a very small club' of people who think about the astounding prospect of human space settlement, which requires certain discipline.

I also thank my students over the last 20 years, for asking good questions about human biology and behaviour that have led me to many lines of research that have contributed to the biocultural approach that shapes this book. My late mentor, Prof. Kenneth M. Ames, did not share my specific fascination with human space settlement, but he supported my research in the field and encouraged me to publish about it in the research journals to shape and legitimize the work. I also thank the many students and interns of my *Pacific Spaceflight* research group (2013–present) who have helped with envisioning very practical aspects of human space settlement as we have designed, built and tested many spacesuits, one of the most immediate and bodily conditioned technological adaptations humanity will use beyond Earth; in particular I thank Michelle Yan, Alexander Knapton, Amy Magruder, Ben Wilson, Winnie Black, Sarah Taylor, Mathew Lippincott, Trent Tresch, Michael Rudis and Peter Dukluyver.

Finally, I thank my parents, Dr. Donald E. Smith and Margit J. Smith, who have always encouraged my research. My mother has given me a lifelong love of books and research deriving from her professions in library sciences and ancient book restoration. My father gave me a NASA publication on space settlement (*Space Settlements—A Design Study*) in the 1970s that has been a source of inspiration for decades, now. And, Dad's *Ranger* series of prints, displayed at the Jet Propulsion Laboratory just as the moon was being explored robotically in advance of humans, are a reminder of the human element of space programmes (*Ranger II* is displayed below). As fascinating and impressive are the technologies, they are meant to ultimately support and perpetuate humanity, which is characterized among all species by its uniquely creative mind. Thanks, Mom and Dad, for your love and encouragement in my project of contributing one little block of stone to the cathedral of permanent human space settlement, a project requiring both research and imagination.

Contents

Chapter 1
An Introduction to Space Anthropology

Abstract This chapter introduces two central concepts: first, human space settlement as adaptive evolution, and second, how evolutionary and anthropological sciences can aid in making the project of human space colonization more likely to succeed. I then continue by defining some basic concepts and terms, and the essential time- and space-thoughtscapes addressed in this book.

> Definitions of adaptation vary across disciplines, but they all capture the idea of adjustments in a system's behavior and characteristics to cope with stress or change, leading to an increased probability of reproduction or persistence… Coined in evolutionary biology, the term broadly refers to the evolution of genetic or behavioral outcomes that enable organisms or systems to cope with externally and internally driven changes to survive and reproduce…Adaptation refers to both the current state of being adapted and to the dynamic evolutionary processes leading to adaptation. Adaptation enhances the fitness and survival of the evolving entities. (Reyes-Garcia et al. 2016:1)

1.1 Introduction

Cultures worldwide and through time have recognized that the Earth is a finite body, and that other such bodies might exist, and might be inhabitable, beyond our own. Most of us know such ideas from the Western tradition; the astronomer Giordano Bruno is remembered for being burned at the stake for such ideas in 1600, and as early as 1620, Englishman Francis Godwin published *The Man in the Moone, or a Discourse of a Voyage Thither*, describing a fantasy of reaching the Earth's moon in a carriage drawn by birds. In an entirely different tradition, Polynesian legends records the following about the female Pacific ocean explorer, Hina:

> [at]…a peninsula called Motu-tapu (Sacred Island), in Ra'iatea, was the canoe station of Ru and Hina…by which they [explored widely across the sea]….After exploring [all the Pacific] Hina's love of discovery did not cease. So one evening when the full moon was shining invitingly, being large and half visible at the horizon, she set off in her canoe to make the moon a visit. On arriving there, she was so pleased with it, that she stepped into it, leaving to the mercy of the sea her canoe, which was never seen again. (Henry 1985 [1928]:462–463)

C. M. Smith, *Principles of Space Anthropology*, Space and Society,
https://doi.org/10.1007/978-3-030-25021-8_1

By the early 20th century AD the essential technologies for bridging the gap between what was recognized as the Earth, and distant, tangible bodies, some of which might be habitable, were being outlined. By the 1920's the Russian teacher and mathematician Konstantin Tsiolkovski (1857–1935) was publishing foundational literature both i the realities of propulsion in a vacuum and fictional works envisioning the colonization of space by humans for peaceful purposes. This early work, while grappling with the realities of space voyaging (Tsiolkovski, indeed, published the rocket equation in 1903) emphasized humanity's central place in space exploration. In the course of the first space race (1959–1972) the concepts of 'space' and particularly 'people in space' were highly exoticized for many reasons. This is not to say that what was being done was not dramatic or technically masterful; rather I am pointing out that many underlying messages of the times were about technological achievement, manufacturing precision and procedural expertise—not, really, about humanity. But humans-in-space has been the concept from the beginning. It has been argued extensively that humanity should not attempt to settle space (e.g. Billings 2018; Stoner 2017); I have addressed this directly elsewhere (Smith and Davies 2006; Smith 2014a, b) and I address the issue in this book only in an Appendix. For the purposes of this book I propose human space settlement as a moral good and a responsibility of our species.

Space was for Tsiolkovski a natural medium to explore and for humans to settle, and in 1926 he proposed 16 steps for human exploration and settlement of space; I have checked the steps so far achieved in space exploration:

1. Design of rocket-propelled airplanes with wings. ✓
2. Progressively increasing the speeds and altitudes reached with these airplanes. ✓
3. Designing of a pure rocket without wings. ✓
4. Developing the ability to land on the ocean surface by rocket. ✓
5. Reaching of escape velocity and first flight into space. ✓
6. Lengthening of the rocket flight time into space. ✓
7. Experimental use of plants to make an artificial atmosphere in spacecraft. ✓ [limited]
8. Using of pressurised space suits for activity outside spacecraft. ✓
9. Making of orbital greenhouses for plants.
10. Building of the large orbital habitats around the earth.
11. Using solar radiation to grow food, to heat space quarters, and for transport needs throughout the solar system.
12. Colonization of the asteroid belt.
13. Colonization of the entire solar system and beyond.
14. Achievement of individual and social perfection.
15. Overcrowding of the solar system and galaxy colonisation.
16. The sun begins to die and the people remaining in the solar system's population move to other solar systems.

Figure 1.1 is one of Tsiolkovski's many drawings from his dozens of publications on space exploration; importantly, while today we commonly associate 'space exploration' with machinery, rockets, and computers, for Tsiolkovski and others at

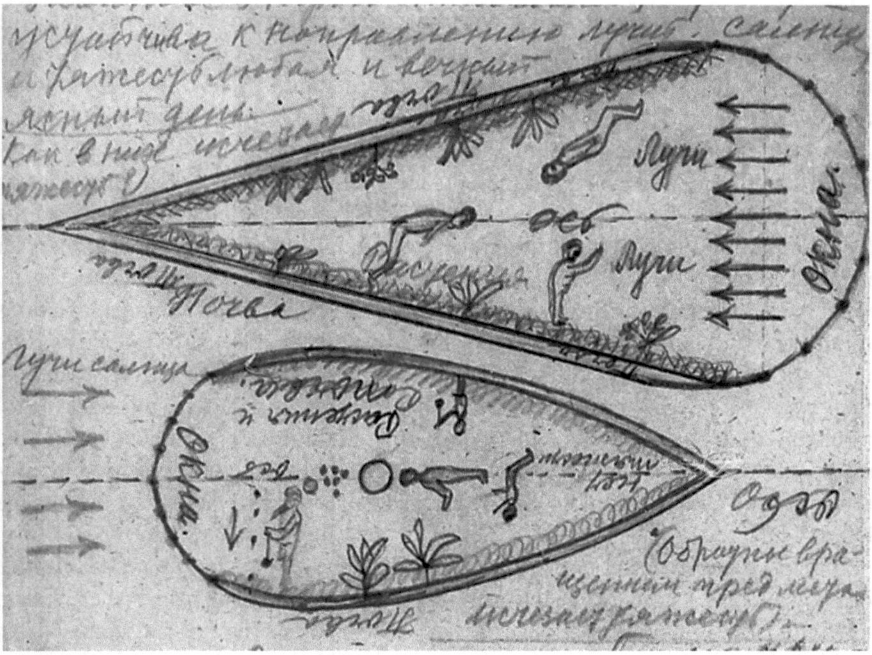

Fig. 1.1 Sketch of humans in space, highlighting people and plants rather than space hardware. From Tsiolkovski (1933). *Album of Space Travel* (Kaluga, Tsiolkovsky State Museum of the History of Cosmonautics)

the beginning of serious thought about space beyond Earth, the focus was humanity, not the technology. Figure 1.1 highlights this perspective, featuring human beings, floating in groups and supplied with vegetation, which Tsiolkovski knew would be central to any long-term human activities beyond the boundaries of Earth.

It has taken only about half a century—just over a single human generation—to operationalize the essential technologies for space travel, and even, as we see currently, for those technologies to migrate from their essentially military and exclusive domains into the public domain. In this first quarter of the 21st century AD we see the proliferation of private space access companies and the rapid decrease of access to space, a prerequisite for space settlement. At this writing, serious plans for increased human activity in space, including steps toward initial space settlement, are underway. I argue that serious planning for space settlement should begin now.

1.1.1 Definitions

If populations of people begin to live off of Earth, they will be adapting to new environments as they have done on Earth for at least two million years, as the genus

Homo dispersed globally, settling environments as challenging as the Arctic and the Pacific. I argue that the foundation for successful space settlement will be to recognize that it will be a continuation of an ancient method of settlement—*biocultural adaptation*—and that we should take lessons from the history of evolution (sometimes evolution at large, and sometimes human evolution specifically) about how best to adapt. And I argue that it is the academic field of anthropology—the scientific study of humanity—that is best equipped to assist space settlement as it studies the coevolution of biology and culture in our species. Other fields are not as broad; for example, psychology is of interest in space settlement studies, but it deals largely with cognition; similarly, flight physiology focuses on individual human adaptations to microgravity, but it is populations of humanity (studied by anthropologists) that will settle environments beyond Earth, not just individuals. In recent years anthropology has increasingly been used in an applied manner, to aid in solving 'real world' problems beyond the normal reach of academia; in this book I introduce an applied anthropology specific to assisting humanity's inevitable attempt to settle environments beyond Earth—also known, broadly, as 'space'. I will begin here with some definitions.

Biocultural adaptation refers to the use of both biology and culture as adaptive tools, by humanity, to adapt to environmental conditions differing from those of Earth, including gravity field, radiation flux, and atmospheric pressure and composition (inside habitats); these are introduced in more detail below. Adaptation in the genus *Homo* has required strong emphasis on learned, cultural guides to behaviour for at least a million years; that is, we survive far more despite our bodies than because of them, and our survival is much more about carefully-tailored behaviour than carefully-tailored genes. Having said this, many genetic adaptations have tailored the human body and its processes to a variety of environments worldwide. Table 1.1 indicates some biocultural adaptations of our species worldwide.

By *space*, *extraterrestrial* and *beyond Earth* I simply refer to environments beyond the surface of Earth, outward from Earth. These effectively begin at Low Earth Orbit (*LEO*, on the order of 100–1000 km above the Earth's surface) and extend to other habitable places beyond Earth, including our moon, the surfaces of other solar system bodies (e.g. Mars) and their moons, 'Lagrangian points' of solar system gravitational equilibrium (suitable for placement of free-floating space colonies; European Space Agency 2018) and even, in a more distant future, such places in other solar systems, now being rapidly discovered and characterized. From the perspective of adaptation, humans (and the many species with which we coevolve, mentioned throughout this book) going to such places 'in space' are simply moving through a medium [e.g. the *interplanetary medium* (Lang 2000)] to another habitat (e.g. the surface of Mars) as they have moved across Earth media (e.g. the Pacific Ocean, the high Arctic archipelago) to certain habitats. The human settlement of space, from this perspective, is focused on going to specific sites. I argue that while technically challenging, the traverse portion of such movements should be de-emphasized in space settlement planning and discussions. It is a premise of space settlement planning that this traverse portion of space settlement will be carried out safely, and so I mean to 'de-exoticize' the 'space voyage' and shift focus to what populations of people will do when they are

Table 1.1 Human biocultural adaptations to earth environments. Adaptations include heritable (genetic) variations favored over time, heritable (genetic) variations consciously selected for by human consciousness, and invented (cultural) variations in tools and techniques that improve fitness of individuals and groups

Biome adaptations	Limiting factors	Biological, cultural and technological
Arctic/Cold	* Extremely low temperatures for long periods * Extreme light/dark seasonal cycles * Low biological productivity	*Biological*: * Increased Basal Metabolic Rate * Increased shivering, vasoconstriction and cold thermoregulation activity and efficiency * Compact, heat-retaining body stature *Cultural and Technological*: * Bilateral kinship = demographic flexibility * Clothing insulates but can prevent sweating * Semisubterranean housing including igloo made of local, free, inexhaustible resource (snow) * Population control methods including voluntary suicide and infanticide * High value on apprenticeship * Low tolerance for complaint; 'laugh don't cry'
High altitude	* Low oxygen pressure * Nighttime cold stress * Low biological productivity * High neonatal mortality	*Biological*: * Dense capillary beds shorten distance of oxygen transport * Larger placenta providing fetus with more blood-borne oxygen * Greater lung ventilation [capacity] *Cultural and Technological*: * Promotion of large families to offset high infertility * Use of coca leaves to promote vasoconstriction and caffiene-like alertness * Woolen clothing retains heat when wet * Trade connections with lowland populations
Arid/Hot	* Low and uncertain rainfall * High evaporation rate * Low biological productivity	*Biological*: * Tall, lean, heat-dumping body * Lowered body core temperature * Increased sweating efficiency * Lower urination rate * Increased vasodilation efficiency *Cultural and Technological*: * Flexible kinship and land tenure system = demographic flexibility matching shifting water resources * Intercourse taboos maintain sustainable population * Loose, flowing clothing blocks sunlight

Compiled from the author from Durham (1991), Alland (1970), Moran (1979) and Frisancho (1993)

at the new site, the new environment for our families and communities. As an analogy, consider the exploration and settlement of the Pacific over the last 3000 years. While much cultural attention was of course given by the voyaging cultures to their sailing vessels, methods of navigation, and methods of staying alive while at sea, the real focus—the point of the voyaging—was arrival at habitable shores. The properties of space between Earth and Mars, or a Lagrangian point, then, are fascinating, and the technology to convey us across this space equally so; but they are not the ultimate point. The ultimate point is supporting communities beyond Earth.

By *humans-in-space* I refer to human activities beyond Earth and therefore subject to conditions almost never replicating those of Earth. Even though all reasonably-proposed space settlement plans at the moment propose living in habitats built, for example, on Mars or the Moon, where we would adjust the habitat temperature to be comfortable, other basic environmental variables will have significant effects. These variables are *gravitational field* (Earth = 1.0 g), *atmospheric pressure* (Earth's human-habitable range is about 14.7 psia to about 9.0 psia), *atmospheric composition* (Earth's roughly 20% oxygen and c. 80% nitrogen and other gases), *temperature* (Earth average habitable range from around −50 F to about +100 F) and *radiation flux* [Earth's varies but for most people is about 620 millirem or 6.2 millisieverts/year (EPA 2018)]. Humans being in space, from this perspective, is not an exotic proposal at the limits of possibility, but a problem of adapting to these novel conditions with technology, biology and behavior. In this book I focus on the settlement of Mars, as it is most likely a first destination, but I also mention a few other possibilities, discussed later in this chapter. Later in this book I argue that *microgravity*, a condition that many commonly associate with 'being in space', is a condition that will *not* significantly structure human space settlement or require significant human adaptation. This is because no current and reasonably-expectable proposals of space settlement envision people spending more than a period of transit (perhaps of some months, but certainly less than a year) in microgravity before reacting a site with a gravity field much more approximating that of Earth.

I use the terms *space settlement*, *space colonization* and *habitation beyond Earth* interchangeably and equally to describe populations of humans attempting to live multigenerationally in communities beyond Earth. The terms 'colonization' and 'settlement' have negative associations and connotations in the Earth's history, particularly in European encounters with the native peoples worldwide. These associations, however, do not apply to the settlement of any site in our solar system, as these other sites are uninhabited by people. The possibility that humanity might encounter and detrimentally affect other varieties of life in our solar system has been formally investigated and is considered today in processes such as sterilization of spacecraft before they are sent to Mars (NASA 2018). By human space 'settlement' or 'colonization', then, I do not propose a project of destructive advance across our solar system. I would not advocate space settlement if I thought that were likely. Rather I think that space settlement, at least as carried out by its earliest generations, will be eminently responsible because there will be a focus on maximising resources, rather than wasting them, when they are not as plentiful as they are on Earth. In any case the terms settlement, emigration, colonization, dispersal and others I use in this book are used (a) interchangeably and (b) broadly to refer to the human occupation of habitats beyond Earth, and attempts to make that occupation multigenerational and, ultimately, to make those settlements independent of Earth.

Biological evolution is often defined as the *change through time of the heritable features of a population*, or, more technically, the *change through time of gene frequencies in a population* (Strickberger 1985). This field has several main scales of understanding and analysis, though they are difficult to disentangle as occurrences at a given level have effects at other levels, and such effects are not necessarily immediate or obvious (e.g. they may be *complex*). Starting with the smallest level, we begin with DNA, the replicating molecule that builds the essential human form; this is largely referred to as the *genotype*. The bodies built by the DNA (and modified by it, and subject to growth as well as environmental influences) is another important scale, that of the *phenotype*. No life form lives in isolation, and individuals form discrete *populations*, which themselves may be partitioned into smaller units, such as families, and larger units such as communities and cultures. Each of these biological scales will be addressed in this book with respect to how they will be shaped by emigration beyond Earth and into extraterrestrial contexts. Another scale is that of the *hologenome*, currently being recognized, defined and investigated as the totality of the genome of a given species and all its symbionts and other co-evolving species, such as the many microbial species that ride upon and coevolve with the human body (Costello et al. 2009). Though fascinating and important, this is beyond my scope in this book, which again focuses largely on humanity.

Traditionally, the study of humanity's adaptations has been the domain of *anthropology* the scientific study of humanity's biology and behavior. Over the last century this field has capably documented our species' remote origins, long and complicated evolution, and myriad manifestations in the present, but it has only occasionally (and then unsystematically) forayed into humanity's distant future (e.g. see Finney 1992; Finney and Jones 1985; Smith and Davies 2012). It is a premise of this book that that future should include the human settlement of environments beyond Earth, particularly for the purposes of safeguarding humanity's apparently unique mode of consciousness, humanity's many domesticates, and the totality of human knowledge—accumulated over about 3000 generations since the origins of behavioral modernity—by the method of establishing populations of humanity culturally and biologically independent of our home planet. I focus on how the resources and expertise of anthropology may be deployed to assist in the goals of human space settlement.

Humans are not just biological, of course, and our evolution is strongly conditioned by complex behavior; this behavior and its products (material and social) are generally referred to as *culture*. A century or so of attempts to study culture in a quantifiable way have not succeeded in any single quantifiable or mathematical model of culture, and a recent review indicates that "it is still too early to claim that every aspect of culture is captured by any single model or even by all existing models taken together," (Taras et al. 2009:362). Nevertheless, culture may be defined operationally here, and in anthropology at large few would argue with the defining statement that *culture is the total body of knowledge of socially-transmitted guides to behaviour shared by a group over multiple generations* (Durham 1991). Culture may also be considered the *extrasomatic means of adaptation* (White 1959:8); that is, culture is (as an action word) deployed knowledge, received by the individual as stated above, that assists in survival by adaptation, in contrast to the biolological means of survival supplied not socially, but genetically. Pragmatically, culture provides the range of options that a person has as their guide to behavior when considering a course of action. This range of options is likely not infinite, or representative of all the physically-possible ways to carry out an action, but is rather the set of ways that the culture's earlier members determined were acceptable and worth transmitting to the next generation. Culture then provides a set of options for behaviour. In the individual mind, played out in the course of daily life, culture:

> informs the knowledge base, the structural organization of cognition, and the individual's hierarchy of values. These various dimensions of cultural influence combine to focus thought by prioritizing the detection and utilization of cues relevant to the performance of a given task in context. The cultural organization of context includes systematic patterns of social relationships, of recurrent activities, and of meanings. Over the course of their socialization, children gradually appropriate this complex system and become a part of it. The distinctiveness of cultures can be explained in part through a historical analysis of their evolution, including ecological, technological, and philosophical dimensions. Each cultural tradition tends to combine many such features and also organizes them into a coherent interpretive scheme, which variously informs the discourse and associated practices of everyday life.... (Serpell and Boykin 1994:369)

Because culture has been used to adapt to a multitude of environments worldwide—and for many other reasons, some linear and others decidedly nonlinear—culture varies across both time and space; there is not just one human language or one human food preference, there are thousands of languages and palates and there are and have been thousands of human cultures. A given culture consists of a set—however loose and fuzzy at the margins—of guides to behaviour, for example in the guides suggesting the appropriate cuisine in Japan or Germany.

Biological features and cultures both change through time by evolutionary processes. A simple change, such an animal losing a limb, or a human being thinking a new thought, is not considered an evolutionary change unless it is heritable in some way or encoded in some structural way in the next generation or among others of a population. This is why evolutionary change specifies change among the *heritable* features of a population, e.g. the genes. Things are somewhat more complex with cultural evolution as ideas (also known more formally as *memes*, or units of cultural evolution) are heritable because they may be transferred to the next generation, and indeed members of the same generation, in ways that are *dissimilar* from sexual replication (for a review see Smith et al. 2018a, b). In short, genes are the units of biological change and memes (ideas) are the units of cultural change, and each is simply a channel of information flow from one generation to the next and, in the case of ideas, from one individual to another within a given generation. Of these biological and cultural evolutionary processes, in this book I am most concerned with *adaptation*, a consequence of both biological and cultural evolutionary processes which we will explore in detail. For the moment, we my think of *adaptation* as a thing or process that allows life where it could not persist before; a space suit is a material adaptation if we wish to stand on the moon just as a polar bear parka and sealskin boots are material adaptations of early native explorers of the Arctic. Biological adaptations are evolved structure or processes that allow the organism to survive, such as gills that separate oxygen from water.

Anthropology's century or so of more-or-less systematic and scientific study of our species has sketched out the essential organizing features of human culture. In the same way, a century or so of evolutionary biology have sketched out the essential organizing features of living systems. Table 1.2 indicates the essential topics covered in modern biology and modern anthropology textbooks; these are excellent guides the central concepts in these fields. Note that no analogy is implied between adjacent topics in the rows, these are simply the way these books introduce their subject matter.

Table 1.1 illustrates, then, that biological evolution is a phenomenon involving variation, adaptation and diversity, heredity and so on. Human cultures involve languages, concepts of kinship and descent, religious traditions and other domains of behavior; later in this book we will see just how these broad behavioural realms are and have been shaped to benefit human beings globally, and how such shaping can continue to better fit our species to environments beyond Earth. In the evolutionary biology column we see the processes and mechanisms of adaptation in Nature and in the cultural anthropology column we see the mechanisms and processes of cultural adaptation.

Importantly, in neither evolutionary biology nor cultural anthropology do we see structuring discussion or topical organization that indicates inevitable outcomes. That is, biologists and cultural anthropologists have searched for, but not found, significant inevitabilities in the processes they have studied. Certainly there are trends, and patterns, and some regularities, but inevitabilities such as the oft-cited 'ladder

Table 1.2 Most regularly-occurring topics in modern evolutionary biology and Anthropology texts, indicating these disciplines' major fields of study. No analogy between the fields is implied in the adjacent column items

Evolutionary biology	Cultural anthropology
Origins of life and DNA	Culture as learned guides to behavior
Variation	Language
Adaptation	Kinship and descent
Diversity	Power relations
Heredity	Sex and gender
Novelty	Equality and inequality
Classification	Religion
Biogeography	Economy and subsistence
Macroevolution and speciation	Myth, ritual and symbol

Compiled by the author from Love (2010), Bonvillain (2006), Lavenda and Schultz (2013) and Ferraro (2006)

of evolution' leading inevitably from one form to the next, or the 'cultural pyramid' in which human cultures are thought to progress from lower to higher stages, are missing from today's texts while just a century ago chapters on "The Progress of Cultures" would have been prominent. That such innate, centrally-controlled or divinely-scheduled tendencies are not found in either of these fields is one of the most significant revelations of the past 150 years of scientific inquiry. Prior to this century and a half of Darwinian approaches [Darwin published *On the Origin of Species* in November 1859 (see Smith 2011)], change in nature and human culture were assumed to be the result of a kind of striving or self-perfection, but this has not been substantiated scientifically. Certainly in biology there is a tendency towards complexity over time (McShea and Brandon 2010), but evolution is characterized by extinction and 'perfection' is an illusion. In culture there are also recurring patterns of complexity, but human cultures have been shaped by historical contingencies and issues of resource distribution, not an unspecified, internal urge or drive towards a particular goal other than survival (note that the goal of cultural evolution, as proposed a century ago, was to evolve through various well-defined stages to accumulate all the features of the Victorian European). This is muddied in humanity, however, as we *do* have the ability to consciously change our own culture, by a proaction that is entirely missing from the entirely reactive processes of biological evolution. While these topics are too expansive to explore here, I do wish to draw from these facts an important point, which is that the common conception of human space settlement as destiny, a product of an oft-cited 'innate will to explore', or a result of exploration 'being in our genes', is erroneous. Anthropologists and historians have found far more contingency in change through time than orderly, goal-seeking behaviour, except on the most basic levels of survival. The lesson for space settlement planners is that successful space settlement have to be the result of *proaction*. Such proaction is seen in all cultures to advance their aims; among South African hunters, adolescents are given ritual scars meant to remind them "to *want* to hunt", and among some Pacific Island cultures, adolescence rituals marking a new stage of life, which would traditionally involve much seafaring, the following dirge was sung (Fried and Fried 1980):

> Busy yourself with the voyage,
> Leap aboard to be heaved up on the sea,
> You carry your family
> Prepare your craft!
> Prepare your paddle!
> Not all will weep at the trail of foam!

Successful space settlement will only occur if it is planned carefully; we cannot take it any more for granted than Polynesian voyagers did their own project. Such projects are puzzles composed of many individual pieces. This book is meant to help us shape some of the pieces of the puzzle of human space settlement.

1.1.2 Objectives

In other writings I have sketched out an adaptive framework for human space settlement (e.g. Smith and Davies 2012; Smith 2016). This book is intended to be a more formalized, specific and scholarly work that delivers more of this framework, and serves as an evolutionarily-informed guide to space settlement planning.

As a starting point we may note that in 1963, Siegfried J. Gerathewohl, NASA's biotechnology chief, wrote the following passage early in his foundation text, *Principles of Bioastronautics*, outlining the need for this new field of study:

> Manned excursions into space require new types of vehicles, machines and hardware which were unknown in conventional flying. They will carry the traveler into such foreign environments as to pose serious problems of health and survival. The new field of medicine, which studies the human factors involved and the protective measures required, has been called *space medicine*. From its cooperation with modern technology, particularly with electronics, cybernetics, physics, and bionics, space biotechnology has branched out as a novel field of bioengineering. (Gerathewohl 1963:5–6).

'Excursions into space' may be characterized as exploration, quite different from settlement, as summarized in Table 1.3.

While bioastronautics was established during the (first) Space Age with tight focus on safeguarding the short-term health of individuals or small crews, space settlement by families and communities raise many other issues; individual physiology is a different phenomenon than, say, population genetics, and individual psychology as short-term adaptation is different from cultural adaptation by reshaping cultural norms in accordance with new circumstances.

In the same way that space exploration required Gerathewohl's *bioastronautics*, then, space settlement planning requires a field of study to ensure that plans are designed and carried out informed by all we know of the adaptive tools and techniques of our species. I propose that new field to be titled *space anthropology*, *exoanthropology* or *extraterrestrial anthropology*.

In the same way that Gerathewohl identified the need for his field in the quotation at the opening of this *Preface*, below I formally outline the need for space anthropology:

Table 1.3 Chief differences between space exploration and space settlement relating to adaptation

	Space exploration	Space settlement
Goals	Specific, short-term	General, long-term
Group size	Small (crews)	Large (communities)
Social organization	Command hierarchy	Civil community
Essential social units	Crews	Families and communities
Adaptive means	Technological, individual behavior, and some reversible acclimatization	Technological, cultural and biological adaptation
Adaptive timescale	Short; weeks to months	Long; multigenerational

> Space settlement will require novel biological and cultural adaptations to support populations of humans, on multigenerational timescales, in environments so far unfamiliar to our species even after 100,000 years of human cultural and biological adaptation to myriad Earth environments. The new field of anthropology that studies such adaptive efforts may be termed *space anthropology*, *extraterrestrial anthropology* or *exoanthropology*, *exo-* referring to beyond Earth, in the same way it is used in the term *exobiology*.

Specifically, I propose space anthropology to have three main functions:

1. **To identify the biological and cultural adaptive suite of humanity globally and to date**, resulting in a catalog of our species' adaptive tools and capacities useful to space settlement planners.
2. **To evaluate the capacities of humanity's various adaptive tools to adapt to reasonably forseeable space settlement plans**, bettering the prospect of productive adaptations to new conditions, e.g. on Mars.
3. **To make recommendations, some broad and some specific, that would assist in human adaptation to environments beyond Earth**, particularly based on evaluations of human adaptive capacities identified in functions 1 and 2.
4. **To assist with direct, actionable space settlement design plans**, for example in the design of human habitats beyond Earth.

The scope of exoanthropology, then, will be broad. I propose it as an applied form of anthropology with the specific goal of evaluating the *adaptive capacities* of our species, both biologically and culturally, so that they may be best deployed to assist in successful permanent space settlement. This will guide space settlement planning in a genuinely adaptive and evolutionarily-informed way, applying the lessons of billions of years of Earth life adaptation to humanity's project to disperse life throughout the solar system and, later, beyond. In fulfilling Function 1, exoanthropology will survey humanity's adaptations through time and across the globe, identifying patterns pertinent to space settlement planners. In fulfilling Function 2, it will review the adaptive competence of many of our species' adaptive tools, allowing us to evaluate our readiness for space settlement and, where we find ourselves unready, suggest courses of action; it will also characterize forseeable space settlement conditions and limiting factors as needed. In fulfilling Function 3, recommendations for space settlement planners will be formulated, varying in specificity, based on the lessons identified in the surveys serving Functions 1 and 2. Finally, in fulfilling Function 4, directly actionable engineering and other design recommendations will be made, materially assisting in space settlement planning. This book focuses on Functions 1 and 2 as stated above, and at this time can only—though I think productively, in the sense of exploring the issues—begin to address Functions 3 and 4.

1.2 Models of Space Settlement

'Space settlement' is too broad a term to reflect the realities of potential permanent human settlement of environments beyond our home planet. Specifically we can identify at least four environments that might be settled by our species in the forthcoming few centuries; these are tabulated below.

Table 1.4 indicates that a variety of conditions exist, with respect to the selective agents that may shape human biological, behavioural and technological adaptation over the next few centuries, beyond Earth. In some cases, technology can mitigate selective agents, such as radiation shielding to mitigate cosmic radiation. In other cases, such as living on the surface of a body with a significantly different gravity field than that of Earth, there will be a combination of *biological* adaptations (just as migration to high altitudes required some biological adaptation on Earth), *technological* adaptations (such as using shielding against Galactic Cosmic Radiation and solar flares at Lagrangian Point settlements) and *behavioural* adaptations (such as new concepts regarding the passage of time on such bodies as a Mars Moon or Asteroid Belt Dwarf Planet, which have unalterable, very different day- and year-durations compared to those of Earth). It is the goal of this book to provide guidelines making such adaptations more likely to succeed, using the principles of evolutionary adaptation and anthropology's body of knowledge regarding humanity's adaptive range. We will revisit some specific properties of settlements such as a Mars or Lagrangian colony below, in the Sect. 1.5 *Human Scales of Adaptation and Space Settlement: Habitats.*

Table 1.4 Gross properties of earth surface and beyond-earth settlement sites conceivable in the next few centuries

Settlement site	Gravity field Earth surface = 1.0	Cosmic radiation flux (millisieverts) (per year without shielding)	Day (hours)	Year (days)
Earth surface	1.0	0.6 msv	24	365
Earth-orbiting settlement	Variable, 0.01–1.0	200–400 msv	Variable	365
Lagrangian-point Settlement[a]	Variable, 0.01–1.0	1.0 msv	Variable	None
Earth's Moon	16	380 msv	672	365
Mars surface	0.38	200–800 msv	26	687
Mars Moon (e.g. Phobos[b])	0.005	c. 200 msv	8	687
Asteroid belt Dwarf Planet (e.g. Ceres)	0.03	200 msv	9	1682
Interstellar Generation Vessel traveling at c. 0.3 light speed	Variable, 0.0–1.0	c. 1.0 msv[c]	Variable	No such measure

Data compiled from Chancellor et al. (2014), Reitz et al. (2012) and Wheeler and Martin-Brennan (2000), Hassler et al. (2013) and simulations run by the author in NASA's online OLTARIS (*On-Line Tool for the Assessment of Radiation in Space*) software (https://oltaris.larc.nasa.gov/)
Note [a]L-4 or L-5, [b]See Wheeler and Martin-Brennan (2000), [c]See Maciel (2013)

1.3 Space and Time Contexts for Human Space Settlement

My earlier book, *Emigrating Beyond Earth*, provided a lengthy evolutionary context for human space settlement. In this section I update this context and sharpen some of its features. Two major variables structuring human adaptations to environments beyond Earth will be space and time.

1.3.1 Space

Regarding *space*, or the distribution of our genes and populations of genes, we may begin with levels of identity including the genome, individual, close associates (such as family members) and extend outwards from these (somewhat variably, by culture) to ever-more-inclusive (but sometimes, paradoxically, increasingly-difficult to define) groups including members of residences, aggregations of residences, political units and—beyond Earth—planetary bodies (for example). How humanity adapts to environments beyond Earth will in part depend on the constitution and arrangement of such human groups across space. For example, a Mars colony may be considered an open system, genetically, as it can continue to receive novel genes in the form of DNA-carrying people arriving from Earth, however its population for the first generation or two or three is likely to be low, so planners will have to consider its small population and the genetic *founder effect*, explored in Chap. 3. A culture beyond Earth, that wishes to be completely disassociated from Earth, however, closes itself, genetically, so that genetic diversity of the founder population would be even a greater concern, no matter how much genetic diversity may be laid in stock in the form of stored sperm and eggs, and/or available through gene editing with advanced methods.

For broad context, let us consider the distribution of Earth life and then humanity specifically. We note that animal life is abundant and widespread on Earth, and that in the past 400 million years animal life has even undergone multiple *evolutionary transitions* [substantial biological reorganizations including evolution of the cell wall, and evolution of multicellularity; see Szathmary (2015)] from one environment or medium to another (e.g. mammals adapting to marine environments over 40 million years ago), requiring adaptation to radically new conditions of temperature, ambient fluid (atmospheric) pressure and composition, IR and UV exposure and even the effect of Earth's 1 g gravity field on the body, as seen in Fig. 1.2 [note that microbial species, not discussed here, are similarly widespread, facing few geographical obstacles to movement (Findlay 2002)].

Looking outward from the surface of Earth, we see that Earth-derived bacteria and fungi have been found to inhabit the atmosphere from 9 km to 77 km in altitude, including UVC-radiation-resistant bacteria (Yang et al. 2005). This work is ongoing [see the 2017 paper *The Biopause Project: Balloon Experiments for Sampling Stratospheric Bioaerosol* (Ohno 2017)] but at present it appears that while mechanisms to

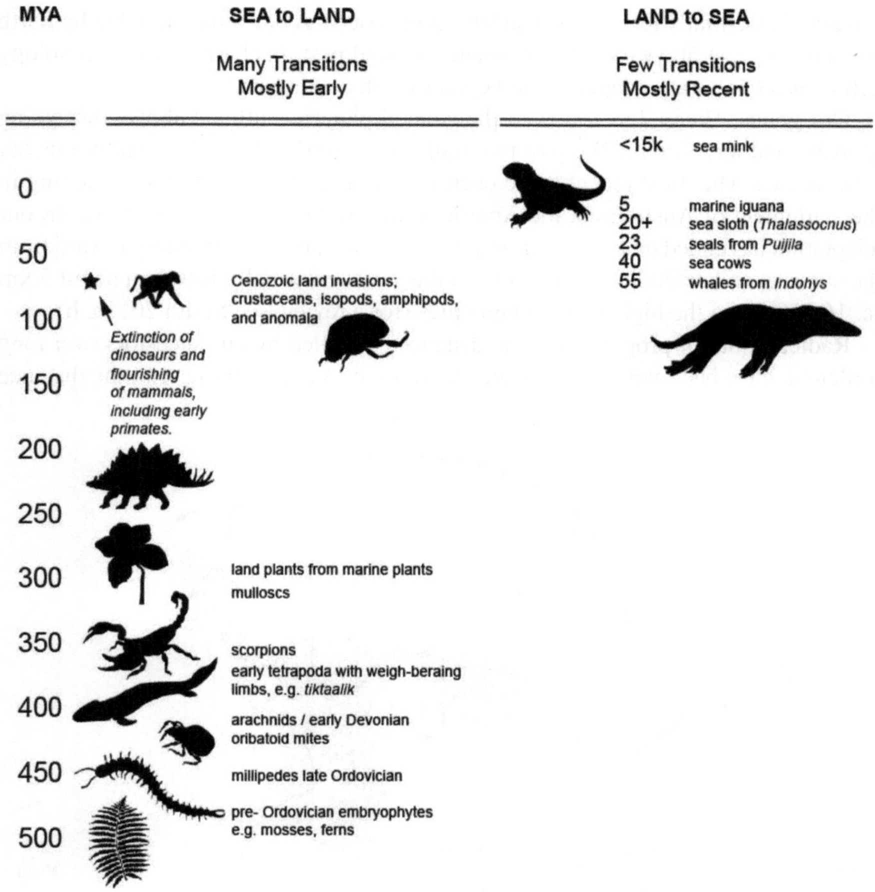

Fig. 1.2 Sample evolutionary transitions from marine to terrestrial (and terrestrial to marine) habitats. Author's diagram is based on figures presented in Vermeij (2000)

carry Earth life upwards include volcanism (to c. 20 km) and gravito-photophoresis (light-activated particle motion; see Cheremisin et al. 2005), this does not appear, to date, to have resulted in broad adaptation to space-like conditions (above 50,000 ft. or 15 km) by terrestrial microbial life. Despite the finding of UVC-resistant stratospheric bacteria, UV radiation appears to remain a substantial barrier: experiments with the aerobic bacterium *B. subtilis* indicate its vulnerability to UV radiation, which is encountered at significantly higher doses in the stratosphere. While in 2011 test bacteria survived space-like low temperatures, pressures and dessication, 99.9% did not survive six-hour UVB and UVC radiation exposure at doses expected at 20 km altitude, the experimenters concluding that "the stratosphere can be a critical barrier to long-distance microbial dispersal and that survival in the upper atmosphere may be constrained by UV irradiation" (Smith et al. 2009). Earth life, then, even when counting the microbial—which is most numerous and evolved earliest—has remained essentially terrestrial. The barrier of space has been impenetrable by Earth life until the last 50 years, when humanity moved past this barrier with technology acting as a temporary adaptation to beyond-Earth conditions.

Our genus, *Homo*, has followed the general patterns outlined above, dispersing globally and terrestrial in the past two million years (Fig. 1.3) in several pulses out from Africa. The most recent pulse occurred in the last 100,000 years, resulting in the settlement of Australasia, the Americas, the Arctic and Oceania. However, our adaptation has ended on the surface of the world's oceans, not extending inward from the surface, and extending outward from the surface normally less than about 5 km (c. 16,000 ft.) in the highest mountain cities (for a review see Smith 2018a, b).

Radically out of proportion to the distances travelled by our ancestors over long millennia have been voyages of space exploration. As seen in Fig. 1.3, the distance

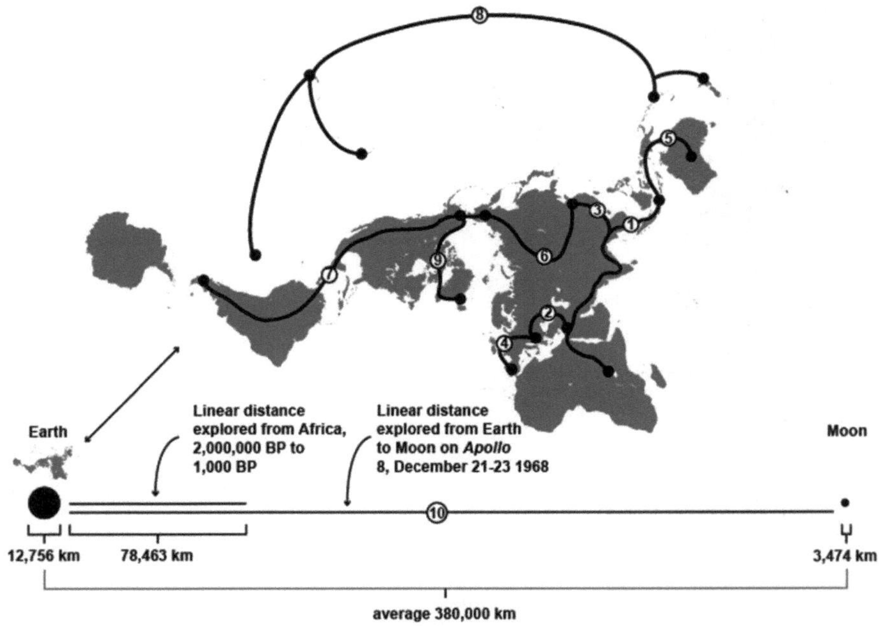

LEG	YEARS BEFORE PRESENT	LEG DISTANCE (KM)
1: Africa - SE Asia	2,000,000	14,247
2: Africa - W Asia	2,000,000	2,099
3: SE Asia - E Asia	1,500,000	2,555
4: W Asia - Europe	1,000,000	4,418
5: SE Asia - Australasia	50,000	6,018
6: E Asia - NE Asia	40,000	7,073
7: NE Asia - Americas	20,000	17,719
8: SE Asia - Oceania	3,000	18,482
9: NW Americas - NE Americas	1,500	5,852
10: Earth - Moon	50	384,400

Fig. 1.3 Human dispersal distance in the last 50,000 years. Note the vast distances covered by the space exploration effort of *Apollo 8* in 1968. Data and visualization by the Author

covered by *Apollo 8*, just one-way to the Earth's moon, in about 70 h, was about 400,000 km, compared to the entire past two million years of dispersal, which covered about 80,000 km linearly. Extending Earth life beyond current space boundaries will of course require just such technology to increase the speed of travel during the life of the individual (multigenerational voyages sending hibernating populations are a fascinating but not yet nearly feasible alternative). Humanity has long experience in increasing locomotion speed with technology; in the past few centuries we have radically decreased Atlantic and Pacific crossing times from months to less than a day, a factor of 180 (Fig. 1.4), with various technologies.

Such technological advances radically increase any individual human's potential *natal dispersal distance*, the distance travelled by an individual between birth and

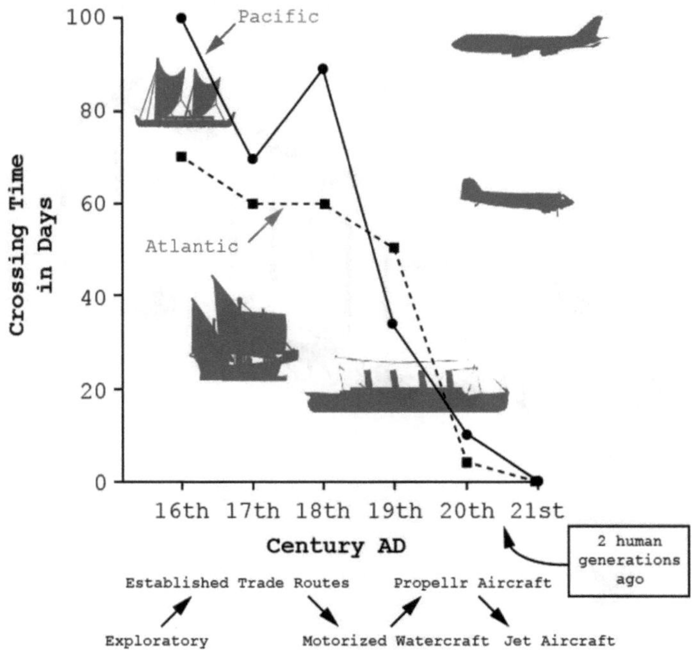

Fig. 1.4 Reduction of Atlantic and Pacific crossing times via transport technologies, 16th–21st centuries AD. Author's visualization. For the Atlantic, 16th Century based on Barlow's sailing voyage from 27 April (departing Canaries) to coast of Virginia, 04 July 1584, 17th Century based on Pilgrim voyages, 18th Century based on Comte de Rochambeau's crossing to Rhode Island from 02 May to 11 July 1780, 19th Century based on common transit times for immigrant sailing vessels, 20th Century based on combustion engine liner ship crossing times and 21st century based on fast commercial passenger aircraft. For the Pacific, 16th Century based on Magellan's Pacific Crossing from Cabo Deseado on 28 November 1520 to the Mariana Islands, landing 06 March 1521, 17th Century based on William Dampier's voyage from Cabo Corrientes (Mexico) to Guam, 31 March to 21 June, with a long layover in Guam (1686), 18th Century based on Rogeveen's 1722 crossing from Patagonia (17 March) to Solomon Islands, 01 July, over 90 sailing days, 19th Century based on clipper ship Challenge, Hong Kong to San Francisco, 1952, and 21st century based on fast commercial passenger aircraft

first mating (Pasinelli et al. 2005). A survey of 311 species of Earth life (Fig. 1.5) demonstrates that natal distance across these life forms averages some meters for much plant life (seeds and propagules often being distributed by animal activity), some tens of kilometers (averaging 46 km in this sample) in most land mammals and a few hundred kilometers in most birds and aquatic species (note these figures do not include seasonal *migrations* from one region and back to a home territory). Natal dispersal distance is limited by several factors including travel speed (itself conditioned by physical terrain properties).

Figure 1.6 indicates the maximum speed, standardized to body lengths per second, for a variety of Earth life (land animals, swimming fishes and birds using wings for propulsion rather than diving at higher speed); these top out at about 10 body lengths per second (Meyer-Vernet and Rospars 2016). While animal life does not travel at maximum speed all the time, the figures are instructive when compared with

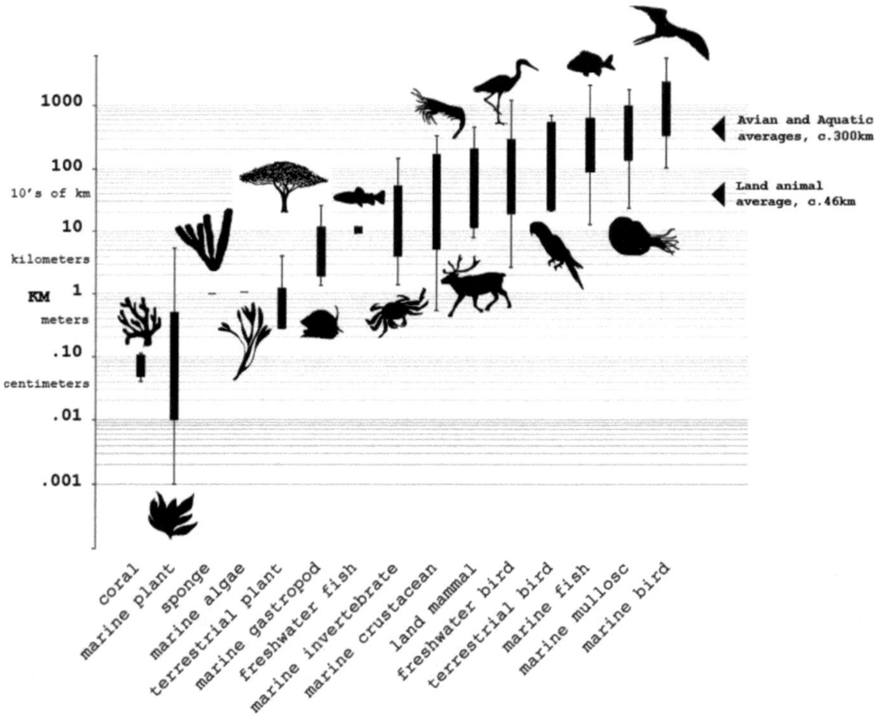

Fig. 1.5 Natal dispersal distances summarized for 311 species of earth life. Raw data were selectively drawn from Daniel et al. (2013), Paradis et al. (1998), Sutherland et al. (2000) and Kinlan and Gaines (2003) after which they were statistically characterized in IBM SPSS 24 and visually summarized by the author. Averages calculated for the sampled species, in kilometers, are Coral = 0.18, Marine Plant = 0.98, Sponge = 0.99, Marine Algae = 1.27, Terrestrial Plant = 3.18, Marine Gastropod = 5.35, Freshwater Fish = 10.97, Marine Invertebrate = 29.47, Marine Crustacean = 37.43, Land Mammal = 42.37, Freshwater Bird = 50.90, Terrestrial Bird = 101.10, Marine Fish = 106.87, Marine Mullosc = 118.99, Marine Bird = 213.93

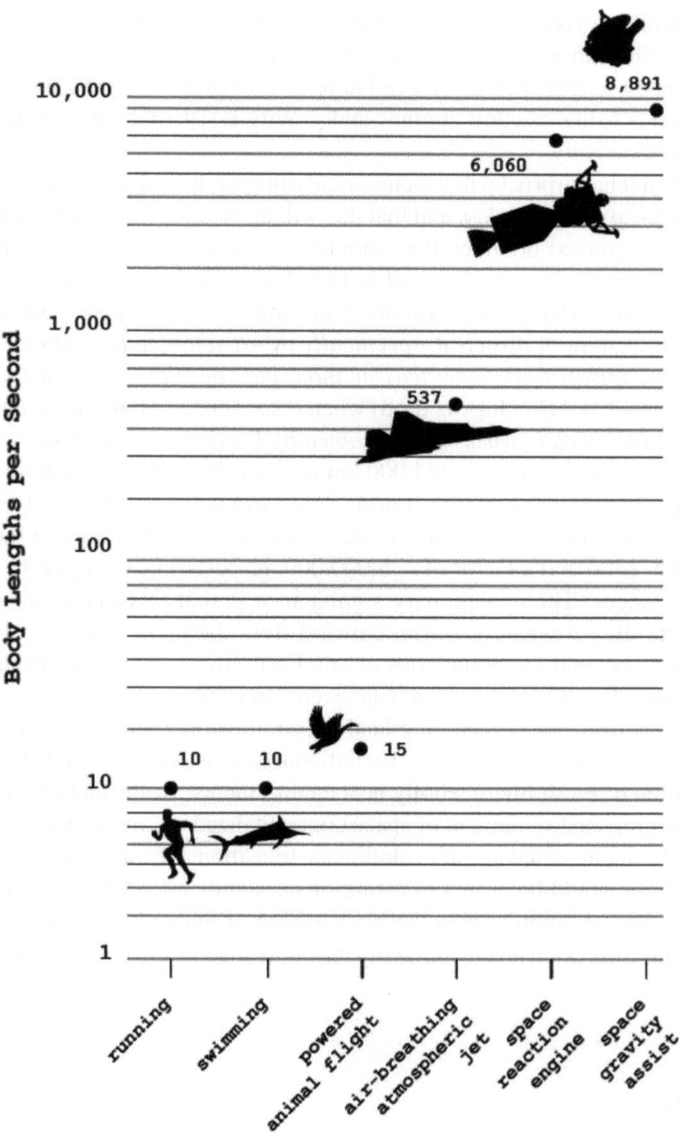

Fig. 1.6 Travel speed in body lengths per second for selected animal life and technological conveyances. Raw data derive from Meyer-Vernet and Rospars (2016), processed by the Author. For Technological Conveyances, body length is 2 m (six feet) rather than vehicle length; for the air-breathing atmospheric jet, I refer to SR-71 traveling at over 2000 mph, for a reaction engine in space I refer to *Apollo IX*'s speed to the moon of c. 25,000 mph, and for a gravity-assist craft I refer to the *New Horizons* probe to Pluto at c. 36,000 mph (gravity assist has yet to be used to transport people). Note new Parker Solar Probe speed, also using gravity assist from the sun, is planned to reach c. 430,000 mph, a speed of about 70,000 body lengths per second for that 3 m-long craft

technologically-devised locomotion. This is also indicated in Fig. 1.6; air-breathing jet engines allow speeds of c. 500 (human) body lengths per second and over 6000 and 8000 body lengths per second (a factor of 800 over biological locomotion) in space vehicles using reaction engines and gravity assist (not yet used to transport humans), respectively.

It should not be surprising that technologies ultimately derived from Tsiolkovski's ideas allow us to move quickly, and that this will allow us to move to habitats beyond Earth. But the context provided here should reveal the evolutionary significance of such technology for our species, and in fact, Earth life itself. Figures 1.4, 1.5 and 1.6 show that humanity can do far more than simply 'explore space', we can engage in an ancient pattern of dispersal, *specifically by extending natal distance of a wide variety of organisms from some tens or thousands of kilometers* (on Earth today, depending on what technology is used) when constrained to Earth *to some hundreds of millions of kilometers when unconstrained by Earth*, accessing potential habitats including the moon (averaging 384,000 km from Earth) within a few days and Mars (averaging 225,000,000 km from Earth). Such extension can be used to disperse members of our species (and other specie we choose to take along) on the order of a factor of c. 8300 and a factor of c. 5,000,000, respectively, over purely naturally-evolved distances. The evolutionary significance is that this makes it possible to extend Earth life, of whatever form, outward from the point of early evolution, to other habitats beyond the boundaries of any Earth life so far known in the past 3+ billion years of Earth evolution. Such an extension would be at least as significant as the transitions from sea to land, and land to sea, illustrated in Fig. 1.2 and, I argue, would constitute the origins of an 8th Evolutionary Transition of Earth Life evolution, the adaptation of Earth life to wholly new environments. At the end of this book we will see the proposed settlement of space contextualized as constituting one more of the currently-documented seven evolutionary transitions of Earth life to date.

While these would be events of evolutionary significance for Earth life at large, they would also be significant in the perpetuation of human life in general. I argue below that just this extension of humanity through space would promote our extension through time.

1.3.2 Time

While the range of Earth life in space may be observed today, for the range of Earth life through time we use the tools of palaeontology and archaeology to look into the past. These disciplines have furnished clear indications that the evolutionary fate of most known Earth species has clearly been extinction. Figure 1.7 (panel b) indicates the timing and percent kill rate of fossil species in the five major extinctions to date, indicating an average marine genus duration of c. 20 million years, and average species duration c. 2–5 million years are reported (Raup 1994). Similarly, most bipedal large primates (the *hominins*, which include our genus, *Homo*) have become extinct over the past few million years and Fig. 1.7a (based on my own calculations)

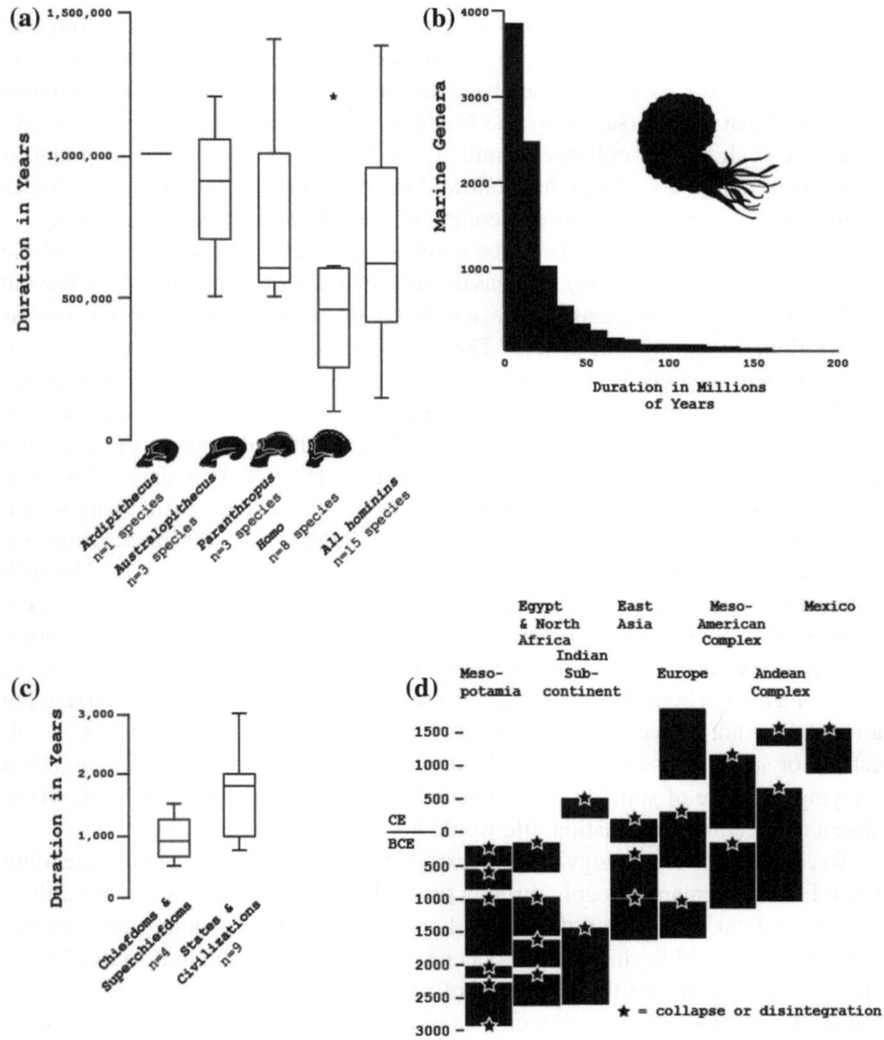

Fig. 1.7 Durations of hominin genera, marine genera and complex social organizations including civilization. Panel **a** boxplots indicate fossil hominin durations for 15 species recognized in Cachel (2015), Day (1986) and Klein (2009), (high outlier in genus *Homo*, indicated by the *, represents *H. erectus*). Panel **b** histogram (10-million year bins) indicates number of marine genera surviving for 17,500 species as reported by Raup (1994). Panel **c** boxplots indicate durations of 13 chiefdoms, superchiefdoms and proper states (sensu Childe 1950) based on chronologies in Trigger (1992) and Wenke and Olszewski (2006). Panel **d** indicates collapses or disintegrations of state organizations in Mesopotamia (Sumeria-Babylon Complex), Egypt & North Africa (Egyptian state), Indian Subcontinent (Harappa and Gupta states), East Asia (Shang, Zhou and Han states), Europe (Greece, Rome and the Modern West), the Mesoamerican (Olmec, Maya) and Andean (Caral through Moche, and Inka) complexes, and central Mexico (Toltec-Aztec) as discussed in Tainter (1990) and Weiss and Bradley (2001)

indicates the average genus duration c. 1.4 million years. From the archaeological record (further explored in Chap. 6) we see (Fig. 1.7c, d) that most chiefdoms and civilizations in the past 6000 years have an average duration of less than two millennia and that 99.9% (save only the Modern West since the post-Medieval) have experienced significant collapse (Smith 2017). In our biology and social organizations, then, our genus, *Homo*, has followed the palaeontological pattern of species extinction, and repeated collapse of complex cultural and economic structures.

We may say that to date, the space boundary for Earth life has been that of the lower atmosphere (c. <20 km) whereas the time boundary for animal life has been in the handful of tens of millions of years and that of the political organization known as civilization, in the millennia at most. The space boundary is set by radically different selective environments from one's origins (e.g. ambient temperatures, pressures and radiation flux) which are beyond the adaptive capacities (conveniently referred to as the *reaction norm* (see Gotthard and Nylin 1995) mentioned again in Chap. 2) of a given life form in the course of a single generation. The time boundary is less-well understood but is sometimes set by stochastic events such as mass extinctions; still, it must be telling that ancient civilizations have all collapsed or disintegrated, a topic we examine more closely in Chap. 6. While Humanity has so far been restrained by such boundaries, we are uniquely capable of proactive adaptation, an evolutionary feature that has given our species the potential to extend life beyond Earth with the technologies discussed earlier. This is perhaps best visually communicated in the astounding photo of a human being on the surface of a solar system body (Earth's moon) having arrived there not by accident (as some life forms are carried to high altitudes by volcanism or sandstorms and other mechanisms) but by *consciously devising* and then carrying a bubble of suitable environmental conditions to that environment, where absent this technical adaptation, life would not be possible (Fig. 1.8).

To consider the technology, time and space together, we may note that extending Earth life and humanity deeply through time—beyond the duration of the Earth as a viable habitat, a timespan that is not known considering random global extinction events—would begin with and depend on the extension through space. This is clear when we consider the properties of the longest-lived of Earth species, again found in the palaeontological record. Examining these, Vermeij (2008) has identified both active survival strategies and the properties of long-lived fossil species, the most salient properties being (a) wide geographical dispersal, (b) great variability or diversity and (c) large populations. On Earth, humanity has a relatively large population of quite variable individuals at present, but because we are all on the one planet we are vulnerable to a single catastrophe (other life forms are as well; and finally, civilization is far more fragile than life itself and could easily collapse given all we know from collapse studies). If we are to learn from the historical sciences we should use the speeds available via aerospace technology to allow us to implement the "survival habits of the longest-lived species" by traveling far from a very reasonably-expectable Earth catastrophe; Mars, many believe, is a viable such habitat for an insurance policy for humanity and its many required domesticates and symbionts. The extension through time, then, will require extension through space,

Fig. 1.8 Earth life form transported to its moon by proactive, adaptive measures. NASA photo

as made possible by technological adaptation. Space exploration may be seen, then, not simply the technological wonder it is often used to epitomize, but a process of genuinely evolutionary significance; the perpetuation of humanity—and the symbionts and domesticates we choose to protect, and all of our cultural knowledge and history—across space, to new habitats. This extension, in turn, can extend us and our lineages through time, by insulating us from any single Earth catastrophe. In short, speed will give us space, and space will give us time.

Table 1.5 places the human settlement of environments beyond Earth in a larger context of human evolution, showing that such settlement would be only another of several major adaptations in the history of our species. There are two lessons in pointing this out. First, it gives the project of space settlement its legitimate significance; settlement would not be a short-term, flag-planting exercise, but a significant aspect of human evolution at large, and recognizing this helps many to understand its importance. Second, the evolutionary and adaptive context here reminds us of the significance of tailoring space settlement as an adaptive and evolutionary endeavor, and in so doing, to learn as much as we can about evolution and adaptation to best help the project succeed.

Table 1.5 The extraterrestrial adaptation in larger context. '−' = characteristic is decreasingly significant, '+' = characteristic is increasingly significant. MYA = million years ago

	Biological change = phenotype's structure and anatomy	Biological change = phentoype's metabolism and processes	Cultural change = behavioral and cognitive change; includes technology
1. Terrestrial *by 5mya* genus *Australopithecus*	+ Bipedal anatomy	+ Thermoregulation	+ Territorial range
	− Arboreal anatomy	Efficiency	+ Ecological plasticity
2. Technological *by 2mya* genus *Homo*	+ finger opposability	+ hand/eye coord	+ Enculturation time
	+ Brain volume	+ Caloric needs	+ Econiche breadth
	+ Body stature	+ Caloric needs	+ Econiche breadth
	− Tooth size	− Digestion requirements	+ Econiche breadth
			= Reliance on technology
			= Reliance on culture from anatomy
3. Cognitive *by 100kya* Anatomically and behaviorally-modern *Homo sapiens sapiens*	+ Brain volume	+ Caloric needs	+ Spatial sense
	+ Neuronal LAT (Lowered Activation Threshold facilitates association of perceptions into more complex representations of the environment in the mind)	+ cog. complexity	+ Temporal sense
	− Tooth size	+ Tool variety and complexity	+ Re-presentation
	− Body robusticity		+ Extrasomatic data storage
			+ Communicative' subtlety and crossing
			= Reliance on complex culture, tech. and data
			= Decoupling data storage/retrieval from biology
			= Increasingly complex data management e.g. mythology
4. Extraterrestrial *Beginning today*	Early: none	← See	+ Data management
	Later: largely unforeseeable and will depend on selective environment, e.g. Mars surface—versus—Mars orbit + biological differentiation		+ Evolution manipulation
			+ Spatial sense
			+ Temporal sense
			+ Cultural differentiation

(continued)

Table 1.5 (continued)

	Biological change = phenotype's structure and anatomy	Biological change = phentoype's metabolism and processes	Cultural change = behavioral and cognitive change; includes technology
			= Reliance on complex culture, tech. and data
			= Reliance on culture of innovation and creative problem solving

Adapted by the author from Table 8.1 of Smith and Davies (2012)

1.4 Human Scales of Adaptation and Space Settlement: The Life Course

If permanent, multigenerational space settlement is the proximate goal that will allow the time- and space-extension of humanity, then we must consider humanity as life is lived by the individual. We do not experience civilizations or millennia, but lifetimes. And evolutionary adaptation, while having long-term effects (*longue duree* is a convenient term indicating the shifting, over time, of the centers of cyclic activity) is played out in the success or failure of individuals to adjust their internal state to new environmental pressures in the course of the lifetime. As an organizing principle, then, I suggest beginning with the human life-course; that, in fact, is the most precious cargo and greatest concern of any activities of humans-in-space. I thus take an *anthropocentric* approach to human space settlement, one which places humanity as the central concern.

The main of stages of the human life-course are identified in Fig. 1.9; note that success is required at different stages of the life course, and that at different stages, success might rest more with biology or culture—that both are required for lifelong survival and success. This highlights the *biocultural* mode of human adaptation mentioned earlier in this chapter. As we will see in the following chapters of this book, the challenge of human space settlement will be to devise some combination of technological, biological and cultural adaptations to ensure the safe development of each of the stages of human life indicated in Fig. 1.9.

1.5 Human Scales of Adaptation and Space Settlement: Habitats

If humanity will have to adapt to protect the life course biologically, behaviourally and technologically beyond Earth, we may consider how these variables will manifest in at least three major types of beyond-Earth settlements currently in discussion; these

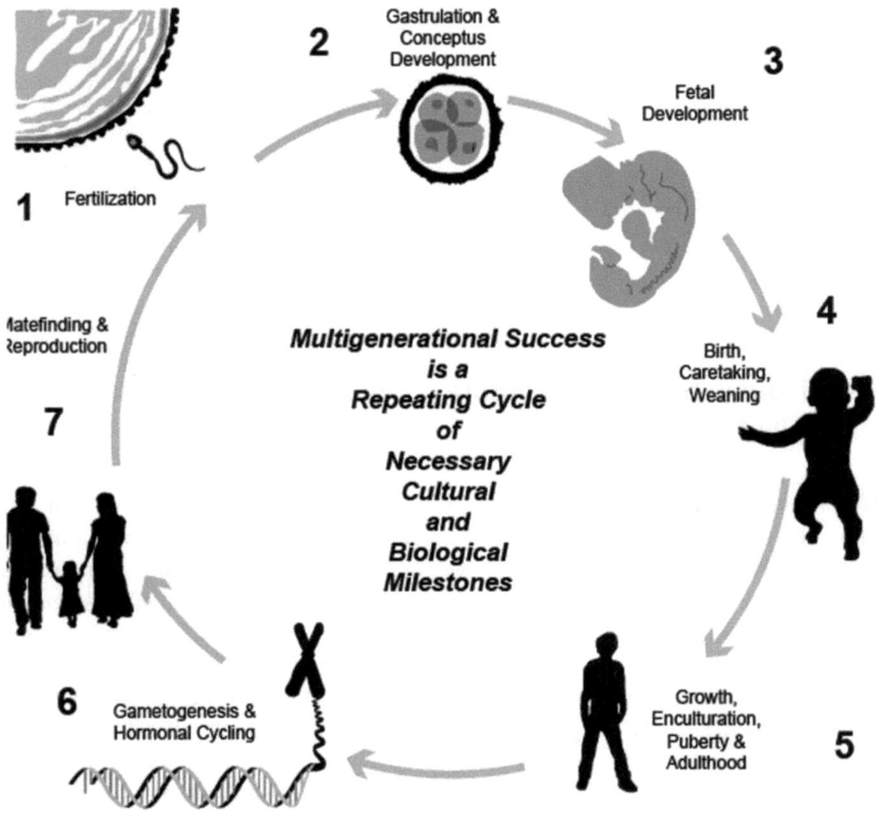

Significant Influences of:

	Biology	Culture
1	sperm & egg viability	-
2	developmental scheduling & oogenesis	-
3	systems development	-
4	early maturation	learning interactions, social & natural history intelligence
5	later maturation & hormonal scheduling	expanding space & time awareness, linguistic & technical intelligence
6	spermatogenesis	-
7	reproductive biology	matefinding & establishment of stable lifestyle

Fig. 1.9 Seven major stages of human life: growth and development, fertilization to matefinding and reproduction. The general stages are largely derived from Jasienska et al. (2017) and Malina (1975) with some elaboration by the Author

are briefly introduced below. Here the discussion is meant to condition how the reader considers the lessons of the rest of the book rather than as thorough explorations of these settlement types, a topic for more specialized studies.

1.5.1 Terrestrially-Tethered Settlements

Terrestrially-tethered settlements will be those that maintain, for several generations at least, substantial (e.g. organizing, structural) biological, cultural and/or political ties with Earth. Included in this category would be colonies in close Earth orbit, colonies placed to orbit Lagrangian points L4 or L5 (about four million miles, or 6.4 million km, about a week from Earth, one way, at *Apollo 10* speed) or lunar settlements that would require some prolonged period of contact with Earth to import soils for agriculture. In such habitats we may expect a slower pace of cultural divergence between the donor (Earth) and recipient (Moon) regions both biologically and culturally, through time. The relatively short travel times would make it easier for gene flow between the populations, and certainly material and ideas would be easily exchanged, such that from the outside, not much would appear to change from Earth conditions. But for people and communities living in such settlements there would of course be change from the experiences of Earth. Figure 1.10 depicts a plausible timeline for what might happen over several centuries in a Lagrange colony. Note that predicting such changes is not the specific function of space anthropology, rather it is predicting which aspects of human biology and behaviour will be involved in such changes that represent adaptation over time. I revisit this topic in Sect. 1.6.1 *Prescription and Prediction in Space Settlement Planning*.

1.5.2 Independent Colonies on Other Solar System Bodies

In the case of settlements or colonies that become structurally independent of Earth, in biological, cultural and/or political connections, we may expect more rapid adaptation and diversification of our biology, culture and technologies; such diversifications, or *adaptive radiations*, are well-known in biology generally [for one of many examples see Elmer (2009)]. From these cases, generally speaking, and more specifically from cases of modern cultural diversification, which includes the phenomenon of the evolution of new identities, or *ethnogenesis* (Fennell 2007), we can learn a great deal that will assist in making a success of attempts of humanity to settle environments beyond Earth. Sometimes we can learn by identifying what human cultures have not succeeded in doing, by the study of cultural maladaptations, sometimes warnings from the past of human adaptive 'dead-ends' (Edgerton 1992).

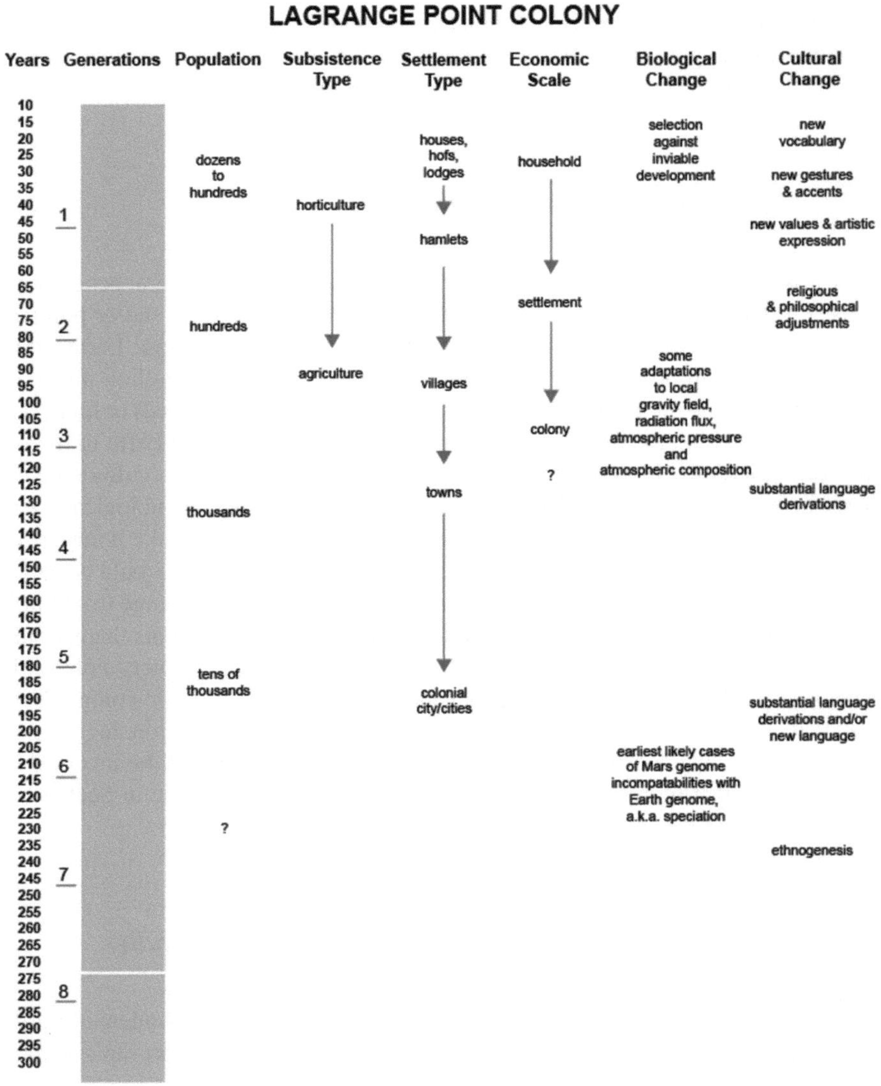

Fig. 1.10 Ballpark timescales in major variables of human populations in an hypothetical lagrange point 5 colony

Such settlements will tend to be materially rather distant from Earth, with significant implications for the genetic health of the independent settlement; for example, it will be desirable for such settlements to grow rapidly because in all vertebrate biology it is recognized that smaller populations are more vulnerable to any conceivable catastrophe than larger population (Vermeij 1978). How do our culture and biology best facilitate rapid but sustainable population growth? Carefully researching and answering such questions, in such a contextual ‘thoughtscape’ as that of the rapidly-growing population of a Mars settlement, will give such efforts a greater chance to succeed than if we do not consider them from the outset (Fig. 1.11).

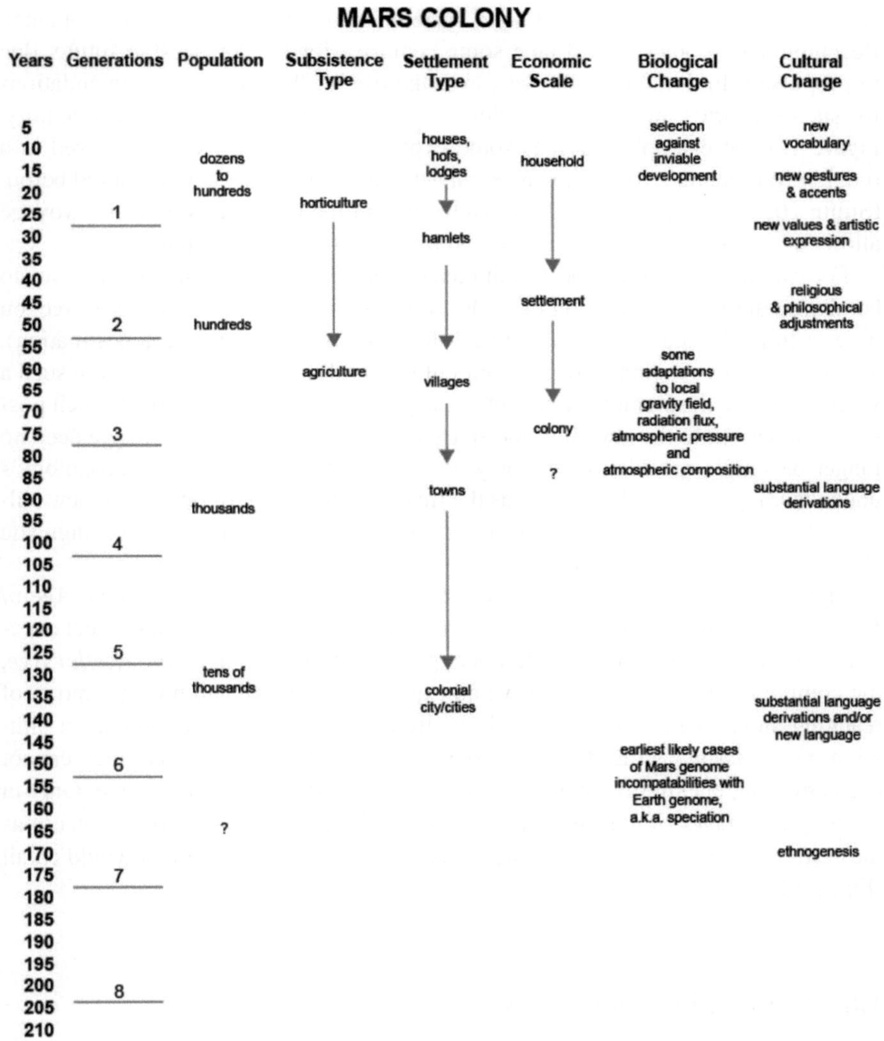

Fig. 1.11 Ballpark timescales in major variables of human populations in an hypothetical permanent Mars settlement. Significant organizational transitions are indicated by gray arrows. Here, one generation = 25 years. Note that the bulk of reasonably-expectable change required to adapt to new conditions occur in the first two generations. The populations making the sacrifices to accommodate such change from Earth conditions will have to be strongly motivated

1.5.3 Independent Colonies Aboard 'Closed-System' Spacecraft

One of the most thought-provoking of concepts of humanity settling beyond Earth is that of settlement of a distant solar system, or, more specifically, an exoplanet. Reaching such bodies would take some centuries for the foreseeable future, due to propulsion limitations. Still, we can make useful, thoughtful recommendations for such scenarios, even as we explore them as 'exotic concepts' in this century. Figure 1.3, for example, outlines some expected 'boundary events' involved in a roughly four-century voyage to an exoplanet, a thoughtscape I have explored before (Smith 2014a, b; Hein et al. 2019). Sketching out the boundaries of such a voyage allows us to research its implications and make useful recommendations.

For example, we may expect both conceptions of *time* and *space* themselves to be altered among a population of such voyagers, who might well begin to reckon time in time elapsed since departing Earth (or years until arrival at a destination), and who might well emphasize, in their cultural traditions, Earth, early on in such a voyage, but the destination exoplanet in the generations closer to arrival. Such possible adaptations to the new time and space issues implied by such a voyage need no longer be simple speculation, but may be responsibly addressed by anthropology's understanding of such phenomena as the timescale for the development of new cultural identities (e.g. ethnogenesis as introduced above) and even the rate of linguistic change (for example, see Nettle 1999).

Figure 1.3 also depicts three 'Ages' plausible in such a scenario. In an *Age of Departure*, the individuals aboard the craft would likely have had some direct experience of Earth, though they would not expect to return to it. In an *Interstellar Age*, the community members would have no experience of Earth and no expectation of reaching an exoplanet, so lives would be lived out entirely in the interstellar situation, not 'missing' an Earth never experienced, or 'lacking' a planet they cannot even envision, but simply living out lives with the worldship as their home. Only in the very last stages of the *Age of Arrival* would 'planetness' again be a concern as the culture began to consider landing on a new world, with all that that would entail (Fig. 1.12).

1.6 Concluding Comments

In this chapter I have outlined a new variety of anthropology, a *space* or *extraterrestrial* anthropology, and specified its function as serving to make the human settlement of environments beyond Earth more likely to succeed; the guiding theoretical principles has been outlined as that of *evolutionary adaptation*, and the chief tools of this approach have been identified as the corpus of anthropological knowledge of our genus' *adaptive range* in space and time to date. Figure 1.4 visually establishes

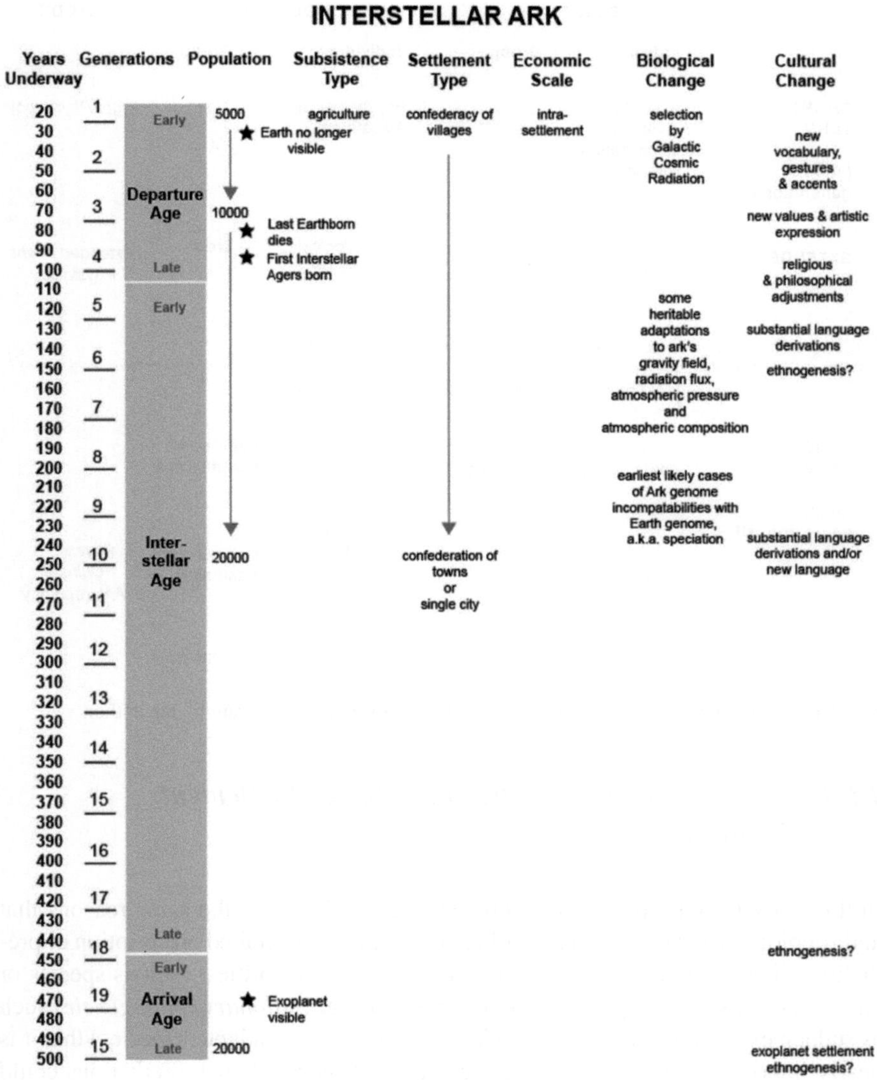

Fig. 1.12 Some hypothetical, but reasonably expected, events of a multigenerational voyage to an exoplanet. Issues introduced here are only to sketch out the 'thoughtscape' and are elaborated upon in later chapters of this book

space or exoanthropology as addressing separate scales of human adaptation than the fields of *flight physiology*, even *extended flight physiology* (Fig. 1.13).

However each of these topics may be expanded in the future, it is sufficient at present to have them established programmatically. We may now proceed to apply these theoretical, research and recommendation tools to the project of human space settlement, as outlined in the following chapters.

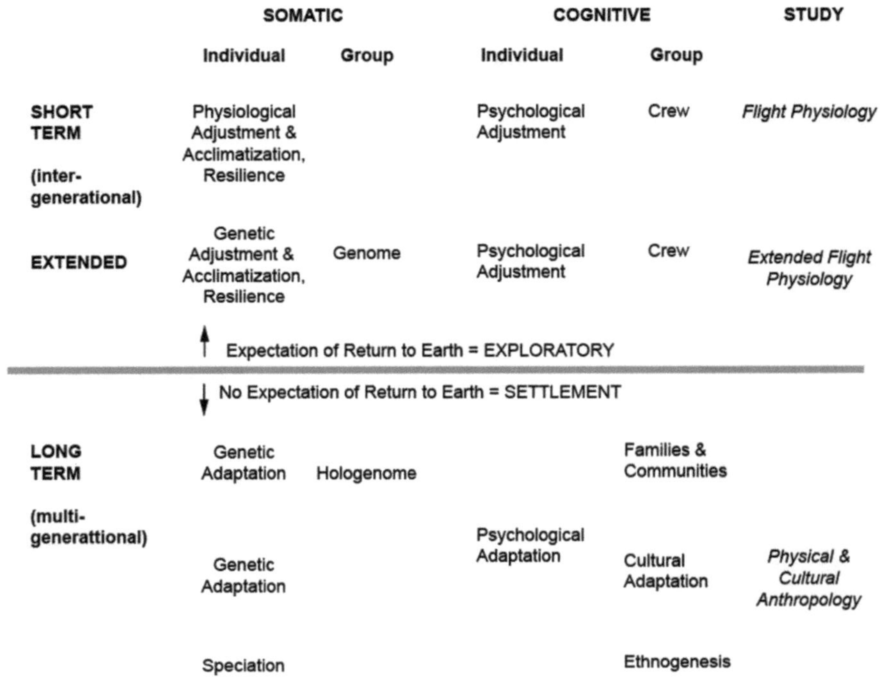

Fig. 1.13 Space anthropology as distinct from flight physiology. Diagram by the author

1.6.1 Prescription and Prediction in Space Settlement Planning

In this book I do not prescribe or predict in great detail for the same reasons that anthropologists and evolutionary biologists refrain from detailed prescription or prediction. First, so much of the evolution of evolutionary entities, such as species or cultures, results from impossible-to-predict factors (*evolutionary contingencies* such as sudden extinction events that knock out otherwise well-adapted species) that it is nearly impossible to make accurate predictions. Second, if such predictions could not reasonably be thought of as accurate, they should not be used in serious planning, and I feel basing details of a policy or plan for space settlement on something as historically inaccurate as predictions about the future would be foolish. Finally, I think it is a reasonable expectation that the people who settle space will not be living out a "master program" written out generations before by relative strangers—they will be the space settlers themselves, people living out their lives and making adjustments to how they live, at every level from daily routines to economic policies. They will make the rules of space settlement, not us. However, in this generation, we can give those eventual space settlement planners the best information possible on which to base the deliberations that will shape their own space-settling cultures. That is really the intent of this book, and the work it represents; the establishment of a way of studying and to a degree planning human space settlement, based on our best understanding of evolutionary adaptation, to make such eventual plans better-informed than they might otherwise be.

For these reasons, while in this book I do comment on such interesting possibilities as cultural change on interstellar arks over some centuries, my main focus is on (a) the first few generations of a given space settlement scenario and (b) the settlement of Mars or orbital/Lagrangian colonies near (within a month or so, travelwise) to Earth. This is because, as I illustrate throughout the book, it is the first few generations of space settlers who will absorb the biological and cultural 'shocks' of settlement. Based on what we know of biocultural adaptation, we shall see that after a century or so the space-settling genomes and cultures will be sufficiently adapted to beyond-Earth conditions to ensure a good likelihood of multigenerational success. Cultures will also have established new words and concepts, and new routines of life, scultping the human experience also to these new conditions.

Finally, I do not rely on advanced technological solutions to the difficulties of space settlement as they are currently understood. Certainly some will be invented and change our priorities, but we must not expect that problems will be solved somehow, someday, by someone, *aside from* a general expectation that technologies will be invented that allow basic human survival in environments beyond Earth (which, I would argue, is an indispensable assumption for this book and its proposed field). On example is the use of exotic technologies, such as human hibernation, to overcome some of the challenges of a voyage between Mars and Earth. While such methods may have merit, they are not yet even in principle proven to be worthy of consideration. I also avoid 'use' of cultural behaviour that would violate basic human cultural traditions. While this book is about how humanity will change as it settles space, there are boundaries to that malleability that I think should be explored by space settlers themselves, rather than us planners in this generation. The reason for my reticence here is that I prefer to propose human adaptive behaviors that are already known to work, or have worked on Earth. Staying somewhat within these bounds is I think the safest way to proceed at present, by suggesting actions that are known and not alien to our cultural traditions. An example that often comes up in my work is the concept of sending not thousands of colonists to an exoplanet, but just a handful of crew accompanied by a large bank of preserved sperm- and egg-cells, to be used in establishing a new civilization. My argument against this is that the cultural shock of beginning with just a small crew would be so large that it might now work as a foundation. Another example comes from a recent suggestion that a crew of less than 500 people could genetically survive a one-way voyage of several thousand years to an exoplanet; but we have no known agricultural village culture that has persisted for so many years in such a small population without any contact from outside. Wherever possible I suggest we plan, at this stage, with reasonable expectations of moderate technologies in the near future, and with cultural adjustments for space settlement that do not create entirely new multigenerational structures or arrangements and problems. Rather, I suggest we remain closer to what has worked for humanity for some time, known and repeatedly-proven patterns of behaviour that may be slightly adjusted for the project of permanent human space settlement.

With the objectives, contexts and thoughtscape of this book now outlined, we may go on to the specific examination of our topic.

1.7 Summary Points

- Permanent human space settlement will be the natural continuation of the ancient trend of human dispersal
- Permanent human space settlement will require multigenerational success of both human biology and culture
- Multigenerational biocultural success can be aided by an evolutionary and adaptive approach
- Human biocultural adaptation is best-known to the academic field of anthropology, the scientific study of our species
- In the same way that bioastronautics and astrobiology were established when they were required in the first space age, the establishment of space anthropology will be required for the space-settlement purposes of the current Second Space Age
- A productive space anthropology may be established as a space science founded in evolutionary and adaptive dynamics
- Human space settlement will require protection of not just humanity but also our hologenome and the genomes of our domesticates
- Human space settlement will safeguard our species by making it more numerous, diverse and widespread
- Humanity's future may be better secured, or extended in time, by technologies that allow rapid extension across space
- Human space settlement is not a luxury but a responsibility
- Human space settlement will be an event of evolutionary significance for all Earth life, the extension and ultimate independence of evolving matter from beyond its originating planet.

References

Alland, A. (1970). *Adaptation and cultural evolution: An approach to medical anthropology*. New York: Columbia University Press.

Billings, L. (2018). Should humans colonize other planets? No. *Theology and Science, 15*(3), 321–332. https://doi.org/10.1080/14746700.2017.1335065.

Bonvillain, N. (2006). *Cultural anthropology*. New Jersey: Pearson.

Cachel, S. (2015). *Fossil primates*. Cambridge: Cambridge University Press.

Chancellor, J. C., et al. (2014). Space radiation: The number one risk to astronaut health beyond low Earth orbit. *Life, 4,* 491–510. https://doi.org/10.3390/life4030491.

Cheremisin, A. A., Vassilyev, Y. V., & Horvath, H. (2005). Gravito-photophoresis and aerosol stratification in the atmosphere. *Journal of Aerosol Science, 36*(11), 1277–1299.

Childe, V. G. (1950). The urban revolution. *Town Planning Review, 21,* 3–17.

Costello, E. K., et al. (2009). Bacterial community variation in human body habitats across space and time. *Science, 326,* 1694–1697.

Daniel, A. S., Schmidt, J., & Hughes, J. M. (2013). Indirect estimates of natal dispersal distance from genetic data in a stream-dwelling fish (Mogurnda adspersa). *Journal of Heredity, 104*(6), 779–790.

Day, M. H. (1986). *Guide to fossil man*. Chicago: Chicago University Press.

Durham, W. H. (1991). *Coevolution: Genes, culture and human diversity*. Stanford: Stanford University Press.

Edgerton, R. B. (1992). *Sick societies: Challenging the myth of primitive harmony*. New York: Free Press.

Elmer, K. R., et al. (2009). Pleistocene desiccation in East Africa bottlenecked but did not extirpate the adaptive radiation of Lake Victoria Haplochromine Cichlid Fishes. *Proceedings of the National Academy of Sciences (USA), 106*(32), 13404–13409.

EPA. (2018). *Radiation sources and doses*. United States Environmental Protection Agency. See https://www.epa.gov/radiation/radiation-sources-and-doses.

European Space Agency. (2018). *What are lagrange points?* See http://www.esa.int/Our_Activities/Space_Science/What_are_Lagrange_points.

Fennnell, C. C. (2007). Crossroads and cosmologies: Diasporas and ethnogenesis in the new world. *African Diaspora Archaeology Newsletter, 10*(3), 16.

Ferraro, G. (2006). *Cultural anthropology: An applied perspective* (6th ed.). Belmont, California: Thompson/Wadsworth.

Findlay, B. J. (2002). Global dispersal of free-living microbial eukaryote species. *Science, 296*(5570), 1061–1063.

Finney, B. (1992). *Space migrations: Anthropology and the humanization of space*.

Finney, B., & Jones, E. (Eds.). (1985). *Interstellar migration and the human experience*. Berkeley: University of California Press.

Fried, M. N., & Fried, M. H. (1980). *Transitions: Four rituals in eight cultures*. New York: W.W. Norton & Co. My paraphrasing of the Tikopia dirge and mention of the !San marking at the time of first killing a large male animal are derived from this work.

Frisancho, A. R. (1993). *Human adaptation and accomodation*. Ann Arbor: University of Michigan Press.

Gerathewohl, S. (1963). *Principles of bioastronautics*. New Jersey: Prentice-Hall.

Gotthard, K., & Nylin, S. (1995). adaptive plasticity and plasticity as an adaptation: A selective review of plasticity in animal morphology and life history. *Oikos, 74*(1), 3–17.

Hassler, D. M., et al. (2013). Mars' surface radiation environment measured with the mars science laboratory's curiosity rover. *Science, 343*(6169), 1244797. https://doi.org/10.1126/science.1244797.

Hein, A., Smith, C. M., & Marin, F. (2019). Worldships: A review. *Acta Future: European Space Agency Advanced Concepts Office*. In press.

Henry, T. (1985 [1928]). *Ancient tahiti: Based on material recorded by J. M. Orsmond*. Bernice P. Bishop Museum Bulletin 48, Honolulu.

Jasienska, F., Sherry, D. S., & Holmes, D. J. (Eds.). (2017). *The arc of life: Evolution and health across the life course*. New York: Springer-Verlag.

Kinlan, B. P., & Gaines, S. D. (2003). Dispersal in marine and terrestrial environments: A community peerspective. *Ecology, 84*(8), 2007–2020.

Klein, R. G. (2009). *The human career*. Chicago: University of Chicago Press.

Lang, K. R. (2000). *The Sun from space*. New York: Springer.

Lavenda, R. H., & Schultz, E. A. (2013). *Core concepts in cultural anthropology* (5th ed.). New York: McGraw-Hill.

Love, A. C. (2010). Rethinking the structure of evolutionary theory for an extended synthesis. In M. Pigliucci & G. B. Muller (Eds.), *Evolution: The extended synthesis*. Cambridge: Massachusetts Institute Press.

Maciel, W. J. (2013). *Astrophysics of the interstellar medium*. New York: Springer.

Malina, R. M. (1975). *Growth and development in man*. Minnesota: Burgess Publishing Company.
McShea, D. W., & Brandon, R. N. (2010). *Biology's first law: The tendency for diversity & complexity to increase in evolutionary systems*. Chicago: University of Chicago Press.
Meyer-Vernet, N., & Rospars, J.-P. (2016). Maximum relative speeds of living organisms: Why do bacteria perform as fast as ostriches? *American Journal of Physics, 83*, 719. http://aapt.scitation.org/doi/10.1119/1.4917310.
Moran, E. F. (1979). *Human adaptability: An introduction to ecological anthropology*. North Scituate, Massachusetts: Duxbury Press.
NASA. (2018). *Brief summary of requirements for landers and rovers*. Office of Planetary Protection. See https://planetaryprotection.nasa.gov/requirements.
Nettle, D. (1999). Is the rate of linguistic change constant? *Lingua, 108*, 119–136.
Ohno, S., et al. (2017). The biopause project: Balloon experiments for sampling stratospheric bioaerosol. *Lunar and Planetary Science, XLVIII*, 1890.
Paradis, E., Ballie, S. R., Sutherland, W. J., & Gregory, R. S. (1998). Patterns of natal and breeding dispersal in birds. *Journal of Animal Ecology, 67*, 518–536.
Pasinelli, G., Scheigg, K., & Walters, J. R. (2005). Genetic and environmental influences on natal dispersal distance in a resident bird species. *The American Naturalist, 164*(5), 660–669.
Raup, D. M. (1994). the role of extinction in evolution. *Proceedings of the National Academies of Science of the United States, 91*, 6758–6763.
Reitz, G., Berger, T., & Matthia, D. (2012). Radiation exposure in the moon environment. *Planetary and Space Science, 74*(1), 78–83. https://doi.org/10.1016/j.pss.2012.07.014.
Reyes-Garcia, V., et al. (2016). Multilevel processes and cultural adaptations: Examples from past and present small-scale societies. *Ecology and Society, 21*(4), 2. https://doi.org/10.5751/ES-08561-210402.
Serpell, R., & Boykin, A. W. (1994). Cultural dimensions of cognition: A multiplex, dynamic system of constraints and possibilities. In R. Sternberg (Ed.), *Handbook of perception and cognition II: Thinking and problem solving* (pp. 369–408). San Diego: Academic Press.
Smith, C. M. (2011). *The fact of evolution*. New York: Penguin Random House.
Smith, C. M. (2014a). Estimation of a genetically viable population for multigenerational interstellar voyaging: Review and data for project hyperion. *Acta Astronautica, 97*(2014), 16–29.
Smith, C. M. (2014b). Starship humanity. *Scientific American, 308*, 39–43.
Smith, C. M. (2016). An adaptive paradigm for human space settlement. *Acta Astronautica, 119*, 207–217.
Smith, C. M. (2017). Evolutionary and archaeological perspectives on estimating the likelihood of civilization collapse. *Nanotechnology Perceptions, 13*(2), 116–122.
Smith, C. M. (2018). *An atlas of human prehistory*. San Diego: Cognella Academic Publishing.
Smith, C. M., & Davies, E. T. (2006). The extraterrestrial adaptation. *Spaceflight, 47*, 85.
Smith, C. M., & Davies, E. T. (2012). *Emigrating beyond earth: Human adaptation and: Space colonization*. Springe-Praxis.
Smith, C. M., Gabora, L., & Gardner-O'Kearny, W. (2018). The extended evolutionary synthesis paves the way for a theory of cultural evolution. *Cliodynamics, 9*(2), Winter.
Smith, D. J., Griffin, D. W., McPeters, R. D., Ward, P. D., & Shuerger, A. C. (2009). Microbial survival in the stratosphere and implications for global dispersal. *Aerobiologia, 27*, 319–332.
Space. (1992) *NASA SP-509: Space resources, Volume 4: Social concerns*. Washington, D.C.
Stoner, I. (2017). Humans should not colonize Mars. *Journal of the American Philosophical Association, 3*(3), 334–353. https://doi.org/10.1017/apa.2017.26.
Strickberger, M. W. (1985). *Genetics* (3rd ed.). New York: MacMillan.
Sutherland, G. D., Harestad, A. S., Price, K., & Lertzman, K. P. (2000). Scaling of natal dispersal distances in terrestrial birds and mammals. *Conservation Ecology, 4*(1), 16. http://www.consecol.org/vol4/iss1/art16.
Szathmary, E. (2015). Toward major evolutionary transitions theory 2.0. *Proceedings of the National Academies of Science of the United States, 112*(33), 10104–10111.
Tainter, J. A. (1990). *The collapse of complex societies*. Cambridge: Cambridge University Press.

Taras, V., Rowney, J., & Steel, P. (2009). Half a century of measuring culture: Review of approaches, challenges, and limitations based on the analysis of 121 instruments for quantifying culture. *Journal of International Management, 15,* 357–373. https://doi.org/10.1016/j.intman.2008.08.005.

Trigger, B. (1992). *Understanding early civilizations*. Cambridge: Cambridge University Press.

Tsiolkovski, K. (1933). *Album of space travel* (Kaluga, Tsiolkovsky State Museum of the History of Cosmonautics). See https://commons.wikimedia.org/wiki/File:Tsiolkovsky_Album_55.jpg/.

Vermeij, G. J. (1978). *Biogeography and adaptation*. Cambridge: Harvard University Press.

Vermeij, G. J. (2000). Why are there so few evolutionary transitions between aquatic and terrestrial ecosystems? *Biological Journal of the Linnean Society, 70,* 541–554.

Vermeij, G. J. (2008). Security, unpredictability, and evolution: Policy and the history of life. In R. Sagarin & T. Taylor (Eds.), *Natural security: A darwinian approach to a dangerous world.* Berkeley: University of California Press.

Weiss, H., & Bradley, R. S. (2001). What drives societal collapse? *Science, 26*(291), 609–610.

Wenke, R., & Olszewski, D. I. (2006). *Patterns in prehistory*. Oxford: Oxford University Press.

Wheeler, R. M., & Martin-Brennan, C. (Eds). (2000). Mars greenhouses: Concepts and challenges. In *Proceedings from a 1999 workshop*. Florida: NASA-Kennedy.

White, L. A. (1959). *The evolution of culture*. New York: McGraw-Hill Book Company.

Yang, Y., et al. (2005). *Investigation of cultivable microorganisms in the stratosphere collected by using a balloon in 2005*. Japanese Aerospace Research Association paper available at https://www.researchgate.net/publication/233387559_Investigation_of_cultivable_microorganisms_in_the_stratosphere_collected_by_using_a_balloon_in_2005/.

Chapter 2
Individual Hominin Biology Beyond Earth

Abstract This chapter covers the biology of the individual human (or, more broadly, hominin) in environments beyond Earth; I consider the essential properties of such environments, such as gas composition and pressure, and gravity field, but I must emphasize that decades of biology research in Low Earth Orbit (LEO) have not furnished humanity with significant information regarding the properties of environments beyond Earth we are most likely to settle for some centuries. Mars or lunar settlements will not be at 0 g, as in LEO (where a great deal of space biology research has been done), but at 0.38 g (Mars) or 0.16 g (Moon), and molecular, cellular and larger-scale physiological processes will certainly differ in such environments. What in fact have nearly three decades of continuous biology research in LEO taught us? Something about what we might encounter if we choose to fly to Mars without rotation for artificial gravitation for about 100 days–but, to be blunt, that is about it. As regards the long-term human settlement of space it is arguable that not much learned from microgravity research is directly applicable. But we must begin somewhere—and I will note that in 2016, NASA began to fund artificial gravity research—and in this chapter I review the issues of the individual human beyond Earth, specifically tracking effects on development through the course of life.

> Our bodies have a story–an evolutionary story–that matters immensely…evolution explains why our bodies are the way they are, and thus yields clues on how to avoid getting sick…[and] it isn't over…we are still evolving.—Dan Lieberman

2.1 Introduction

The effects of off-Earth conditions on the biology of the individual organism have been studied for over four decades and are published in a vast corpus of data. Table 2.1 indicates the most common effects of *spaceflight* (largely indicating microgravity with roughly Earthlike atmospheric gas pressure and composition).

While these effects on the individual are known, the focus in this chapter is not so much on these specific effects on the individual, but on the biological reasons for

C. M. Smith, *Principles of Space Anthropology*, Space and Society,
https://doi.org/10.1007/978-3-030-25021-8_2

Table 2.1 Common medical conditions of spaceflight so far observed in the past quarter century

Common and anticipated	Occasional instances	Incidence-prioritized space exploration medical conditions
Space motion sickness	Renal stone formation	Radiation sickness
Nasal/sinus congestion	Acute urinary retention	Severe decompression illness
Constipation	Cardiac dysrhythmia	Barotrauma
Headache	Urinary tract infection	Osteoporosis
Back pain	Gastroenteritis	Seizure
Upper respiratory tract infection	Prostatitis	Anaphylaxis
Minor abrasion	Serous otitis media	Anxiety
Musculo-skeletal trauma	Contact dermatitis	Depression
Corneal irritation	Decompression sickness	Medication overdose/misuse
Insomnia	Near drowning after spacesuit failure	Palliative treatment
	Aspiration of foreign body	Diverticulitis
		Appendicitis
		Sepsis
		Herpes reactivation cellulitis
		Otitis media/externa
		Dental
		– Cavity
		– Pulpitis,
		– Toothache,
		– Avulsion, Loss
		Eye penetration
		Limb amputation (lifesaving)
		Chest trauma/Pneumothorax
		Obstructed airway
		Haemorrhage
		Burns (thermal or chemical)
		Smoke inhalation

Adapted by the Author from Table 3 of Hodkinson et al. (2017)

these effects, and what they can tell us about the likelihood that these effects will shape human adaptation to beyond-Earth conditions over multigenerational time. Therefore I am less interested here in the field of *flight physiology*, which deals with characterizing, understanding and mitigation of medical issues of the individual in short timespans, but rather in the underlying mechanisms of spaceflight symptoms, on levels from the molecule to cell and physiology, and on the timescale specifically of the human life-course, including stages such as infancy, childhood and so on into reproductive age.

In this chapter I summarize what is best known from these data, organized chronologically by the age of the organism, from sex cell formation to the end of the life cycle. I also mention areas that are poorly-known and make recommendations for both future research and space settlement in both the near- and long-term. I deemphasize 'simulated microgravity' research on multicellular life, such as the tail-suspension and leg-suspension experiments carried out on mice; I agree with critiques that they are so disruptive to the normal functions of the organism that effects observed cannot surely be attributed to microgravity conditions; indeed such experimentation is nearly absent from the recent 'International roadmap for artificial gravity research' (Clement 2017).

Many studies of the effects of non-Earth-normal conditions use 'reference species' or 'model species', such as mice, rats and fishes, as proxies of human beings; this is, in part, because these animals have short natural lives, so their many life stages may be examined over a relatively short (to the human) timespan, and in part because they share many developmental genetic patterns with humans (Wohlgemuth and Murashov 1995). These developmental similarities in genetics are critically important and have been discovered, in fact, in the period since human spaceflight really began; they involve the HOX gene family, which I have introduced elsewhere in this way: "there are a few hundred "immortal genes," shared among all animals, that are responsible for the essential shape of a given animal. It used to be thought that, generally, different kinds of animal bodies would need very different genes, which intuitively makes sense. But, as evolutionary biologist Sean B. Carroll has pointed out, all animals "no matter how different in appearance, share several families of genes that regulate major aspects of body pattern" (Smith 2011: 97).

In preparing this chapter I have surveyed the research literature largely (but not exclusively) dating to the last two decades, during which genomic and developmental biology have so broadly improved our understanding of biological systems. This is not to say, of course, that space biology from times prior are all invalid, but only that improvements in understanding are real and significant in the genomic age of biology. This was forseen by microbiologist Carl Woese (1928–1012) who introduced his 2004 essay *A New Biology for a New Century* by writing:

> Science is an endless search for truth. Any representation of reality we develop can be only partial. There is no finality, sometimes no single best representation. There is only deeper understanding, more revealing and enveloping representations. Scientific advance, then, is a succession of newer representations superseding older ones, either because an older one has run its course and is no longer a reliable guide for a field or because the newer one is more powerful, encompassing, and productive than its predecessor(s). (Woese 2004: 173)

Indeed, the research article *Mechanisms of Gravity Effects on Cells: Are There Gravity-Sensitive Windows?* published in 1992, asks questions about cellular structure and processes that are radically better-understood with a modern understanding of the cellular cytoskeleton and microtubule architecture, a field in its infancy in the early 1990s. Today the cellular cytoskeleton is known to be actively reshaped in response to microgravity exposure (Louis et al. 2015), a process controlled on the molecular level of signalling proteins that regulate intercellular processes.

The literature I have searched in this period has appeared in many journals, but few books. The main journals I have searched have been *Gravitational and Space Research*, *Microgravity*, *Microgravity Science and Technology*, *Cell*, *Astrobiology*, *Life Sciences in Space Research*, *Advances in Space Biology and Medicine*, the *Biophysical Journal* and *Biological Sciences in Space*. The bulk of studies in these journals address the effects of *microgravity* (sometimes denoted μg in the literature) on living systems, mostly in Low Earth Orbit conditions aboard various space stations. *Hypergravity* (>1 g) is also studied, on occasion, but much less so than microgravity. Studies in microgravity have been carried out to better understand living systems once they are relieved of much of the Earth's gravity field, and hypergravity to better understand the effect of increased g load on living systems accelerating to leave a planetary body (normally taking less than half an hour) or decelerating when one is entering a planetary body gravity field (a similar timeline). I comments on this at the end of this chapter.

Whatever the gravity field, and other environmental variables, of a given beyond-Earth environment (see Chap. 1), these will be novel to the essential life course of the human body at all its stages of development and maturation. Figure 2.1 illustrates the timing of the main human life-course stages and some differences in health risks between individuals of the male and female sex. These chronological and biological developmental stages, however occasionally fuzzily-bordered, are important organizing units for considering human futures beyond Earth. For example, note that in both the male and female sex, health risks are earliest developmental-genetic, then shift to communicable disease during early socialization, when children interact more often, then include more violence in male adolescence and more risks to women during pregnancy. Depending on the demographic plans and outcomes, and radiation concerns of a given habitat beyond Earth, pregnancy might well be more carefully scheduled, and more resources put into ensuring the health of the mother and offspring, than here on Earth. In the same way, while males remain reproductively viable normally for some decades after females, increased radiation damage to male sex cells, simply as a function of time, might demand a more formalized cessation of childbearing in males, beyond Earth. The point here is not the specifics, but the recognition of these life stages and examination of how they may be protected with careful, adaptive planning.

Figure 2.2, for example, considers just the development and growth of a human infant in a Mars gravity field. Based on current understanding of two variables—*locomotion* (habitually and obligately bipedal in humans) and *instinctual physical actions* (conditioned by the phenomenon of Central Pattern Generation)—it is entirely plausible that significant (though superficially small) differences between Earth and Mars

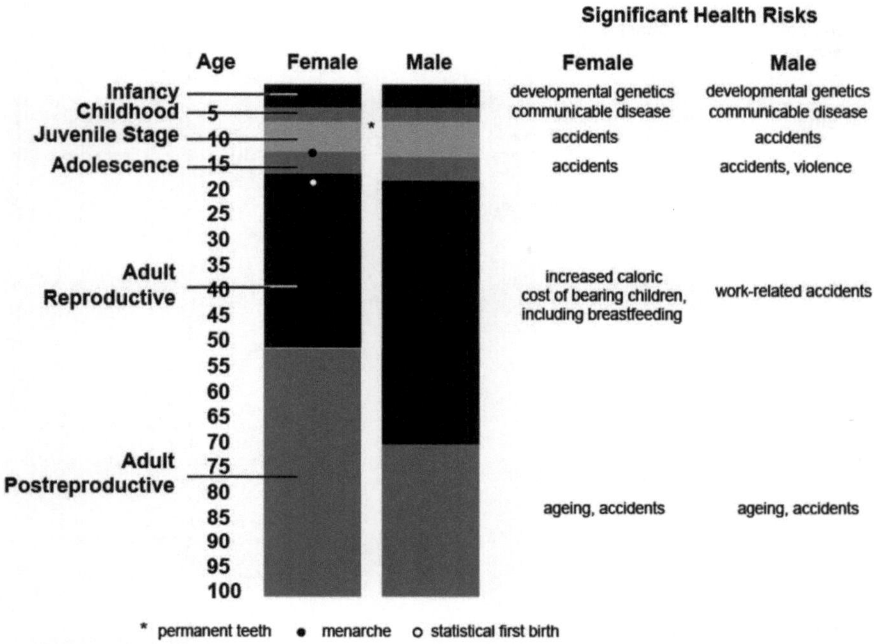

Fig. 2.1 Human biological life stages. Derived from Bogin and Smith (1996) and data from the *US Health Resources and Services Administration*. Data are aggregate and do not reflect ethnic, income-level or other variables. Section 3.5, in the next chapter of this book, details some of the genetic health risks in the earliest stages of life

environments will have effects on the life course, in behaviour and anatomy, of the *phenotype* (the born, developed individual). If Mars conditions, in this case, drive small, reversible adjustments of the body (*acclimatizations*) we would not, biologically speaking, call this structural change or adaptation (though technically this discounts the reality that the person will unlikely encounter Earth conditions and 'revert' to an Earth phenotype). If the effects are strong enough to cause selection in the population such that some individuals are less likely to pass on such altered traits to the next generation, then we may say that biologically significant adaptation is underway; but it might well then be interrupted by cultural and technological methods, complicating any answer to the common question "what will adaptation be like on Mars?" If Mars-driven traits become widespread or *fixed* in a population, we may say that adaptation is well underway; and if such adaptations become structurally significant, such that Earth-born and Mars-born mated individuals could not have biologically fertile offspring of their own, then we would be able to say that *speciation* has occurred.

A catalog of what these specific effects will be, however, remains beyond our grasp—and considering the complexity of biological systems undergoing adaptation, will likely remain so for generations, even as space settlement is commenced. A 2014 review of radiation- and microgravity-biology research since 1962 indicated that,

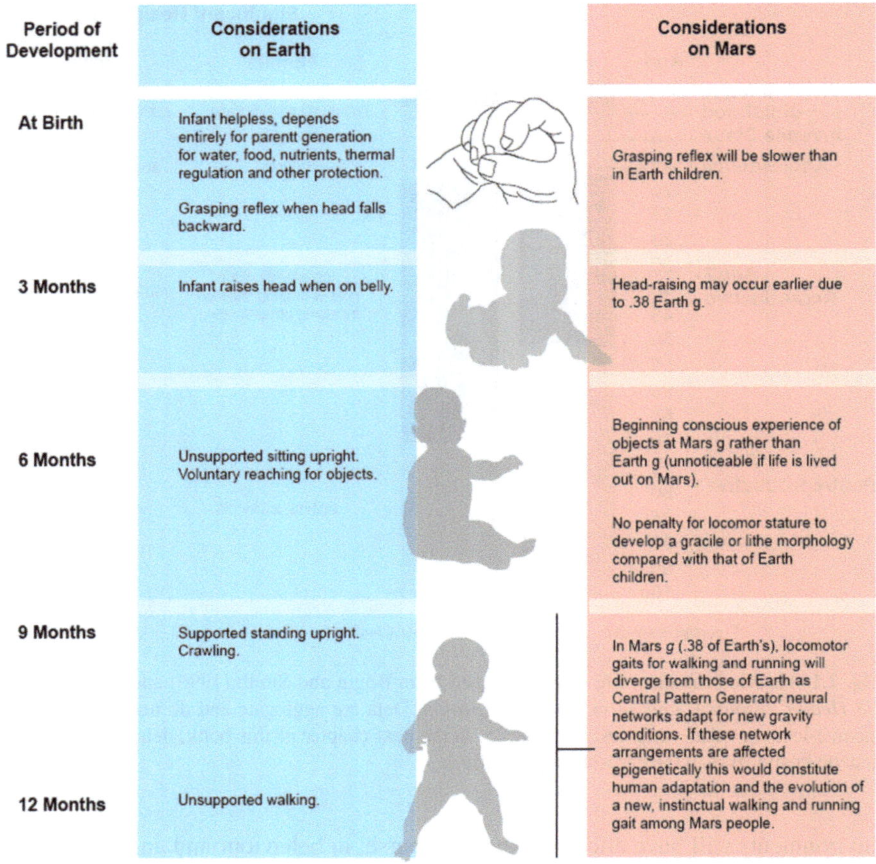

Fig. 2.2 Physical considerations in the first 12 months of life on Earth and Mars specific to locomotion and central pattern generation. Immediate considerations seen here will have multigenerational effects discussed in the text; such effects, when detected, might well be offset by technological and behavioural methods, muddying any answer to the common question '*how will we adapt to conditions on Mars?*'

broadly speaking, issues of causes of observed effects "…remain not well understood [and that] further improvements in space experimental technologies are required for future studies" (Yatagai and Ishioka 2014: 76). Considering this state of affairs after 40 years of research analysis regarding microgravity conditions, it seems that the effects of the Mars environment—0.38 g, not 0 g—are unlikely to be known in detail any time soon. Rather, I suspect they will be learned in situ on Mars itself. What we can do at present is (a) survey the literature on the effects of some variables important to development and growth on the human body and behaviour and from this (b) make some recommendations for biocultural adaptation to such conditions. These recommendations serve one of the functions of Space Anthropology as established in Chap. 1, namely "*To make recommendations, some broad and some specific, that would assist in human adaptation to environments beyond Earth*".

2.2 Beyond-Earth Mammalian Reproduction

Mammalian, and in particular, human, biological reproduction beyond Earth will be required of course for multigenerational success. The issues of reproduction have not been extensively researched in microgravity, but they have been sketched out in broad principles. In this chapter the main stages of reproduction, as part of a larger cycle mammalian life, are detailed, highlighting (a) their essential features, (b) research done so far and (c) recommendations for supporting the healthy continuation of these in non-Earth environments, using both the adaptive tools of biological evolution and cultural modifications of behaviour in a synthetic effort of *biocultural adaptation.*

Note that effects of 'space' on development is a misnomer, as off-Earth conditions are found in many variables, chiefly gravity field, atmospheric pressure, atmospheric composition, and temperature range. In this way we must distinguish between microgravity as experienced in Low Earth Orbit and reduced gravity as we would find on Mars, at 0.38 that of Earth's 1.0 g field. This chapter focuses on the effects of a gravity field different from that of the Earth's surface 1 g field which has conditioned many variables of Earth life for some billions of years.

Similarly, what are the effects of these conditions on animal life? They may be measured in many ways. One thing to consider is their longevity; some are *transient*, some are *permanent* and some are *reversible*, such that after alleviation of microgravity conditions, the body or process returns to an Earth-normal state. Such subtleties highlight the fact that blanket statements such as 'humanity cannot settle space' are over-generalized. It is precisely by understanding the subtleties that we will identify solutions to adapting to environments beyond Earth.

2.2.1 Spermatogenesis and Sperm Health

Mammalian spermatogenesis, or sex cell generation, begins 'at puberty after a long preparatory period of *prespermatogenesis* in the fetus and the infant" (Holstein et al. 2003); these early periods are so poorly-known that they cannot be discussed usefully here (Di Agostinio et al. 2004), but it is important to note that pre-puberty sex cell generation in the male must be considered in conditions beyond Earth-normal. A normal time for generation of mammalian sperm is about 90 days and is controlled largely by a small number of genes (Yamauchi et al. 2016). Development may go awry, producing structural and mechanical abnormalities (e.g. malformed sperm cell tails) and such abnormalities as low sperm counts (less than c. 20 million after 72 h' sexual abstinence). One study of sperm morphology and performance measured the following important variables, indicating some of the chief anatomical

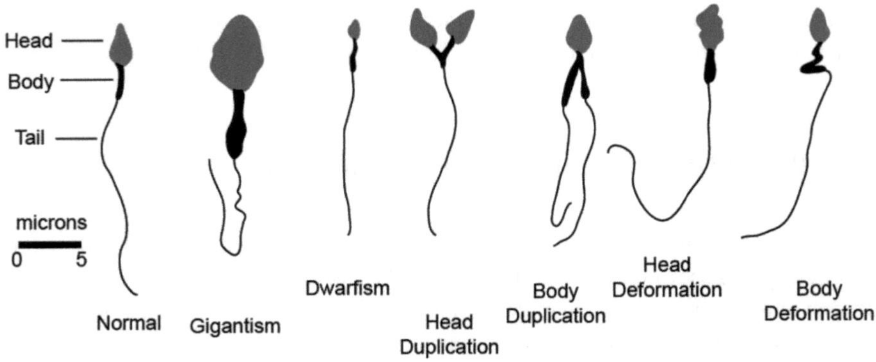

Fig. 2.3 Healthy human sperm cells (left) and Sperm cells exhibiting physical abnormalities (R). Diagram by the author

features of these sex cells: *total sperm length, head length, midpiece length, flagellum* (tail) *length* and *straight line velocity* (Tourmente et al. 2011). Figure 2.3 illustrates healthy human sperm cells and some of the most common sperm cell abnormalities.

The effect of 30 days of simulated microgravity on sperm cell generation has been studied, the authors concluding that "Testicular weights and spermatogenesis were significantly reduced…such that no spermatogenic cells beyond round spermatids were present and epididymides were devoid of mature sperm. In many tubules, loss of all germ cells, except a few spermatogonia…was observed…[demonstrating that spermatogenesis] is severely inhibited by long-term simulated microgravity, whereas testicular androgen production is not. These results have significant implications regarding serious effects of long-term exposure to microG on the reproductive capability of scrotal mammals, including humans" (Tash et al. 2002: 1191). Note that I emphasize the word simulated in this quotation, and that simulation by hindlimb suspension or other methods has yet to be shown to be significantly analogous to actual microgravity conditions.

The effect of actual (rather than simulated) microgravity on the male gamete and its production have been studied, and several observations are worth noting here. In the mouse, one study as shown that the process of sex cell generation (meiosis) is not negatively affected by simulated microgravity, despite reduced testis size; in fact in that case entry into meiosis seems to have been promoted by microgravity (Pellegrini et al. 2010). Also, both mammalian testosterone levels and sperm production have been documented to increase in microgravity (Ricci et al. 2004), and sperm production is not particularly inhibited by microgravity conditions (Serova 1989). The produced sperm have been observed to swim at higher speed than on Earth (Englemann et al. 1992; Tash and Bracho 1999), but what specific effects this has on fertilization is as yet poorly-known. It is known that sperm cell morphology can effect the likelihood of the cell being able to fertilize the egg (Ombelet et al. 1995), and that cell morphology is supported by the cytoskeleton, which itself can be altered by microgravity (Crawford-Young 2006); but actual spermatogenesis has not been studied in such conditions and the topic is at present poorly-understood.

The viability of developed sperm cells to LEO-level radiation has been observed. Mouse spermatozoa were exposed the radiation flux of LEO for nine months; on return to Earth the sperm cells, which had been subjected to c. 0.03 Msv/day (about 10 times the Earth surface normal) "could not be distinguished from that of the ground control samples, at least at the light microscopy level…[S]ome spermatozoa showed breakage between the head and tail or fragmented tails, but this is typical for freeze-dried spermatozoa…" and following this examination, "the spermatozoa were used for intracytoplasmic sperm injection (ICSI) of fresh oocytes…[resulting in]fertilized and formed normal-appearing pronuclei irrespective of mouse strain [control—vs.—space-exposed]…" (Wakayama et al. 2017: 5990). Furthermore, the offspring of the space-exposed spermatozoa were healthy and were themselves mated, and their off-spring proved also to be healthy and viable; so far as this is a gauge for human reproduction, exposure to microgravity and the radiation flux in LEO over about seven months seem reasonable. Another experiment exposed Medaka fish *(Oryzias latipes)* to 'beyond Low-Earth Orbit' radiation flux; specifically, early spermatogonia, immature spermatids and mature sperm were subjected to HZE (high energy ions) simulating GCR (galactic cosmic radiation) somewhat beyond Earth, the result being that while mutation rate increased with HZE dose, the number of "viable mutations from all exposed germ cell stages are very low and would not be expected to yield significant numbers of mutations above background levels…the hazard to male [sperm cells and their production to GCR beyond LEO over a short time] is probably temporary sterility but not significant numbers of mutagenic effects observable in progeny." (Shimada et al. 2005: 6066).

In sum we may say that some aspects of sperm health appear to be conditioned by the gravity field, but also that vertebrate sperm can develop effectively healthily even in microgravity. Sperm development (meiosis) is an important and delicate phase, but so far exposure to microgravity has not been seen to statistically significantly alter sperm cell development. At this writing (in the last quarter of 2018) frozen human sperm have been sent to the International Space Station for further study regarding viability in the *Micro-11* experiment.

2.2.2 Oogenesis and Egg Health

In mammals, many female gametes (*oocytes*) are generated in the fetal stage, and it has long been believed that these formed the life supply of the individual; however recent research strongly suggests that in some mammals (including humans) the individual can continue to grow eggs after birth due to ovarian germline stem cells (Dunlop et al. 2014). If this phenomenon is confirmed it would extend the period of greatest susceptibility of oocyte *development* from just the fetal period some years into, presumably, sexual maturity and beyond, a significant development of obvious importance to space settlement.

The development of oocytes in the fruit fly (*Drosophila melanogaster*) have been shown to be effected by exposure to microgravity, including (a) increase in oocyte production rate and size of oocyte and (b) yolk differences from those of Earth-conditions oocytes, though the incidence of these was rather low and the authors conclude that "No significant accumulation of lethal mutations in any of the experimental conditions was detected as measured through the male to female ratio in the descendant generation. It seems that *Drosophila melanogaster* flies are able to sense and respond to the absence of gravity, changing several developmental processes even in very short space flights. The results suggest an interference with the distribution and/or deposition of the maternal components involved in the specification of the anteroposterior axis of the embryo" (Vernós et al. 1989: 213). A more recent set of experiments observed multiple generations of *Drosophila* oocyte formation in 1 g and space-station microgravity conditions, specifically studying gene activity regarding cytoskeletal proteins, which give the egg its essential shape. The complex results of the experiments to date include a better understanding of egg-formation genetics, and in sum concluded that:

> exposure to weightless conditions [...] increased the transcription of metabolic genes and cuticle components and decreased the transcription of genes involved in morphogenesis, cell differentiation, cytoskeletal organization and genes associated with the plasma membrane...[but that these did not so disrupt oocyte function significantly, and that in sum] The preservation of fertility implies that human exploration of other planets (such as Mars) and outer space, with alternating stays under weightless conditions and different gravities, may be possible in terms of species preservation. (Ogneva et al. 2016: 18–20)

Among vertebrates, however, oogenesis has not been observed to be significantly affected by microgravity in some fish, toads and mice (Tan 2006), although *simulated* microgravity is claimed to increase the frequency of 'cytoplasmic blebbing'—a deformity of the cell wall—in mouse oocytes (Wu et al. 2011). The effect of *hypergravity* (greater than 1 g) on egg cell development has been observed to be detrimental (Ning et al. 2015), indicating at least that gravitation fields can effect such development; this is largely unimportant for concepts of space settlement unless hypergravity is endured during the period of egg cell development itself (for many cells, during the fetal period, but as noted above, this may extend into adulthood), as in high-g launches from Earth and high-g landings on, say, Mars. Egg development in the ovarian tissues of mice have been observe in actual microgravity, resulting in the observation that "no gross morphological changes occur with regard to follicular development and the formation of *corpora lutea* in the ovarian tissue of mice exposed to the spaceflight environment for a period of 13 days or less." (Smith and Forsman 2012) (Fig. 2.4).

The viability and health of developed oocytes has not been significantly studied in microgravity conditions, but in Terrestrial conditions oocytes are known to age after the ideal period of fertilization, which in humans is within 24 h of ovulation. Unfertilized oocytes "undergo a...deterioration [including] decreased fertilization rates...chromosomal abnormalities...and hardening of the [ovum cell wall]..." (Miao et al. 2009). Such, and other, results of aging eggs result in lower fertility rates of the eggs themselves and increased rates of malformation of the embryo if they are

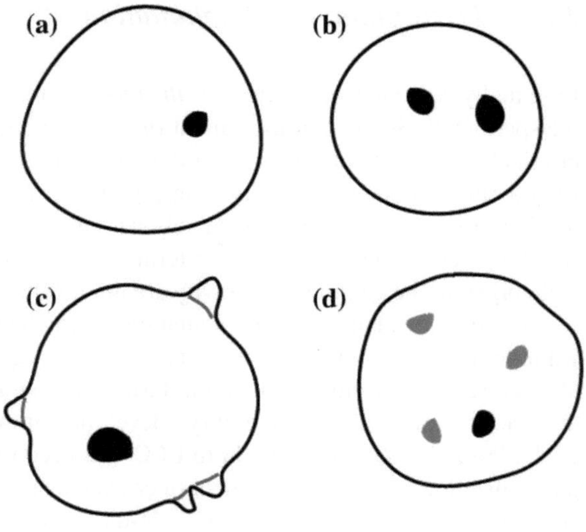

Fig. 2.4 Schematic diagram of human oocytes. **a** healthy egg cell, **b** egg cell with two nuclei (normally fail to develop normally), **c** egg cell exhibiting non-cell-death-related blebbing (cell wall protrusions) observed in some microgravity experiments, **d** egg cell exhibiting excess and large lipofuscin bodies (gray blobs), associated with severe developmental abnormalities. Diagram by the author. Figure adapted by the author from Smith and Forsman (2012)

fertilized. Clearly, if oocyte ageing is somehow accelerated in off-Earth conditions, it will be important to understand. Oocytes of African clawed frogs have been subjected to microgravity in parabolic flights, revealing that transmembrane ion currents are susceptible to gravity field alteration from Earth-normal, though to what effect is still unknown (Richard et al. 2012). Frogs of the genus *Xenopus* were observed in the microgravity of a space shuttle mission; the eggs developed normally, ovulated (being released from ovaries) normally (Black et al. 1996), and produced healthy and sexually-viable offspring after the eggs were artificially fertilized (for practical reasons), though the offspring had some developmental variations from ground control animals. Fruit fly (*Drosophila*) oogenesis was observed on another space shuttle flight, the authors concluding that

> No significant accumulation of lethal mutations in any of the experimental conditions was detected as measured through the male to female ratio in the descendant generation. It seems that Drosophila melanogaster flies are able to sense and respond to the absence of gravity, changing several developmental processes even in very short space flights. The results suggest an interference with the distribution and/or deposition of the maternal components involved in the specification of the anterioposterior axis of the embryo. (Vernós et al. 1989: 213)

Overall we may say that some oocytic abnormalities have been observed in microgravity conditions, and egg cells are generally susceptible to gravity effects on their development and overall health; but in some cases egg development has proceeded in a healthy way in non-Earth gravity conditions. Processes such as ovarian follicle development and ovulation, while apparently affected by gravity field differences from that of the surface of the Earth, are generally successfully achieved by vertebrates in microgravity.

2.2.3 *Fertilization and Implantation*

In sexually-reproducing species, *fertilization* refers to the penetration of the egg by the sperm cell. Sperm cell movement (*motility*) is required to reach the ovum, and physical differences in sperm cell morphology (e.g. size or size of different parts, such as the midbody or the tail) can give some sperm advantages over others in reaching the ovum, which immediately hardens its outer shell, the *zona pellucida*, which prevents other sperm from entering. Details of the sperm's journey from the male organ to the egg in microgravity are unknown.

One study of mammalian implantation in simulated microgravity indicated that while fertilization could occur well enough in microgravity conditions, live birthrate of mice resulting from implantation during microgravity was reduced, suggesting that "normal preimplantation embryo development might require 1 g" (Wakayama et al. 2009). A 2005 Soyuz flight to LEO was used to observe microgravity effects on fertilization and the development of cricket neurons (which are particularly large and easy to observe in insects); an eight-day exposure to these conditions showed that fertilization proceeded normally in microgravity, resulted in healthy offspring with neural architecture statistically identical to control individuals fertilized and allowed to develop in 1 g conditions on the ground, but that microgravity-fertilized eggs hatched significantly earlier than the ground control population; why has not yet been identified (Kirschnick et al. 2006). Fertilization of salamander eggs in microgravity has been observed to occur in about 56% of opportunities, slightly higher than in salamanders on Earth (Aimar et al. 2000). Space shuttle experiments in the early 1990s demonstrated the same pattern, an 88% amphibian fertilization rate in microgravity compared to 73% in Earth-normal 1 g conditions (Souza et al. 1995).

A review of animal fertilization experiments in microgravity reveals that successful fertilization has been achieved in Mekada fish (which subsequently produced healthy offspring), the fruit fly, nematodes (which generated thousands of offspring), rats, tropical freshwater snails, African clawed frogs, sea urchins, salamanders, newts and crickets (Dayanandan 2011). There does not appear to be a strong effect of microgravity on this phase of sexually-reproducing life, with sperm cells apparently entirely able to move effectively in microgravity, and find and fertilize the female sex cell.

2.3 Beyond-Earth Mammalian Growth

2.3.1 *Gastrulation*

Gastrulation, a three-day period in the human, results in (a) cellular differentiation including the origins of skin and nerves (*ectoderm* cells), gut and other organs (*endoderm* cells) and muscle, bone and heart tissues (*mesoderm* cells), and (b) spatial organization of cells, establishing the essential geometrical axes of the body (ventral-dorsal and vertical). It is a period of such significance for the organism that developmental biologist Lewis Wolpert (b.1929) has remarked ""It is not birth, marriage, or death, but gastrulation which is truly the most important time in your life" (Wolpert 2008: 12). Figure 2.5 indicates the zygote, blastula and final gastrula irrespective of the chronological timeline, which varies by species; in humans the development from zygote to gastrula takes about two weeks.

Fertilized salamander eggs have been observed to gastrulate abnormally in microgravity condition, where they are larger and less numerous than those forming in the 1 g gravity field of the Earth's surface (Clement 2006; Danilchik and Savage 1994). In some fish and newts, however, there was little to no effect of microgravity on gastrulation (Ijiri 2003). Late-stage African clawed frog gastrulation (after about six days of the process) has been subjected to the observed in real microgravity, resulting in "considerable derangement of structure…disproportionately long tails and short bodies, a failure to inflate their lungs, and a reduced, misshapen branchial (anterior limb) apparatus…the sudden onset of [microgravity, on entering LEO] may have caused or contributed to these irregularities…. If the process of cell migration is sensitive to gravitational cues, possibly mediated through the cytoskeleton…then this stage of development may be especially vulnerable to [microgravity]" (Olson et al. 2010: 167).

As far as it has been observed for mammals and close model species, gastrulation is sometimes seen to proceed normally, while other times it has been interrupted, resulting in disorder. In humans, this three-day period, typically commencing about 17 days after fertilization, may be particularly sensitive to conditions beyond Earth. One social response to such sensitivity may be to carefully track the reproductive timeline to give the pregnant mother particularly ideal life conditions at this time;

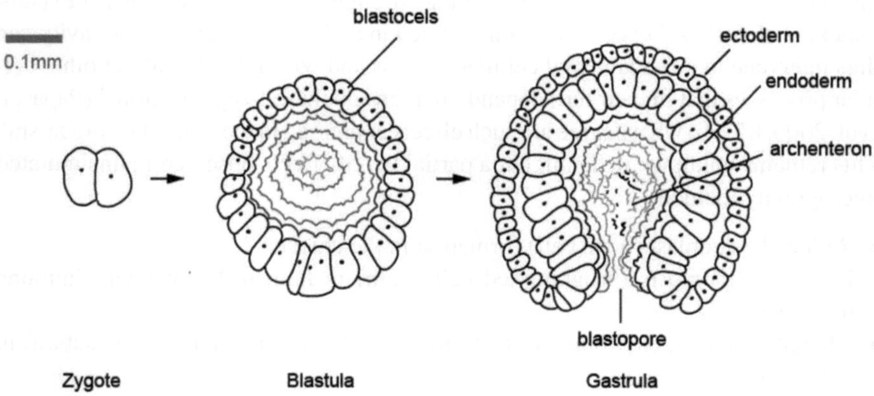

Fig. 2.5 Phases of vertebrate gastrulation. Diagram by the Author. The zygote is the fertilized egg, here undergoing first division. The blastula is the hollow ball of cells that further differentiates into the *ectoderm* cells (which will develop into skin cells), *endoderm* cells (which will develop into tissues and structures inside the ectoderm) and structurally forming also the *archenteron*, a cavity rudiment of the digestive anatomy; note the 0.1 mm scale, about the thickness of a human hair. Human gastrulation begins in the third week

this would require a close understanding of, and attention to, the c. 30-day period of early development leading to fetus formation.

2.3.2 Embryogenesis

The process of embryogenesis includes cellular *differentiation* in which cells take different positions relative to where they were formed (for example, forming a spinal cord foundation) and different functions according to three basic cell types: *stem* (the source of other types of cells), *germ* (egg and sperm cells) and *somatic* (cells that form tissues). The process can be affected by environmental stimuli, but is largely genetically regulated, some of the regulation from well-known gene families including several hundred of the homeobox gene family (Duboule 1995).

Vertebrate embryos assembled by these genes appear to be somewhat and sometimes variably effected by gravity fields significantly different from those of the Earth's surface; in one experiment vertebrate embryos were shown to be lethally damaged by mechanical deformation under increased-g conditions, the researchers concluding that: "…preimplantation embryos sense shear stress, chronic shear stress is lethal, and the *zona pellucida* [embryo wall] lessens the lethal and sublethal effects of shear stress." (Xie et al. 2006: 45).

A 2016 review of microgravity effects on embryogenesis might suggest the reason for this demonstrated susceptibility; it indicated that *microtubules*, tubular polymer 'scaffolds' that form physical structures inside cells (and provide cellular compression resistance), are (as we might expect) gravity-sensitive, resulting in a range of effects including random rather than organized inter-cellular microtubule assembly and arrangement which reduced cellular strength (compression resistance) (Janmaleki et al. 2016). This effect is summarized in a 2000 statement that "Gravity can thus intervene in a fundamental cellular process and will indirectly affect other cellular processes that in their turn depend on microtubule self-organization." (Papaseit et al. 2000: 87). To what degree are such effects expressed in the individual organism? This remains unknown in detail, but a partial list of effects observed in unsimulated microgravity include:

- Reduced osteoblast (bone cell) formation in the chicken
- Increased *apoptosis* (programmed cell death) in human T-lyphocyte (immune response) cells
- Altered (from 1 g environment) tylcholine receptors (related to muscle action) in African frogs
- Disoriented microtubule arrangements in human breast cancer cells
- Altered (from 1 g environment) inner-ear hair form in rat pups.

More recently it has been reported that growth abnormalities (compared to Earth-conditions) "in the proliferation rate of mouse embryonic stem (ES) cells cultured in [simulated microgravity] were shown to be transient, subsiding after the 2nd day, with no difference in cell cycle kinetics compared to 1 g controls" (Wang et al. 2011). Still, other effects of microgravity have been more lasting; among amphibians, abnormal cellular division by about day 5 has been noted (resulting in abnormalities of tissue thicknesses), and among avians there has been a high mortality rate among eggs likely due to "yolk failure to float [properly] in albumen" (Finn et al. 2017) leading to mechanical distancing between nutrient sources and the fetus itself.

We may say that while microgravity does not essentially or unequivocally impede embryogenesis, it does condition a wide variety of embryogenic structures and processes, including intracellular architecture (which can lead to physically weak cells), mutated cell-surface receptors, cell fate scheduling and cellular division through time. The timescales of such effects are not terribly well-understood on the scale of human embryogenesis: of 26 studies of cellular processes during embryogenesis reported in in one review the average time of exposure of an organism's cells to microgravity (in LEO or simulated) is 70.7 h (just under three days), or about 5% of the 192-period (c. eight week) period of embryogenesis (Crawford-Young 2006). Much remains to be learned regarding this critical period, but at present no obvious barriers to successful human embryogenesis have been observed.

Protecting the embryo from non-Earth-normal g and other conditions seems to be a worthwhile cultural mitigation; this would mean, for example, some form of prohibition or limitation of certain activities during embryogenesis, such that women in the first eight weeks of pregnancy might decide not to travel to- or from- space, where altered-g conditions occur. This would not be more culturally disruptive, I would suggest, than current protection of the young is carried out by pregnant women not smoking or drinking alcohol.

2.3.3 Development

The physical animal body is assembled and transformed after embryogenesis in several stages of *development* that depend "on the embryo's ability to maintain a programmed temporal and spatial coordination of morphogenetic [tissue- and organ-constructing] events" (Wohlgemuth and Murashov 1995: 63). The vertebrate developmental schedule includes *critical periods* during which perturbation of the organism by environmental or genetic factors can severely negatively affect the developing organism. Horn (2006: 209) indicates what constitutes a 'critical period' in development: it must be *susceptible to a specific environmental modification at a well-defined developmental stage* and *must persist through postnatal life*. By 1995 a literature review concluded that "several reports [document] the apparently normal development of several vertebrate species, including mammals, under conditions of exposure to space flight during various periods of the development process. Evidence to the contrary also exists and it is therefore likely that some alterations in morphology do occur in a microgravity environment. Although subsequent development may appear overtly normal, more subtle abnormalities result" (Wolgemuth and Murashov op cit). Since that time a certain amount has been learned, as will be summarized below, but let us first look at some specific cases of development to better appreciate its significance.

An example of developmental disturbance in microgravity is found in the development of rats flown on the 16-day Neurolab mission (Miyake et al. 2004). Rat pups in postnatal days 7–13 were found to develop differently (some organs being larger than normal) than rat pups developing after 13 postnatal days' age; such organ size variations may be trivial or catastrophic to health, but only recently have "developmental biologists…started to dissect signaling pathways dedicated to [organ] size control…[including the] the Hippo (Hpo) pathway, which controls organ size by coordinately regulating cell growth, proliferation, and apoptosis [cell death]" (Pan 2007). The point here is that while even a decade ago organ-size differences between Earth-developed and space-developed individuals might well have had an explanation based on gravity effects on higher-order processes like cell division, today they may well have an explanation based on such signaling pathways. At this time we in many such cases cannot be certain, the importance of which is returned to at the end of this chapter. One more example from the biological data is of specific interest here. A comprehensive study of ribbed newt reproduction, from egg fertilization through embryogenesis and development, was carried out on the *Mir* space station in the late 1990s, during which space-born and space-developed individual newts were even later mated with Earth- born and Earth-developed newts. The results are complex but in essence it was found that "During all of development up to adulthood, we detected no abnormalities that could distinguish these animals from standard animals…[among their offspring] percentages of fertilization and development were in accordance with the control animals. The duration of development until metamorphosis was in accordance with that of control offspring, as well as the duration of development of the larvae deriving from [control animals observed on Earth] No genetic abnormalities were detected during the analysis of offspring…" (Dournon et al. 2001: 324). In this case the results indicate that while some differences between ground-control and space-raised individuals did occur—just as we would expect due to microgravity at the least—such differences were not lethal to the first generation of space-born individuals or even the offspring born (though on Earth) of these individuals.

This fits a larger pattern in space biology that has been assembling in the course of this chapter, and which will be returned to at the conclusion; space conditions clearly perturb the course of reproduction and development, but are not clearly and irrevocably lethal to it. That microgravity is not immediately and lethally-disruptive to the vertebrate sex cells, fertilized eggs, or the developing individual may be related to the fact that for a billion years of sexual replication in Earth life—and certainly for the past few million years of hominin biology—early animal development has occurred "in a condition of neutral buoyancy…due to similar densities of the embryo and the [amniotic fluid]…" (Andreazolli et al. 2017). Having mass, all animal tissues (including the individual cell's interior support structures of *microtubules*) are mechanically affected by gravitation, and while hypergravitation has been shown to disrupt microtubule structure (Searby et al. 2005), and thus for cellular structure and function, the relaxation of the 1 g gravity field we feel on the surface of Earth is apparent not a significant 'shock' to the early animal life form. Indeed it is historically 'familiar' to animal life developmental processes and tissues as it is in such a similar gravity field that early animal life developmental stages evolved in the first place, in aquatic life following the Precambrian explosion of animal forms. We will revisit this fact and some of its implications at the end of this chapter.

The operation of genetic-developmental 'programmes' must play out for the development of the biologically-healthy individual human being. Consider the development of the fetal brain. Human fetal development occurs in weeks 9–37 of pregnancy. By the 22nd week, the fetus is beginning to significantly develop brain tissues that are, in their volume compared to body size, uniquely *encephalized* (larger than one would expect for the body size) in the animal kingdom. The biological investment in the brain in our species is evident in the fetus, whose head is greatly enlarged compared to the rest of the body through development and even through infancy. The brain's cells or *neurons* make many important connections in weeks 21–38, during which the essential *connectome*—the array of neural cell connections in the individual brain—takes shape, with some patterns shared among all brains but also exhibiting individual variations; the similarities (sometimes termed the *default mode network*) are likely responsible in part for structurally-similar (across humanity) reflexive and instinctual actions, whereas the variations must have some contribution to our highly-variable personalities, the physical, neural basis of the 'disorderliness' one might say, or creativity, of individual human behavior. Such personality variation is not trivial, in fact it may well be a large component of humanity's tremendous range of behavioral variation, a key element of the evolutionary process which demands replication, variation and selection of information. Recent research has shown that neural architecture development in the human fetus proceeds regionally, with the hindbrain (occiput) forming connections earliest and the lateral brain regions (the parietal lobes) occurring latest (see Fig. 2.6).

In this case of development, since we know that all cells (including neurons) have mass, they are all subject to deformation in gravity conditions different from those of Earth-Evolved Life (EEL); what effect such conditions will have on such critical periods is essentially unknown and I think will not be known for some time. But we may say that a prudent recommendation would be for pregnant women to not experiences NEN gravity fields (or radiation flux) as would be felt in departing Earth or arriving at Mars. This is just one of many methods we may propose to actively protect the processes of human reproductive biology from features of space environments.

In a 2006 review of animal development in microgravity, Horn listed five principal findings up to that date;

- Microgravity conditions do not prevent development
- Microgravity conditions do not radically alter development that begins compared to Earth development
- Healthy individual organisms with a good chance for a healthy life can develop in microgravity

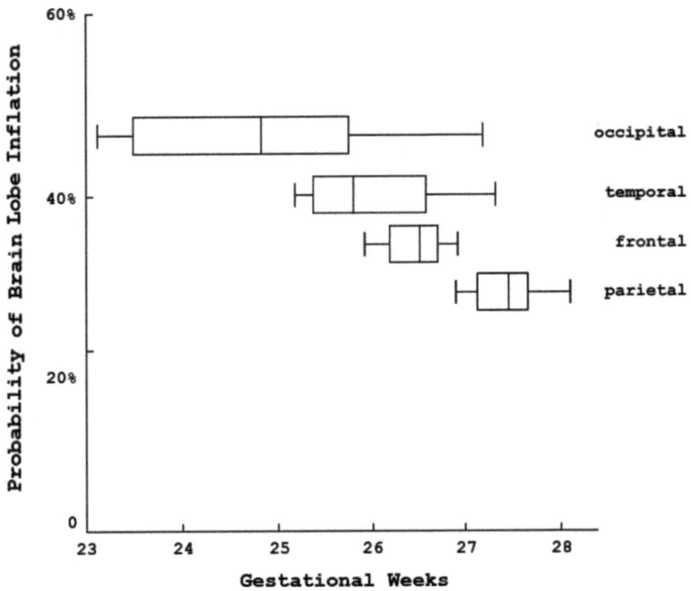

Fig. 2.6 Human fetal brain connectome development in gestational weeks 23–30. See important details in the text. Adapted by the author from Fig. 8 of Jakab et al. (2014)

- Individual organisms developed in microgravity appear essential the same as Earth developed organisms
- It is likely that after development, differences between microgravity and Earth developed organisms would occur, e.g. in the musculo-skeletal system of mammals that on Earth have adapted to 1 g conditions for over 100 million years (Horn 2006).

Regarding the final point above, we have empirical evidence that even during early life the earth-gravity-field sculpted mammalian developmental schedule can be affected by microgravity: skeletal development has been shown to be particularly affected by microgravity conditions; fetal rats on the Cosmos-1514 mission exhibited a "13–17%" arrest in "nearly every areas of the fetal skeleton" (Denisova 1986).

Figure 2.7 indicates an example in development, that of the human forearm. Note that altered-g and breathing gas compositions can effect various substages of the development. Many such details will probably not be understood until they are studied while occurring in beyond-Earth conditions.

Functional development refers to the development of physiological processes. These include bone growth, which in vertebrates commences rapidly after several weeks of pattern establishment in cartilaginous tissues. Table 2.2 indicates the effects of microgravity on bone development in various species; it shows that while microgravity has some effects in each case, in most cases there is also some normal development. Note that these effects are all among non-fish and non-amphibian life forms; it may well be that this reflects the evolutionary history of those life forms in a rather neutral-buoyancy environment.

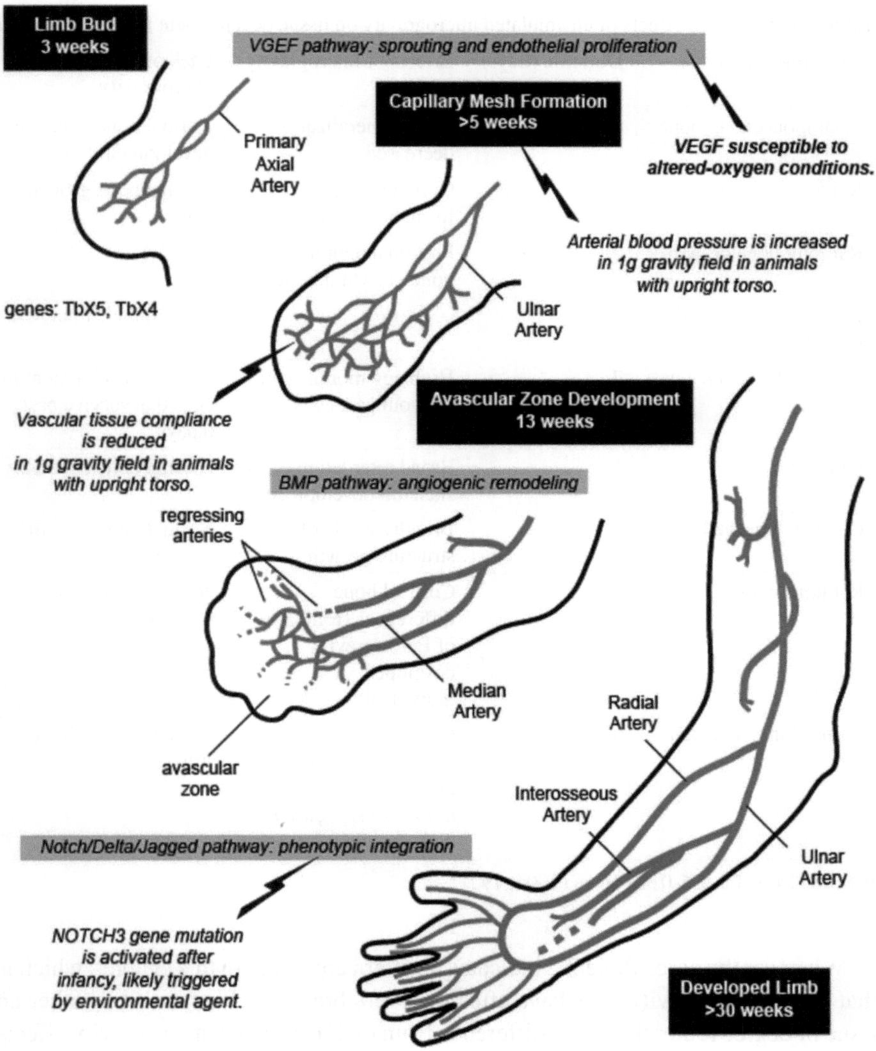

Fig. 2.7 Development of the human forelimb. Depiction by the Author from data Barham and Clarke (2008) and Chappell and Bautch (2010). Notes regarding the susceptibility of various gene functions and pathways to altered-oxygen conditions and gravity field indicate the thoroughgoing susceptibility of developmental genetics to beyond-Earth conditions. When new such conditions result in developmental failure, it is inevitable that there will be selection against such susceptibilities, no matter how advanced our therapies

Table 2.2 Observed effects of unsimulated microgravity on tissue development in various species

Experimental model	Duration (days)	Effects of microgravity	Lack of effects of microgravity
Fetal mouse long bone	4	Bone mineralization decreased	Normal bone collagen development
Rat long bone growth	4	Cortical bone formation retarded	Normal bone growth rate
Rat long bone growth	7	Proximal region of long bone growth inhibited	–
In utero rat pup head growth	9	–	Normal bone thickness.
Fertilized chicken eggs	7	High egg mortality in the youngest eggs	Normal development in eggs that survive past 6 days
Fetal rat	5	Broad retardation of skeleton development	Normal bone chemistry
Chicken bone cells	11	Broadly weaker bone structure growth	Normal bone growth energetics
Rat bone cells	13	Cortical bone differences from those of Earth-gravity-developed individuals	Normal overall bone structure
Mouse long bone cells	6	Complex variations in bone growth *depending on developmental stage at time of exposure to microgravity*	Normal overall bone structure

Adapted from Table 5-01 of Horn (2006: 192)

Whatever the case, Table 2.2 indicates a pattern emerging in this chapter, which is that while microgravity does have effects on vertebrate development, it is rather an issue of degree rather than kind difference from on Earth. Many microgravity effects seem often to be the result of simple mechanical response e.g. cellular structure and growth patterns differing when exposed to non-Earth-normal conditions and seem to be accommodatable by such responses. A life form's genetically-determined (as opposed to physiologically-determined) capacity to respond to such conditions as microgravity is termed *phenotypic plasticity*, defined as "the ability of individual genotypes [genetic instructions] to produce different phenotypes [expressed genes, or 'bodies'] when exposed to different environmental conditions" (Fusco and Minelli 2004: 548). The human phenotypic plastic range (sometimes referred to as a 'reaction norm' as mentioned in Chap. 1) has been sketched out but remains under investigation (Bateson et al. 2014). So far it appears sufficient to accommodate the adult human body in microgravity for many months. Nothing is known about its ability to support human young in microgravity, but, as I will mention again towards the end of this chapter, I see no need in the immediate future for developing young to be subjected to LEO-like gravity conditions. Many microgravity effects on the adult human appear to be reversible (acclimatizations) when the individual is returned to Earth-normal conditions. Currently, under the exploration paradigm it is assumed that the organism will always be returned to such Earth-normal conditions, but in the case of permanent space settlers, this issue will be of little interest structural to their goals of establishing populations independent of Earth.

In addition to *physical* development and *functional* development there is *behavioral development*. In nonhuman life, the bulk of behavior is governed by a variable combination of *instinct* (biologically-transmitted guides to behavior) and *learned* guides to behavior; in humans, and some of our primate relatives, there is a significantly larger component of *socially-transmitted guides to behavior*, which we sometimes refer to as *culture* (the entirety of Chap. 3 deals with culture and culture change). For human offspring an extended period of behavioral development is required to learn the rules of social interaction required to survive independently of the parent generation; in space studies such enculturation has not been studied, but the topic is explored in Chap. 3.

It is important to note that early human development all takes place in heavily-shielded circumstances (see Fig. 2.8). Biological shields include the amniotic fluid

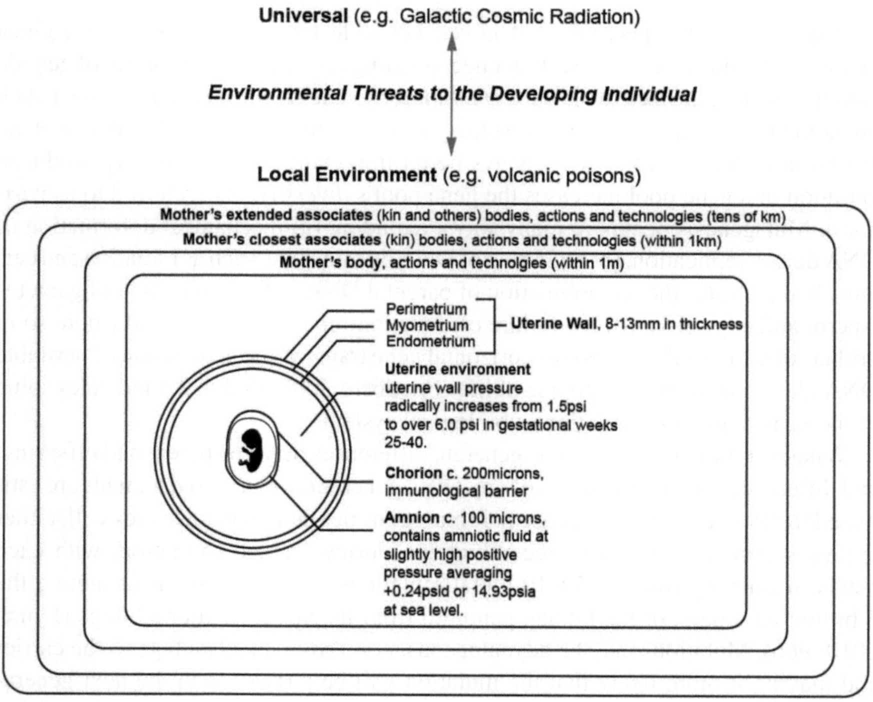

Fig. 2.8 Some of the layers of shielding protecting the human fetus through development and birth. Figure by the Author. For uterine pressure see Woodbury et al. (1938). See text for important details

that suspends the fetus, protects it from shocks and temperature variations (e.g. see, and provides nutrients for growth (and is contained by a thin *amnion* tissue envelope), the *chorion*, a quarter-millimeter thick protective tissue envelope particularly effective in preventing infection of the fetus, and the mother's three uterine walls, averaging 10 mm in thickness; beyond this are the hard tissues of the pelvic bones and the bones, connective tissues, muscles and skin of the mother's body. But there are many other barriers protecting the fetus from such environmental threats as radiation and even physical harm from other individuals; they include technologies (such as a sunshade or clothing) and complex behavior (e.g. maintaining good social relationships with kin who might provide extra food, water, nutrients and/or thermal regulation), such that even from conception, many and complex barriers are arranged to protect the developing individual. Protecting these barriers and supplementing them for new conditions beyond Earth—for example with radiation shielding for habitations on Mars, in various space suits for various space environments, and in various spacecraft mechanisms while travelling through interplanetary space—will be an extension of an ancient evolutionary story, not a novelty.

2.4 Radiation and Mutagenesis Beyond Earth

Mutagenesis is the appearance of novel, heritable traits—mutations—in a given genome. Mutations A recent shift in understanding mutagenesis, the origin of genetic novelty, has emphasized the *failure of ubiquitous gene repair mechanisms* over such phenomena as 'zap mutations' resulting from cosmic radiation (Friedberg et al. 1995), though the former will always be an important mutagen. Newly-produced variation in a gene pool increases the gene pool's *diversity*, also referred to as *variation*. Mutagenesis occurs in many ways, including (a) mechanical deformation of DNA during replication, (b) alteration of DNA by mutagens such as radiation and certain chemicals, (c) the recombination of parental DNA in the production of gametes (sperm and egg cells), (d) the failure of mutation-repair processes, and (more so in prokaryotes than eukaryotes) (e) horizontal gene transfer, the acquisition of heritable DNA alteration during the course of life (Friedberg 2006) (Table 2.3 indicates some of the main sources of variation in biological systems).

Whatever the cause, mutations generate differences between parent and offspring, and differences among members of any given generation. Currently, humans are estimated to have about 38 DNA base pair mutations per human gamete (sex cell, either egg or sperm) as a result of recombination during gamete formation, with each newborn carrying roughly 50–100 total mutations (genetic differences among the 3 billion base pairs of the human genome) from its parents and/or siblings (Lynch 2010: 966). Mutations may be advantageous, conferring an advantage to the carrier and making it more likely that the mutation will be passed on to the next generation, disadvantageous, conferring a disadvantage (e.g. poorer health) to the carrier, making it less likely that the mutation will be passed on to the next generation, or neutral, having no measureable effect on the likelihood that it will be passed on to the

Table 2.3 Sources of variation in biological systems

Source	Mechanism	Example
Mechanical properties of DNA	Temporary mechanical deformation of DNA molecule disturbs the base pair sequence	'Wobbling', 'buckled' or 'slipping' DNA strand may be altered, producing variation in the offspring
Chemical or environmental Mutagen	Energy or chemicals directly encounter and alter DNA	Radiation energy can break an entire DNA strand in two. Alkaline conditions can separate DNA 'rail' strands
Recombination	'Shuffling' of genes before production of next generation	Male DNA is shuffled before formation of sperm cells; female DNA is shuffled before formation of egg cells; offspring inherits shuffled characteristics
In- Migration	New genetic material is introduced into population	A population begins mating with newly-discovered population of same species, 'importing' new genetic material into the 'gene pool'
Horizontal gene transfer	New DNA is taken into the organism from the environment	Unrelated or distantly-related species transfer DNA back and forth; largely known in microbial species (today)

All variations result from some *mutagen*, or source (column 1). This chapter is largely concerned with row two, *Chemical or Environmental Mutagens.* Table by the Author

next generation; this does not mean, however, that neutral mutations never become more or less common in populations. Mutation rate is quite variable among species, subpopulations of species, families, individuals, sexes and even bodily tissues (Duret 2009) so no single mutation rate can be used in all calculations. Having said this, an average nuclear mutation rate per generation for modern humans has been estimated at 1.3×10^{-8} per generation (Leffler et al. 2012).

Mutation rate has long been known to increase with exposure to certain kinds of radiation (Russell et al. 1960). The Low Earth Orbit (c. <2000 km) altitude of the International Space Station is well outside the Earth's dense and radiation-buffering atmosphere but well within the Earth's radiation-deflecting Van Allen Belts (extending to c. 50,000 km); however it and other crewed station orbits pass through the South Magnetic Anomaly, where they receive most of their abnormally-high radiation doses, on the order of 20 millisieverts (mSv) monthly, about 1000 times that of the 0.02 mSv monthly dose experienced by life on the Earth's surface (Fig. 2.9 indicates the main varieties of radiation pertinent to our solar system). It is clear that while microgravity effects on human biology are mixed, space radiation hazards are potentially quite serious; empirically-observed effects of GCR on DNA include chromosomal deletions, DNA strand breakage, increases in mutation frequency, increases

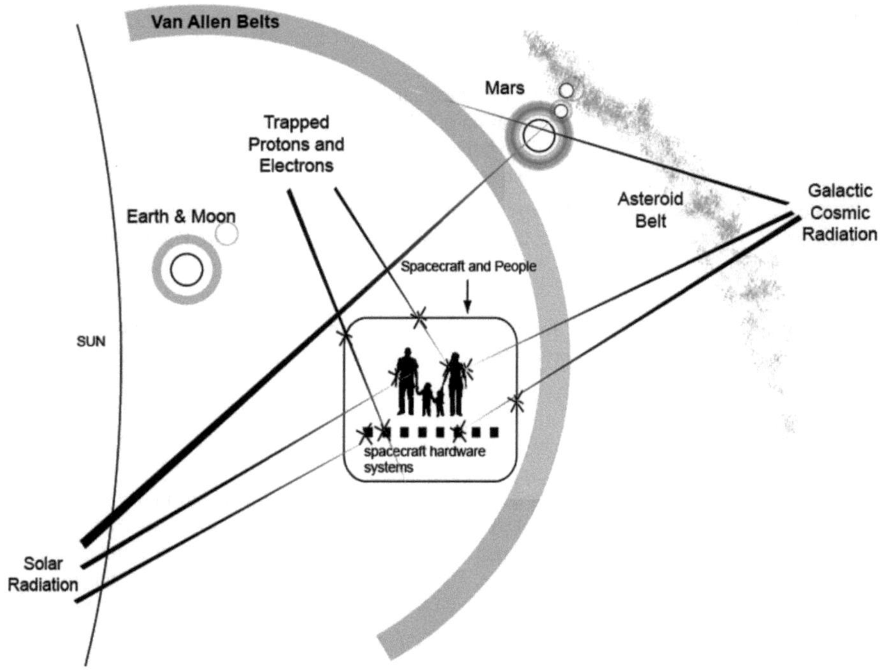

Fig. 2.9 Radiation sources and energies in our solar system. Galactic Cosmic Radiation (GCR) is a concern to all off-Earth environments, whereas trapped protons and electrons are mainly a concern within the boundaries of the Earth's Van Allen Belts (largely extending less than 50,000 km beyond Earth). Significant solar radiation reaches Mars in the form of Solar Particle Events (SPE's) for which at most 20 min' warning could be given if first detected on Earth. Figure by the Author

in chromosomal aberration frequency and interruption of gene expression; it has also been indicated that there may be an interplay of microgravity and radiation hazards to DNA, and that "further experiments need to be performed in the true space environment in order to address this critical question" (Villanueva et al. 2017).

A vast literature resulting from half a century of theory and study addresses the radiation biology of the vertebrate in 'space', divided among environments (e.g. Low Earth Orbit, interplanetary space, the lunar surface and the surface of Mars) but also animal tissues, developmental life stages and so on. In 2006 the National Academies of the United States published *Space Radiation Hazards and the Vision for Space Exploration* (National Academies of Science 2006), reporting on a workshop convened to address space radiation hazards to living systems in light of the then-recently-announced *Vision for Space Exploration*. The radiation risks to the human in *exploratory* (author's emphasis) space missions were tabulated in this publication as follows:

- Carcinogenesis (chiefly leukemia and solid cancers)
- Degenerative Tissue Effects (e.g. heart disease and cataracts)
- Damage to the Central Nervous System (decay of motor skills, accelerated ageing)

- Acute Risks (e.g. immediate vomiting, nausea or death in case of ultra-high doses)
- Reduced Lifespan.

Note that these are, as highlighted by me, issues of concern for *exploratory* voyages, rather than issues of *multigenerational* concern; this is to be expected as NASA has an exploratory function at present, not one of human space settlement. In this book I presume that space settlement will be attempted, which by default assumes that various behaviors and technologies will be used to mitigate the known short-term risks over the long term. Having said this, I do not wish to treat lightly the concerns of radiation, but rather shift our thought from radiation as a concern to the individual on a relatively short *exploratory* voyage (with subsequent return to Earth) to radiation as a concern for *multigenerational communities* beyond Earth, who may never encounter Earth conditions.

To begin we may review the main conclusions of the 2006 survey noted above; they include the identification of the following pressing issues:

- Improving basic understanding of governing physical processes of the space radiation environment;
- Developing models that can form the basis of greatly improved predictive models;
- Validating these models with improved observations;
- Developing the conceptual and hardware infrastructure for an operational space weather system.

By 2015 a report on the mitigation of GCR hazards to human space crews identified a variety of *passive* shielding materials, including polyethelene and aluminum, as being effective in reducing prospective astronauts' exposure to GCR hazards when over 40 g/cm^2 but concluded that because of weight restrictions on foreseeable voyages, researchers should also investigate *active* shielding, including "mini-magnetospheres, electrostatic shielding, and confined magnetic fields using superconducting magnets" (Simonsen 2015). At the time of writing, such shielding specifics are not yet ready.

A 2016 report suggests that new GCR modelling is required to study GCR hazards beyond Earth because while most "…accelerator-based space radiation experiments have been performed with single ion beams at fixed energies…the space radiation environment consists of a wide variety of ion species with a continuous range of energies" (Norbury et al. 2016: 38). Because a new instrument constructed at Brookhaven National Laboratory allows such multiple-species radiation bombardment of samples, a new research effort is suggested to address fundamental questions in this field.

By 2018, a mid-decadal progress report was also produced by the National Academies of the United States, *A Midterm Assessment of Implementation of the Decadal Survey on Life and Physical Sciences* Research (National Academies of Science 2018). Of the study's roughly 100 high-priority recommendations are the following specific to radiation hazards:

- Conduct longitudinal studies of astronauts for cataract incidence, quality, and pathology related to radiation exposures to understand both cataract risk and radiation-induced late tissue toxicities in humans.
- Continue ground-based cellular studies to develop end points and markers for acute and late radiation toxicities, using radiation facilities that are able to mimic space radiation exposures.
- Expand the use of animal studies to assess space radiation risks to humans from cancer, cataracts, cardiovascular disease, neurologic dysfunction, degenerative diseases, and acute toxicities such as fever, nausea, bone marrow suppression, and others.
- Expand understanding of gender differences in adaptation to the spaceflight environment through flight- and ground-based research, particularly potential differences in bone, muscle, and cardiovascular function and long-term radiation risks.
- NASA should define requirements for thermal control, micrometeoroid and orbital debris impact and protection, and radiation protection for EVA systems, rovers, and habitats and develop a plan for radiation shelters.

These are so nebulous ("assess space radiation risks to humans from cancer, cataracts, cardiovascular disease, neurologic dysfunction, degenerative diseases, and acute toxicities such as fever, nausea, bone marrow suppression, and others") that it is difficult to grasp what is known or well-understood about space radiation hazards after decades of study.

During a *Future-in-Space Working Group Telecon* sponsored by the University of Texas-Houston on 06 June 2018, one of the authors of the report, Dan Dumbacher of Perdue University, gave a *precis* of this report. I was fortunate to listen to this most up-to-date evaluation of the work so far. Radiation biology remained a field of particular interest as NASA continues to consider crewed exploration into interplanetary space. Of the life sciences of most interest, Dumbacher indicated "space radiation risk to humans" as a topic of 'Highest Priority'. However, Dumbacher then mentioned that after nearly half a century of theory, data collection and analysis of space biology data, "A fundamental understanding of microgravity and radiation effects on biology needs to be had." (FISO 2018). Following the optimistic tone of the 2006 report, this is a rather underwhelming state of understanding at this point in time. Currently it may be said that the US space agency has sketched out mission- and lifetime-specific radiation doses and uses these guidelines to make astronaut assignments for spaceflight. But beyond this there is as yet no practical radiation solution for exploring a variety of off-Earth environments, much less settling them.

As of the early 2000s the European Space Agency (ESA) had also carefully addressed space radiation dangers with nearly 20 study programmes including:

- Radiation Exposure and Mission Strategies for Interplanetary Manned Missions
- Interactions of Radiation with Biological Systems at the Cellular and DNA Level
- Spacecraft Radiation Shielding and Effects Toolkit.

A more practical approach than many of these reviewed has recently been outlined, highlighting solutions, rather than problems; in *Vive la radiorésistance!: converging research in radiobiology and biogerontology to enhance human radioresistance for deep space exploration and colonization* (Cortese et al. 2018) which presents a a multi-method approach to GCR hazards including DNA-repairing regenerative therapies, hibernation-in-transit, novel shielding methods, careful selection of radioresistant individuals as crew and careful mission planning, among others.

A literature review of Mars-related radiation-mitigation strategies dating after 2010 yields roughly 5000 results, including articles titled *A Preliminary Study on Radiation Shielding Using Martian Magnetic Anomalies* (Emoto et al. 2018), *Nanomaterials for radiation shielding* (Thibeault et al. 2015), *Shielding biomolecules from effects of radiation by Mars analogue minerals and soils* (Ertem et al. 2017, which deals with the issue indirectly) and *Adaptive, Readily Morphing, Optimized Radiation Shielding for Transit Habitats: Flyby Mars Mission* (Barthel and Klijn 2017) among many others proposing various radiation mitigation schemes.

At this time, then, we may make the following substantive comments:

- Various forms and effects of radiation energy beyond the shelter of Earth are potentially significant barriers to successful human adaptation beyond Earth because they actively damage DNA at all stages of human reproduction; however careful scheduling of reproductive biology could substantially protect delicate reproductive stages from major extraterrestrial radiation risks.
- Currently a wide variety of methods to mitigate space radiation hazards are being investigated, including passive and active shielding.
- Such shielding—when combined with exposure scheduling, active measures such as radiation repair therapy—will *have to be devised* if the premise of this work (that humanity will attempt to permanently settle space) is to be accepted.
- Whatever mitigations for radiation risk are devised, a behavioral component, in the increased cultural acceptance of risk (in this case risk of radiation hazards to reproductive biology), will likely be required in any realistic space settlement plans (see an excellent discussion in Schimmerling 2010).
- Thus for our purposes, the precise radiation doses and solutions are of less interest or practical utility than knowing that mitigation will include:
  - *passive* technologies such as built- or in situ shelters (e.g. habitats and caves, respectively)
  - *active* technologies such as magnetic shielding (e.g. natural magnetic field anomalies recently identified on Mars (e.g. Emoto et al. 2018), and battery-powered field generation for interplanetary voyages)
  - *exposure scheduling* (e.g. dose recording per space traveler and assignment of space travel based on accumulated dose, behavioral rules on Mars preventing the very young from being exposed to too much radiation)
  - *radiation therapies*, while promising (see Baarends et al. 2001), are largely unexplored at present and are still dependent upon discoveries in biology.

From this perspective it is easier to understand the potential contribution of anthropology to the broad issue of shielding humans from environmental hazards. Humanity has survived multigenerationally for over 100,000 years (about 10,000 of these years being sedentary farming people) by devising a wide array of behavioral and technological barriers between the body and natural environmental hazards. Figure 2.10 demonstrates just a few varieties of households globally and through time in non-industrialized cultures to remind the reader that shielding the body with technology is

Inuit *iglu*, Canada.

Mandan lodge, North Dakota.

Cliff dwelling, Colorado.

Yurt, Mongolia.

Tufa dwelling, Turkey.

Tallensi homestead, Gahna.

Fig. 2.10 A variety of human shielding solutions globally and through time. Such shelters have shielded humanity from extremely high- and low temperatures, wind, precipitation and other environmental hazards for tens of thousands of years. Space settlement will require new designs and technologies but may be seen simply as extension of a familiar method rather than exotic. All figures are in the public domain except author's drawing of the Tallensi household, lower right

a familiar rather than novel experience for ours species; one may also recall that plant life shelters bulbs and other resources under the Earth's surface, and many varieties of nonhuman animals shelter, for example crabs in seashells, or meerkats in underground warrens. Anthropology, as will be seen in the rest of this book, will assist in better selection of such shielding methods and forms for viable communities of people living permanently beyond the Earth. Some specific anthropologically-informed recommendations regarding off-Earth radiation hazards are discussed in the conclusion of this chapter.

2.5 Beyond-Earth Mammalian Ageing, Senescence and the End of the Life Cycle

The field of evolutionary ageing science, also known as geroscience or biogerontology identifies an array of the effects of human ageing; specifically several variables are positively and (purportedly or demonstratively) correlated with chronological ageing (the passage of time) (see Table 9.3 of Holmes 2017):

- Probability of disease, deterioration and death
- Accumulation of mutations
- Accumulation of cellular damage
- Genetic disregulations
- Failure of processes including autoimmune
- Alterations of lipid and energy metabolism
- Failure of body's homeostasis and cellular-repair mechanisms.

Biological ageing (rather than, of course, chronological ageing) in the flatworm appears to be noticeably slowed by altering the function of seven genes that "encode proteins [in much animal life] and that are likely related to neuronal or endocrine signaling" (Honda et al. 2012). Thus some ageing is genetically-determined, while other ageing effects are the result of biomechanical errors. Almost no research on ageing in microgravity or as it is effected by beyond-Earth radiation conditions has been carried out. It is likely such will only be understood by early space settlers in their own specific environments, whether orbital, lunar, Martian or otherwise.

One way that ageing will be an issue of successful space settlement is in its effect on average life expectancy. Currently it may be equivocally argued that ageing will be accelerated, or slowed, by conditions beyond Earth. Whatever the case we may see that average lifespan can change radically in just decades; Fig. 2.11 indicates the general increase of life expectancy both at birth and at 50 years old for US citizens, plotted over time from 1840 to the present. Such changes have not significantly occurred by natural selection or biological evolution, but nearly entirely through technological improvements in health care and improved nutrition. Whichever effect life beyond Earth has on the lifespan, it will have to be considered in the demographic structure of the settlement population; a topic reviewed in Chap. 3, which concerns populations of humans rather than the individual.

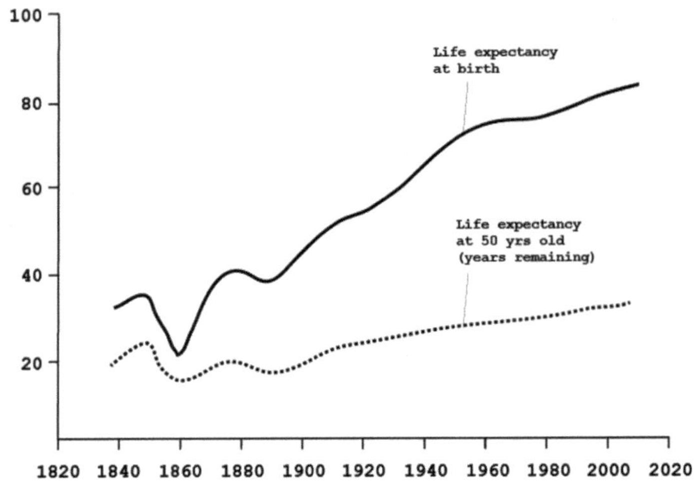

Fig. 2.11 Life expectancy both at birth and at 50 years old for US citizens, 1840–Present. Data from World Bank (2018) and visualization by the Author

2.6 Adaptive Recommendations for Individual Health Beyond Earth

It is clear that the biology of the individual human will be affected by exposure to environmental conditions beyond Earth. This is what we should expect and what would be healthy in an adaptive, plastic life form, whereas fixity—an inability to adapt—results in the biological phenomenon termed *canalization*, in which a life form conserves a particular biology, does not adapt successfully to new conditions, and often becomes extinct. Humanity's adaptive plasticity is both biological and technological. Both should be allowed to fit humanity to its new environments, even if they differ only slightly from those of Earth. For example, on Mars, a 0.38 Earth-normal gravity obtains. Here, expecting and *accepting* that our biology will be affected, expecting people to differ from Earth people over time, would be culturally adaptive. Resisting such change to 'push' an Earth phenotype would be counter-productive in the long term.

How would humans beyond Earth accept and promote rather than resist such adaptation? We may look at the course of life for some examples. It is clear that earliest stages of human development are the most delicate (Ross and Tomko 1998; Neubert et al. 1998; Singh et al. 2010). Much of this early development takes place in the heavily biologically-, socially- and technologically-shielded environment of the uterus. However once the essential phenotype (the infant) is born and loses many of its prior protections, it is rather adaptable in the short- and long-term relative to the life of the individual. A culture promoting and accepting change would be particularly careful to protect these early stages of life and allow natural selection to tailor individuals to their new environments. This means that for some generations there might well be more malformed embryos and unsuccessful pregnancies than on Earth—but note that even a century ago, stillbirth rates were radically higher than today in the developed world (Fig. 2.12). Beyond Earth, for the first few generations, real biological adaptation will manifest in higher such rates, though they should not return to levels of the early 1900s AD. A culture that accepts this painful transition will find itself better-adapted after some generations than one that resists it and tries to maintain an Earth genome, Earth growth pattern, and Earth biology, beyond Earth. How does a culture shape itself to accept such changes from those familiar to people on Earth?

Cultivating a culture of space settlement will require the communication and societal internalization of some key concepts. These may be thought of as guiding principles, not shackles; gusts guiding a sailboat in a general direction, rather than the rails of a railway, determining every move of a train. These principles should be communicated or reinforced, or at least not violated, in the settling culture's essential socialization and educational activities. Ensuring this does not require a draconian communication of commandments, but is better achieved as we today teach the principles of civics, ethics and philosophy in our educational institutions (more formally) and daily socialization (less formally); they may be incorporated into the arts and literature of a culture as well, as we have from Greek plays communicating parables to modern art, music and other entertainment. When the founding documents of a space-settling culture are established, the 'framers' would do well to incorporate these concept as foundations if they wish space settlement to be a multigenerational success; they are based on the evolutionary process that has resulted in billions of species of well-adapted forms of life on Earth over the last several billion years:

- Space settlement is not conquest but adaptation
- Adaptation is change through time that allows survival; it is positive

Fig. 2.12 Stillbirth rates in the US, UK and Sweden, 1900–2000. Figure adapted by the Author from *Reducing Maternal and Neonatal Mortality in Indonesia: Saving Lives, Saving the Future* (National Academies of Science of the United States of America 2013)

- Adaptation is an evolutionary process
- Evolution requires three successful processes: each must be nurtured
  - *Replication*: protection of reproductive health individually and as a population
  - *Variation*: differences from the 'Earth-Normal' human may be beneficial
  - *Selection*: acceptance of change through time from an 'Earth-Normal' human
- Multigenerational survival favors larger populations⁺
- Multigenerational survival favors diverse populations⁺
- Multigenerational survival favors geographically-widespread populations⁺

⁺ = discussed more in Chap. 3.

More specifically, these guiding principles can be applied to each stage of the course of life. I will proceed below in the sequence established in this chapter, from sex cell genesis through development, maturation and the end of the life cycle. The recommendations made for each stage of life are derived from our current evolutionary understanding and are meant to each contribute somewhat to the overall goal of multigenerational success of human populations entirely independent of Earth's resources, biological, material and even philosophical.

2.6.1 Spermatogenesis, Oogenesis and Fertilization

The development and protection of sex cells must be consciously considered rather than taken for granted. Just as we maintain general health by voluntary exercise, various preventative and therapeutic reproductive health measures should be devised and made integral to cultural health. In examples of concrete cultural actions, we may think in terms of rights and responsibilities per age and sex cohort (here I am not concerned with *gender*, a social identity category, but the biological identity category of *sex*). Responsibilities of the *prefertile* would include maintaining reproductive health, of the *fertile* to mate consciously and with commitment to lifelong support of the offspring, and of the *post-fertile* to curtail mating. All human cultures currently have rules to such reproductive behaviour, whether more- or less-strictly enforced. In cultures where resources may be scarce, enforcement of low reproductive rate is particularly important to keep the population in line with what can be supported by the environment. I suspect much the same will manifest in the earliest settlers of space. They will not necessarily face calorie shortages—in fact I think children born in early space settlements will at birth likely be more guaranteed adequate calories and nutrition than many humans ever born on Earth, because no realistic space settlement scenario would begin with extremely small margins for error but rather wider margins. But risks at first will be high; *if* there is a crop failure, for example, early settlers will face acute food shortage if their population is not closely tailored to the immediately-available resources until large stocks of stored calories and nutrients are ensured, and that might take decades, a generation or more.

Should a settlement culture not wish to so carefully control their population—though reproductive health will have to be consciously attended in any space settlement scenario—they might avoid such by ensuring, from the start, large caloric, nutrient, water and other staple resources. However, this work would likely require larger working populations and I think in the earliest settling populations there will be more conscious population control than many are accustomed to in the first world on Earth today. There is no reason to think that such 'social engineering' would be inimical to a settler culture. That settling culture will have to be composed of people highly dedicated to their goal of multigenerational success. Plenty of cultures have had such fortitude in the past and may be models to be admired and venerated by early space settlers; they include the Pacific-exploring Polynesians, who, during long voyages, carried small food margins and curtailed reproduction, and the Arctic-exploring peoples of Canada and Alaska, who carefully kept their populations in line with what the land could support. Only in the last century in the developed world has cultural restriction on reproduction relaxed almost entirely.

But there is a paradox here. It is well-known that small populations of any life form are relatively vulnerable; a single catastrophe could kill them all in one episode (e.g. communicable disease or physical catastrophe) or could so upset the demographic age- and sex structure of a population that viable reproductive members are no longer available, the population entering an *extinction vortex* discussed more in Chap. 3. For these reasons the earliest generations of space settlers might well attempt to grow their population quickly, to avoid such issues as the extinction vortex, but then level the population off as surpluses are built up and can be used to support a given population in the case of catastrophe. In either case more risk will be taken on as the genome and human preventative and therapeutic technologies are adapted to and are sculpted by new conditions beyond Earth. Settling cultures will have to accept social conditions somewhat different from those of first-world Earth in the Post-War era, and they will have to embrace change at all stages of the life course.

A list of practical adaptations to safeguard spermatogenesis, oogenesis and fertilization (each of which has been observed to proceed largely unhindered in microgravity in a model organism, the Japanese killifish (Ijiri 1998) in a generic, beyond-Earth environment are presented in Table 2.4. Some methods proposed, for example in vitro *gametogenesis*, or gametogenetic health therapy (Baarends et al. 2001), are quite new, and may not be viable if they are not developed.

Biological adaptations could include gene therapies to ensure healthy gametogenesis, and nutritional support specific to sperm and egg health. Technological adaptations could include spermatogenesis and oogenesis in vitro to allow for close monitoring of the sperm and egg form (Ibtisham et al. 2017) and/or fertilization in vitro, again to more carefully monitor the process and better guarantee the results. Cultural adaptations to these actions could include simple scheduling rules for reproductive activity; male gametes should be replenished after undergoing hypergravity deformations (at high-g lift-offs and landings, for example) by a period of reproductive absence; similarly, eggs produced or malformed by hypergravity exposure could be identified with close monitoring and removed from the potential fertilization pool.

Table 2.4 Practical recommendations for safeguarding spermatogenesis, oogenesis and fertilization in beyond-Earth habitats

	Biological adaptation	Technological adaptation	Cultural adaptation
Spermatogenesis and oogenesis	Promote healthy gametogenesis via gene therapies	Lower radiation exposure by remaining under shielding while most reproductively active. Possible in vitro gametogenesis	Reproductive sex abstinence in immediate pos-hypergravity period (launch, landing) to allow generation of new spermatozoa and oocytes; monitor to ensure gamete health before reproductive efforts
Fertilization	Promote sperm and egg health via gene therapies	In vitro fertilization in Earthlike (1 g, normal atmospheric pressure and gas composition) facility	Reproductive sex abstinence after a determined cumulative lifetime radiation exposure

See text for important details

Cultural adaptation to such regulations would include enculturating the young to an acceptance of close monitoring and scheduling of reproductive health and actions.

Most important is that while these measures would begin as simple behavioral and technological modifications from tradition, *as they are refined and formalized over time for the specific conditions of they would become adaptations.*

2.6.2 *Gastrulation, Embryogenesis and Development*

As mentioned, it is in these early stages of life that most can go awry; once the individual is born, they rapidly become quite robust and adaptable. In human gastrulation, early cellular differentiation disorders most commonly result from "genetic factors, such as mutations or chromosomal aberrations, but additional factors including environmental agents, such as drugs, viruses or hypoxia also have an effect" (Vaquer and Hadjantonakis 2013: 427). All of these are likely to be involved in new selective environments beyond Earth, as, for example, gene activity is altered by microgravity, mutations occur with new exposure to certain kinds of radiation, and different habitat breathing gas compositions shape the selective environment of the early individual life.

Culturally and technologically, then, it is in the early development and pregnancy, the *prenatal* period, that space-settling people will be most attentive to the health of the next generation. Special provisions will be taken to protect this stage of life. Examples may be found in Earth-based cultures; in cultures of the high-altitude-adapted Andes of South America, pregnant women often migrate down to lower altitudes in the last prenatal weeks; here there is more oxygen for the infant and mother during the birthing process. Afterwards, however, they return, with the infant, to higher altitudes; this is a clear biocultural adaptation to the conditions. A clear cultural adaptation among these populations is a higher cultural tolerance for unsuccessful pregnancies. In settlements beyond Earth, it may be that in the last prenatal weeks a similar adaptation is made, with birthing women being given more care to replicate Earthlike conditions. For example if a Mars habitat has a gas composition and pressure (due to engineering issues) rather different from that of Earth (for example, on Skylab atmospheric pressure was 5.0 psia with an atmospheric composition of 70% oxygen and 26% nitrogen, whereas on Earth at sea level pressure is 14.7 psia with a composition of about 78% nitrogen and 20% oxygen), some prenatal period might be spent in special facilities, built into the colony architecture, with more Earth-like atmospheric composition and pressure. In the case of Mars colony, to replicate the full gravity of Earth, it might even be that late-term mothers would migrate up to an orbital facility, spinning for a full 1.0 g, rather than Mars' 0.38 g field, for delivery. Whatever the specific adaptations, technically, biologically and culturally, I think it is in the prenatal period that most will be done differently than is commonly done to protect the next generation than is done in much of the developed world today. Table 2.5 indicates some of the most common health issues during pregnancy, for mother and fetus. We see that in some cases, conditions beyond Earth may require adaptation. It is not too early to begin thinking about these critical issues.

Table 2.6 indicates some practical ways in which Gastrulation, Embryogenesis and Development could be safeguarded in a generic beyond-Earth context. Some methods, including DNA repair by artificially inducing the body's own ancient DNA repair mechanisms (Nikolaev and Yang 2017), are just becoming available. The recommendations in Table 2.6 are therefore plausible if not yet demonstrated and are comparable to World Health Organization guidelines for general support of reproductive health (World Health Organization 2006).

Biological adaptations include promotion of health in the three noted stages, sometimes using genetic therapies only now being invented. Essentially, while these are biological adaptations, they are consciously directed by people. For example, embryogenesis may be monitored and, if necessary, treated with *gene activation therapy* (Servick 2015) that could promote healthy embryogenesis (Latham and Schultz 2001). The technological adaptations include shielding from radiation during the critical early phase of gastrulation, and possibly even support of the pregnancy in facilities that approach Earth-normal conditions (mentioned in the table). Though this seems reasonable, I do wonder how long it would be reasonable or even ethical to 'force' an Earth genome and Earthlike development in conditions beyond Earth. Such would be refusing to change, rather than changing with the circumstances, the antithesis of adaptation or evolution. Such moral questions require investigation.

Cultural adaptations suggested include a very conscious monitoring of reproductive health among people preparing to have children. There would be careful scheduling of attempts to fertilize, and monitoring might well extend even into preg-

Table 2.5 Common health issues during pregnancy and comments regarding beyond-Earth conditions

Condition	Description	Treatment	Implications/questions for beyond-Earth settlements
Iron Deficiency Anemia	Iron deficiency	Iron and folic acid supplements	Ensure secure iron and folic acid supplies
Gestational diabetes	Elevated maternal glucose levels	Dietary adjustment; medicine	–
Depression and anxiety	Increased stress and depression	Psychological well-being	–
Decreased Fetal Movement	Fetus decreases movement in 3rd trimester	Possible early delivery	How will fetal activity, e.g. rotation to head-down posture, be affected by non-Earth g conditions?
Infections	Mother or fetus can be infected during pregnancy	Wide variety	–
Hyperemesis Gravidarum	Morning sickness'	Wide variety	–
Placentia Previa	Placenta blocks uterus at pregnancy	Prevention and C-section delivery	How will placenta position at birthing be affected by non-Earth g conditions?
Placental Abruption	Placenta dislodges from uterine wall	Early delivery	How will placenta position at birthing be affected by non-Earth g conditions?

Adapted by the author from data at the US National Institutes of Health (National Institutes of Health 2019)

nancy, as the earliest stages, e.g. gastrulation and embryogenesis, are particularly vulnerable; within a day of successful conception, the mother might well be moved to facilities where the conceptus could be monitored in its most vulnerable, early few weeks.

Table 2.6 Practical recommendations for safeguarding gastrulation, embryogenesis and development in beyond-Earth habitats

	Biological adaptation	Technological adaptation	Cultural adaptation
Gastrulation	Promote healthy gastrulation via gene therapies e.g. DNA repair enhancement	Lower radiation exposure to particularly-radiation hypersensitive early gastrulation[a], by remaining under shelter	Careful scheduling of reproductive activity to take advantage of technological shielding
Embryogenesis	Promote healthy embryonic gene activation (EGA) with gene activation therapy	Movement of childbearing mother to Earthlike facilities with c. 1 g gravity field, normal atmospheric pressure and gas composition	Cultural acceptance of reproductive monitoring
Development	Promote fetal development with monitoring and development-specific nutritional and caloric support	As above	Cultural acceptance of reproductive discipline

See text for important details
[a]See Heyer et al. (2000)

2.6.3 *Infancy*

As mentioned, once successfully born, many of the potential dangers to the individual will be relieved; according to US actuarial tables, the death probability at birth is currently 0.0063, whereas just 12 months later it is 0.00045 and by 10 years of age it is 0.000094 (we return to these figures in the following section).

Still, infancy has its dangers, as introduced in Table 2.7. As in the prior table we see that in some cases, conditions beyond Earth will differ from those of Earth and require adaptation. For example, it is suspected that altered gravity conditions effect liver function (Jonscher 2016), which is the root cause of infantile jaundice, which can be fatal. We do not at present know how, for example, the 0.38 g field of Mars would affect liver function, and this is one of many questions we may begin to address in serious space-settlement planning.

2.6.4 *Childhood and Adolescence*

Beyond infancy, early childhood health issues are often nonlethal and result from communicable conditions (e.g. infections) encountered during increased socializa-

Table 2.7 Common health issues during infancy and comments regarding beyond-Earth conditions

Condition	Description	Treatment	Implications/questions for offworld settlements
Distended belly	Gas bloating, lack of bowel movement	Induce bowel movement	How will intestinal gas production and retention be affected by non-Earth atmospheric composition and pressure?
Birth injuries	Infant may break bones at delivery	Promote bone healing	How will mother's pelvic opening size and shape, which alters before birthing, be affected by non-Earth g conditions?
Hypoxia	Infant not properly oxygenated	Wide variety	How will infantile respiration and oxygenation be affected by non-Earth habitat gas composition and pressure?
Jaundice	Accumulation of blood poisons due to liver problem	Wide variety	How will infantile liver function–known to be affected by microgravity–be affected by non-Earth g conditions?

Adapted by the author from *Caring for Your Baby and Young Child: Birth to Age 5* (American Academy of Pediatrics 1993)

tion and schooling. In 'developed world' countries "The main factors implicated in child deaths in developing countries include zinc, iron and iodine deficiencies, poor sanitation, suboptimal breastfeeding patterns and poor nutrition more generally" (Delaney and Smith 2012: 2). None of these seem particularly likely to be increased in frequency in conditions we may expect of early, off-Earth settlements (though they should be reviewed). Note that both nutrition and immune function have been studied in long-term microgravity exposures, showing generally "depressed cellular immunity after space flight and…latent virus reactivation and a decreased ability of phagocytic cells to kill bacterial invaders," (Cogoli 1993). Naturally, adequate nutrition will have to be attended to in settlements beyond Earth. Table 2.8 indicates the established nutritional requirements currently set for the International Space Station. Note that such a table is ideal for 30–50 year-old ISS astronauts and would be adjusted for individuals in the pre- and post-reproductive stages of life, each of which has differing nutritional requirements, as broadly characterized in Table 2.8.

How human beings care for and enculturate children varies widely, globally, in details, but all cultures have systems of action to at least attempt to provide for

Table 2.8 Nutritional and other dietary requirement standards for International Space Station Astronauts

Nutrient	Units	Requirement
Energy	KJ (kcal)	Established as per World Health Organization, per astronaut, according to age, sex and body weight for moderate activity level
Protein	% of Total Consumed	12–15%
Carbohydrate	% of Total Consumed	50%
Fat	% of Total Consumed	30–35%
Fluid	mL/MJ consumed or mL/kcal	238–357 or 1.0–1.5 or 2000 mL/day
Vitamin A	microgram	1000
Vitamin D	microgram	10
Vitamin E	mg	20
Vitamin K	microgram	80
Vitamin C	mg	100
Vitamin B12	microgram	2
Vitamin B6	mg	2
Thiamin	mg	1.5
Riboflavin	mg	2
Folate	microgram	400
Niacin	NE or mg	or
Biotin	microgram	100
Pantothenic	acid	mg
Calcium	mg	1000–1200
Phosphorus	mg	1000–1200
_1.5	times	Ca
Magnesium	mg	350
Sodium	mg	1500–3500
Potassium	mg	3500
Iron	mg	10
Copper	mg	1.5–3.0
Manganese	mg	2.0–5.0
Fluoride	mg	4
Zinc	mg	15
Selenium	microgram	70
Iodine	microgram	150
Chromium	microgram	100–200

Adapted from Table 1 of Lane and Feeback (2002)

the material and socio-cultural-mental health of their offspring. A relatively secular, moderate approach—which I feel is not drastically shaped by religious, ethnic or socio-economic biases, recently proposed "A New Framework for Childhood Health Promotion" which suggested that "[childhood health] programs and policies address multiple capacities so as to be mutually reinforcing and support families to comprehensively provide the experiences and environments for ensuring that children have strong foundations for health and development" (Mistry et al. 2012: 1689–1690). Here we are getting more specific than I would like for the purposes of this book, and we are shifting away from the health of the individual (this chapter's concern) to life stages more shaped by socialization (which we will return to later in this book). The point is that the health of children will have to be as carefully seen to, and perhaps more carefully-so, than on Earth today.

As we move towards adolescence, socialization becomes intense and the likelihood of individual death increases somewhat, as seen in Table 2.9.

However carefully-constructed their habitats, it seems inevitable that beyond Earth people will be exposed to more risk than many of us accept in most aspects of daily life today.

One of the most widely-discussed risks to the individual is the radiation flux beyond Earth's protective magnetic fields, as discussed above. As we have seen, those risks may be 'unpacked' from the general statement that they 'will be higher than on Earth'. Figure 2.13 illustrates that a spectrum of risk obtains in the world of cumulative radiation exposure; lower exposures of course normally entail lower

Table 2.9 Death probability by age

Age	p of Death
0	0.006383
1	0.000453
2	0.000282
3	0.000230
4	0.000169
5	0.000155
6	0.000145
7	0.000135
8	0.000120
9	0.000105
10	0.000094
11	0.000099
12	0.000134
13	0.000207
14	0.000309

Note reduction of probability after birth and through childhood, but slight increase again after age 11. From US Social Security Administration Actuairal Timetables (2018)

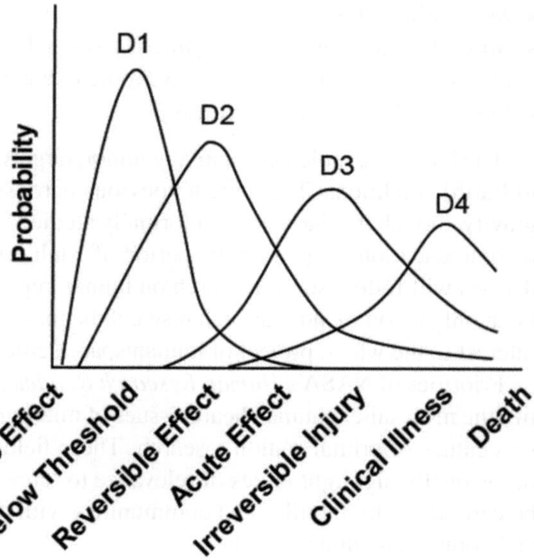

Fig. 2.13 Probability of various levels of effects of radiation with dose. Note that the curves may schematically represent any risk-and-result scenario. Figure adapted by the author from Fig. 1, page 326 of Schimmerling (2010)

adverse effects but may have sever effects, and higher doses of course are normally going to produce more severe results, though sometimes they do not. The point here is that risk itself may be unpacked to better understand the hazards and tailor our mitigations.

Such risks may be more sensitively incurred by spreading them out over a given life course in environments beyond Earth. For example, the cultural acceptance of radiation exposure to the young—people of reproductive age—might well be very low, to protect their reproductive biological health and that of their young. But for postreproductive people, who have had children and do not or are not permitted to have more, the radiation dose acceptance might well be much higher.

2.6.5 *Reproductive Adulthood*

Most astronauts studied for long-term health issues have been adults of reproductive age. These have recently been reported in the *Journal of Women's Health*, listed below as health effects on women *as compared to men* (Mark et al. 2014):

- Less hearing loss
- Less hearing bias (left—vs.—right ear)
- Less visual impairment
- More accurate in tests of alertness
- Better immune response
- More orthostatic intolerance (discomfort at moment of standing, relieved at reclining)

- More urinary tract infections
- More struvite kidney stones (mineral struvite less commonly causes kidney stones on Earth) than in males, who have more oxalate kidney stones
- Less speedy in tests of alertness.

The list above indicates relatively minor, largely reversible (on re-acclimatization to Earth) conditions. There are no obvious barriers noted to reproduction in microgravity, though this has not been formally studied, and it is too soon to make specific recommendations regarding the period of adult reproductive age. Such recommendations will follow space research on human reproductive biology. At this moment I can only recommend that such research begin as soon as possible as it is of central interest to the whole project of human space settlement.

Priorities of NASA's *Human Research Roadmap* are instructive in that they identify the most salient human health issues of microgravity exposure over the last quarter century of orbital station research. These fields are listed below; note that they occasionally highlight issues of relevance to 'crews', whereas space settlement will be carried out by families and communities with different individual psychological and communal cultural needs.

- Concern of Clinically Relevant Unpredicted Effects of Medication
- Concern of Intervertebral Disc Damage upon and immediately after re-exposure to Gravity
- Risk of Acute (In-flight) and Late Central Nervous System Effects from Radiation Exposure
- Risk of Acute Radiation Syndromes Due to Solar Particle Events (SPEs)
- Risk of Adverse Cognitive or Behavioral Conditions and Psychiatric Disorders
- Risk of Adverse Health and Performance Effects of Celestial Dust Exposure
- Risk of Adverse Health Effects Due to Host-Microorganism Interactions
- Risk of Adverse Health Event Due to Altered Immune Response
- Risk of Adverse Health Outcomes and Decrements in Performance due to Inflight Medical Conditions
- Risk of an Incompatible Vehicle/Habitat Design
- Risk of Bone Fracture due to Spaceflight-induced Changes to Bone
- Risk of Cardiac Rhythm Problems
- Risk of Cardiovascular Disease and Other Degenerative Tissue Effects From Radiation Exposure and Secondary Spaceflight Stressors
- Risk of Decompression Sickness
- Risk of Early Onset Osteoporosis Due To Spaceflight
- Risk of Impaired Control of Spacecraft/Associated Systems and Decreased Mobility Due to Vestibular/Sensorimotor Alterations Associated with Spaceflight
- Risk of Impaired Performance Due to Reduced Muscle Mass, Strength & Endurance
- Risk of Inadequate Design of Human and Automation/Robotic Integration
- Risk of Inadequate Human-Computer Interaction
- Risk of Inadequate Mission, Process and Task Design
- Risk of Inadequate Nutrition

- Risk of Ineffective or Toxic Medications Due to Long Term Storage
- Risk of Injury and Compromised Performance Due to EVA Operations
- Risk of Injury from Dynamic Loads
- Risk of Orthostatic Intolerance During Re-Exposure to Gravity
- Risk of Performance and Behavioral Health Decrements Due to Inadequate Cooperation, Coordination, Communication, and Psychosocial Adaptation within a Team
- Risk of Performance Decrement and Crew Illness Due to an Inadequate Food System
- Risk of Performance Decrements and Adverse Health Outcomes Resulting from Sleep Loss, Circadian Desynchronization, and Work Overload
- Risk of Performance Errors Due to Training Deficiencies
- Risk of Radiation Carcinogenesis
- Risk of Reduced Crew Health and Performance Due to Hypobaric Hypoxia
- Risk of Reduced Physical Performance Capabilities Due to Reduced Aerobic Capacity
- Risk of Renal Stone Formation
- Risk of Spaceflight Associated Neuro-ocular Syndrome (SANS).

Included in these concerns is the well-documented effect of microgravity on the hard tissues (e.g. bone) for which there are currently the following countermeasure therapies: *anti-resorptive drugs, nutritional supplements, exercise programs, resistance training* and *pulsed electromagnetic fields* (Loomer 2001: 257) all of which we may expect to be deployed as needed, and supplemented, for people living out entire lives beyond Earth.

2.6.6 Postreproductive Adulthood

Little is known of the individual biology of humans of postreproductive biological age in beyond-Earth conditions (most spaceflight participants being in the 35–50 year old age range). The well-known reduction of bone mass after long-term microgravity exposure, as noted in Sect. 2.6.5, might be in part due to intracellular switching activity inhibition, due to cellular mechanical reshaping as a result of microgravity, leading to a "a rounder cell shape with disorganization of microtubules, stress fibers, intermediate filaments, and focal adhesions. Transcription may be also altered as nucleus shape is changed….both RhoA and Rac1 might be inhibited explaining decreases in osteogenesis" (Louis et al. 2015). What has been learned about individuals over about 50 years of age exposed to LEO conditions has focused on short-term effects and adaptations including J. H. Glennn, 77 years old at the time of a spaceflight, and a commercial spaceflight participant (Jennings et al. 2006). Glenn fared well in his flight and there is no reason to suspect that postreproductive adulthood in environments beyond Earth would be of any less significance culturally—as elders are often revered for possession of accumulated wisdom—than anywhere on Earth today.

2.7 Conclusions and Recommendations

The individual human body, which all human-in-space plans must accommodate and build outward from, is affected by exposure to non-Earth environmental conditions. The effects occur on every scale, from molecule to cell to physiological systems, and on all phases of life, from fertilization to reproductive adulthood and beyond. The body accommodates some of these effects in the immediate term, adapts to some over the longer term and in time will to a degree be sculpted by them, even as we apply technological, behavioral and cultural insulators. Each scale of influence and each life-course stage must be protected by these insulators. We must also recall that the genetic health of individual life forms needed for human life will have to be as carefully investigated and protected as human life. A review of individual plant growth in microgravity conditions recently indicated that:

> The growth and morphogenesis of plants are entirely dependent on the gravitational acceleration of earth. Under microgravity conditions in space, these processes are greatly modified. Recent space experiments, in combination with ground-based studies, have shown that elongation growth is stimulated and lateral expansion suppressed in various shoot organs and roots under microgravity conditions. Plant organs also show automorphogenesis in space, which consists of altered growth direction and spontaneous curvature in the dorsiventral (back and front) directions. Changes in cell wall properties are responsible for these modifications of growth and morphogenesis under microgravity conditions. Plants live in space with interesting new sizes and forms. (Hoson 2014: 205)

It must be emphasized that what is known so far of LEO conditions on the human body feature largely the effect of microgravity and radiation flux on the whole body and some of its developmental systems. If we think about permanent human space settlement, however, we may note that it is unlikely that microgravity conditions will be a large part of most peoples' life experience. In a lunar habitat, people will experience 1/6th Earth g, while on Mars it will be about 0.38 g; in rotating orbital or Lagrangian colonies, the larger part of life will be lived out in something like Earth-normal gravity conditions. As an example, if we assume a 100-day transit to Mars (a plausible speed in the near future) in a spacecraft that does *not* rotate to provide earth-normal gravity, or something like it, the individual settler's exposure to microgravity conditions known from many LEO studies is very short, about 3 months of an 840-month (c. 70 year) lifespan, or about 0.38% of the lifespan (Fig. 2.14).

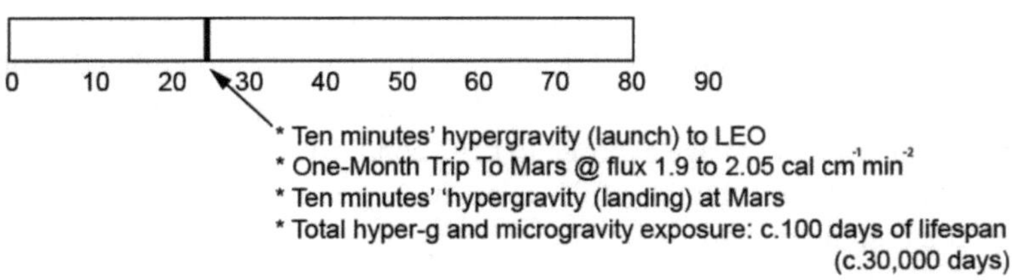

Fig. 2.14 Segment of lifespan spent in microgravity and hypergravity conditions from Earth to Mars. Illustration by the Author

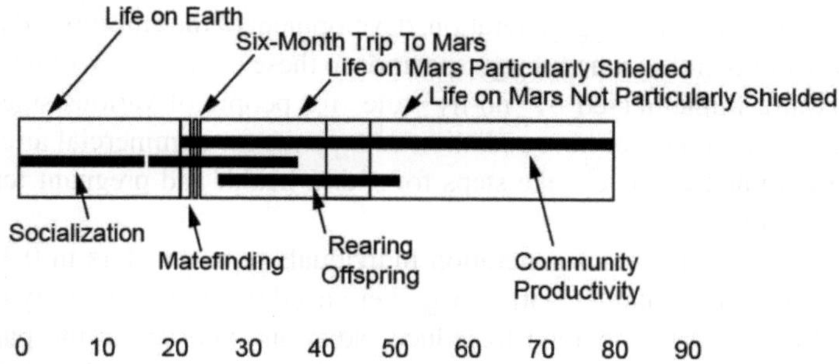

Fig. 2.15 Segment of lifespan spent in microgravity and hypergravity conditions From Earth to Mars: details. Illustration by the Author

We may expand this thinking to the course of life, and include the radiation exposure. Figure 2.15 indicates the exposure of the lifespan of a Mars-Earth settler. Leaving Earth in the time of matefinding (the 20s or so) the individual has experienced only Earth conditions. A brief exposure to hypergravity—a few minutes at 2–3 g's at most—might affect egg and sperm cell development, and developing embryos. To counter this, it would be wise to abstain from reproductive sex until (a) male sperm cells are replenished (c. 90 days) and (b) female egg cells dating to this exposure have either been passed on by *ovarian follicle atresia* [programmed egg cell expiration (Kapia and Hsueh 1997)] or new egg cells have developed (if current studies suggesting that this is a lifelong process turn out to be correct). Due to the delicate embryonic and fetal development processes, it would also make sense to abstain from launch during pregnancy. Several hours or days may be spent reaching transportation to some other habitat, e.g. Mars or a Lagrangian colony, a period unlikely to cause any significant reproductive issues. If the multiple-month voyage to Mars (for example) includes artificial gravitation, the exposure in this period would most likely be to increased radiation flux of the interplanetary region, though for this many mitigations are being proposed. After this time a brief exposure to hypergravity occurs on landing, with again some minutes of exposure to up to 4–6 g's, close to twice that of leaving Earth; it would make sense to apply the same prescriptions for reproductive health here as were given for leaving Earth, that is abstinence from reproductive sex for some time, and a prohibition on hypergravity exposure to the developing infant. Then, life begins on Mars, and for prereproductive people this might be spent particularly shielded until children are born. After giving birth, it may be that parents are willing to take a bit more radiation exposure, out beyond the sheltered communities. The young will be carefully tended to in order to ensure they reach self-sufficiency, also perhaps sheltering them more than they would experience on Earth.

The point in these diagrams is not to underestimate the significance of the space radiation and microgravity-conditions we may expect in certain parts of the space-emigration timeline. Certainly, non-Earth-normal g and radiation conditions could severely effect some of the most delicate and vulnerable processes of the human life course, such as sperm- and egg-generation, development of the embryo and fetus, and pregnancy. But I feel that scheduling exposures to these conditions will be relatively straightforward, implemented by 'no-fly' rules for people of various stages of the life course, just as today we have radiation dose limits for commercial aircrews and just as prereproductive males take steps for sperm health and pregnant females do not drink alcohol.

I also note that for second-generation individuals born on Mars in 0.38 g or in orbital or Lagrangian colonies with c. 1 g, there need be no microgravity exposure, and again life may be scheduled to reduce radiation exposure in the particularly delicate life stages around the time of reproduction. But beyond this time greater exposures may well be acceptable.

Such scheduling mitigations of risk would ultimately be cultural adaptations to conditions beyond Earth. Establishing such rules for norms of behavior has sculpted human culture and biology for the past 100,000 years on Earth, and should be expected. They will not be imposed by 'strangers' or 'mission planners', but encoded into and by the culture of settlement itself. This is because space settlement will not be a mission with the objective of return to Earth, but to establish and propagate new human cultures entirely independent of Earth.

The conception that humanity's biological features changing once extended beyond Earth is somehow problematic is common in science fiction and perhaps rooted in the concept of 'going native'. But humanity is always changing, whether the changes are observable or not. In a reduced gravity field, it is clear that different morphologies take shape as life forms develop, due to that lessened gravity effect (as noted above regarding plant growth). It is entirely possible that such will result, on Mars or anywhere else we experience a different gravity field, in a different stature for humanity over some decades or centuries, but we have had such change before, and it should not be considered problematic. Figure 2.16 indicates the appearance of a modern human stature in Europe since Medieval times, largely in the last few centuries and largely resulting from improved nutrition.

Following evolutionary principles to make this project more successful indicates that on the level of the individual human being, space-settling cultures should expect and accept a slow divergence of space-born people from Earth-born people. This will represent the adaptation to new environments that has allowed humanity to thrive on Earth for millennia. Having surveyed how this is likely to play out in the case of the individual human, we may now increase the complexity by examining adaptation of entire human populations over multigenerational time, in Chap. 3.

2.8 Summary Points

- Human biology is affected at all scales, from molecule to the organism, by exposure to beyond-Earth conditions, the best-known of which is microgravity.
- Although microgravity studies have been important to understanding humanity's future beyond Earth, it is not currently expectable that human populations will

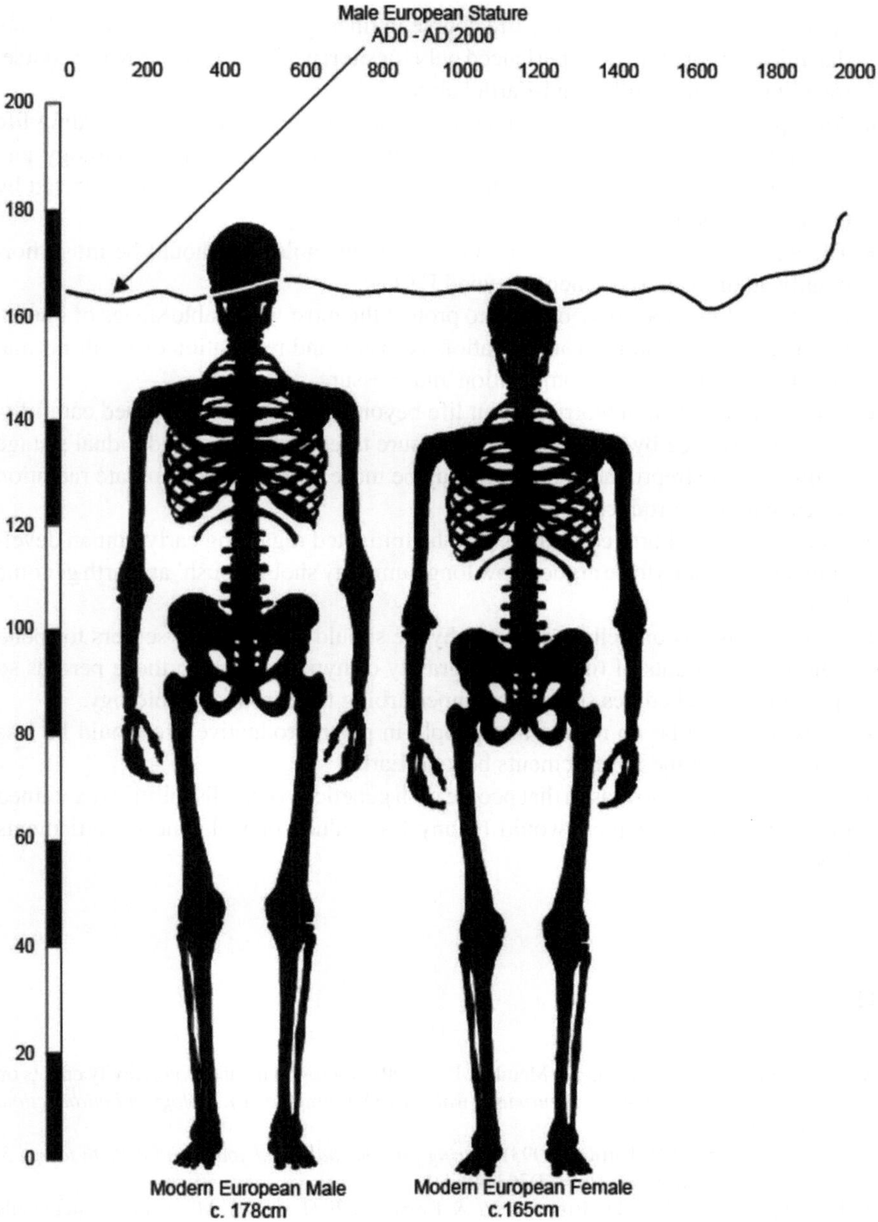

Fig. 2.16 Human stature differences, modern and Medieval times. Illustration by the Author

attempt to actually live out the life course in microgravity; rather it is most likely that microgravity will be experienced only for short periods of the entire life course, traveling to and from beyond-Earth habitats.

- All aspects of the phases of human life must be better-understood regarding life in conditions beyond Earth; likely most important is reproductive biology and the protection of offspring until they mature somewhat and may be protected by technology and behavior.
- The earliest stages of human life are most vulnerable and should be most thoroughly attended in settlements beyond Earth.
- A variety of methods may be used to protect the most vulnerable stages of human life, including shielding from radiation exposure and replication of Earth-normal gravity and atmospheric composition and pressure.
- Some common health worries about life beyond Earth, when examined carefully, may be mitigated by adjustment of exposure tolerances by the individual's stage of life; e.g. postreproductive people may be more willing to accumulate radiation doses than prereproductive people.
- Though beyond-Earth conditions may be mitigated regarding early human development, it remains to be argued how long humanity should 'push' an Earth genome beyond Earth.
- There seems no compelling reason why we should expect space settlers to spend significant amounts of time in microgravity or hypergravity; an those periods so spent may be schedules so as to be unperturbing to reproductive biology.
- There seems to be no reason that people in postreproductive life would be less valued or welcome in settlements beyond Earth.
- There seems to be no reason that people with genetic or other disabilities (examined more in the next chapter) would be any less valued or welcome in settlements beyond Earth.

References

Aimar, C., Bautz, A., Durand, D., & Membre, H. (2000). Microgravity and hypergravity effects on fertilization of the salamander. *Pleurodeles waltl* (Urodele amphibian). *Biology of Reproduction, 63,* 551–558.

American Academy of Pediatrics. (1993). *Caring for your baby and young child: Birth to age 5.* Online at https://eric.ed.gov/?id=ED363446.

Andreazolli, M., Angeloni, D., Broccoli, V., & Demontis, F. G. (2017). Microgravity, stem cells and embryonic development: Challenges and opportunities for 3D tissue generation. *Frontiers in Astronomy and Space Sciences, 4*(2), 1–7. https://doi.org/10.3389/fspas.2017.00002.

Baarends, W. M., Van der Laan, R., & Grootegoed, J. A. (2001). DNA repair mechanisms and gametogenesis. *Reproduction, 121*(1), 31–39.

Barham, G., & Clarke, N. M. P. (2008). Genetic regulation of embryological limb development with relation to congenital limb deformity in humans. *Journal of Children's Opthopaedics, 2*(1), 1–9. https://doi.org/10.1007/s11832-008-0076-2.

Barthel, J., & Klijn, N. S. (2017). Adaptive, readily-morphing, optimized radiation shielding for transit habitats: Flyby Mars mission. *Journal of Spacecraft and Rockets, 54*(6), 1196–1201. https://doi.org/10.2514/1.A33935.

Bateson, P., et al. (2014). Developmental plasticity and human health. *Nature, 430,* 419–421. https://doi.org/10.1038/nature02725.

Black, S., Larkin, K., Jacqmotte, N., Wassersug, R., Pronych, S., & Souza, K. (1996). Regulagive development of Xenopus laevis in microgravity. *Advances in Space Research, 17*(6–7), 209–217.

Bogin, B., & Smith, B. H. (1996). Evolution of the human life cycle. *American Journal of Human Biology, 8,* 703–716.

Chappell, J. C., & Bautch, V. L. (2010). Vascular Development—Genetic Mechanisms and Links to Vascular Disease. *Current Topics in Developmental Biology, 90,* 43–72. https://doi.org/10.1016/S0070-2153(10)90002-1.

Clement, G. (2006). Introduction to space biology. In G. Clement, & K. Slenzka (Eds.). *Fundamentals of space biology: Research on cells, animals and plants in space* (pp. 1–50). El Segundo, California: Microcosm Press; New York: Springer.

Clement, G. (2017). International roadmap for artificial gravity research. *Micrograviṫy,* 3(29). https://doi.org/10.1038/s41526-017-0034-8.

Cogoli, A. (1993). The effect of space flight on human cellular immunity. *Environmental Medicine, 37*(2), 107–116.

Cortese, F., et al. (2018). Vive la radiorésistance!: converging research in radiobiology and biogerontology to enhance human radioresistance for deep space exploration and colonization. *Oncotarget, 9*(18), 14692–14722.

Crawford-Young, S. J. (2006). Effects of microgravity on cell cytoskeleton and embryogenesis. *International Journal of Developmental Biology, 50*(2–3), 183–191.

Danilchik, M. V., & Savage, R. M. (1994). Cytoskeletal organization in advancing cleave furrows of Xenopus laevis. *Molecular Biology of the Cell, 5,* 101.

Dayanandan, P. (2011). Gravitational biology and space life sciences: Current status and implications for the indian space programme. *Journal of Biosciences, 36,* 911–919.

Denisova, L. A. (1986). Effect of weightlessness on the skeletal development of the rat fetus. *Kosmicheskaia Biologiia I Aviakosmicheskaia Meditsina, 20,* 60–63.

Di Agostinio, S., Botti, F., DeCarlo, A., Sette, C., & Geremia, R. (2004, August). Meiotic progression of isolated mouse spermatocytes under simulated microgravity. *Reproduction, 128,* 25–32. https://doi.org/10.1530/rep.1.00184.

Dournon, C., Durand, D., Tankosic, C., Membre, H., Gualandris-Parisot, L., & Bautz, A. (2001). Effects of microgravity on the larval] development, metamorphosis and reproduction of the urodele amphibian Pleurodeles waltl. *Development, Growth & Differentiation, 43,* 315–326.

Duboule, D. (1995). *Guide to the homeobox genes*. Washington, D.C.: U.S. Department of Energy Office of Scientific and Technical Information.

Dunlop, C. E., Telfer, E. E., & Anderson, R. A. (2014). Ovarian germline stem cells. *Stem Cell Research and Therapy, 5,* 98. https://doi.org/10.1186/scrt487.

Duret, L. (2009). Mutation patterns in the human genome: More variable than expected. *PLoS Biology, 7*(2), 217–219.

Emoto, K., Takao, Y., & Kuninaka, H. (2018). A preliminary study on radiation shielding using Martian magnetic anomalies. *Biological Sciences in Space, 32,* 1–5.

Englemann, U., Krassnig, F., & Schill, W. G. (1992). Sperm motility under conditions of weightlessness. *Journal of Andrology, 13*(5), 433–436.

Ertem, G., Ertem, M. C., McKay, C. P., & Hazen, R. M. (2017). Shielding biomolecules from effects of radiation by Mars analog materials and soils. *International Journal of Astrobiology, 16*(3), 280–285.

Finn, L. S., Maccarrone, M., & Blaber, E. (2017, April 25). Microgravity, stem cells and embryonic development: Challenges and opportunities for 3D tissue generation. *Frontiers in Astronomy and Space Sciences*. https://doi.org/10.3389/fspas.2017.00002.

Friedberg, E. C. (2006). Mutation as a phenotype. In L. H. Caproale (Ed.), *The implicit genome* (pp. 39–56). Oxford: Oxford University Press.

Friedberg, E. C., Walker, G. C., & Seide, W. (1995). *DNA repair and mutagenesis*. Washington, D.C.: ASM Press.

Fusco, G., & Minelli, A. (2004). Phenotypic plasticity in development and evolution: Facts and concepts. *Philosophical Transactions of the Royal Society B, 365,* 547–556. https://doi.org/10.1098/rstb.2009.0267.

Future in Space Working Group. (2018). Teleconference audio archived online at the *Future in Space Working Group* archives at http://fiso.spiritastro.net/telecon/Ferl-Dumbacher_6-6-18/.

Heyer, J. M. B. S., MacAuley, A., Behrendtsen, O., & Werb, Z. (2000). Hypersensitivity to DNA damage leads to increased apoptosis during early mouse development. *Genes & Development, 14*(16), 2072–2084.

Hodkinson, P. D., Anderton, R. A., Posselt, B. N., & Fong, K. J. (2017). An overview of space medicine. *British Journal of Anaesthesia, 119*(S1), i43–i53.

Holmes, D. J. (2017). Women's health in post-menopausal age. In G. Jasienka, D. S. Sherry, & D. J. Holmes (Eds.), *The arc of life: Evolution and health across the life course* (pp. 139–157). New York: Springer.

Holstein, A.-F., Schulze, W., & Davidoff, M. (2003). BioMed Central document at http://www.rbej.com/content/1/1/107.

Honda, Y., et al. (2012). Genes down-regulated in spaceflight are involved in the control of longevity in *Caenorhabditis elegans*. *Nature Scientific Reports, 2*(487), 2–7. https://doi.org/10.1038/srep00487.

Horn, E. (2006). Animal development in microgravity. In G. Clement & K. Slenzka (Eds.), *Fundamentals of space biology: Research on cells, animals and plants in space* (pp. 171–226). El Segundo, CA: Microcosm Press; New York: Springer.

Hoson, T. (2014). Plant growth and morphogenesis under different gravity conditions: Relevance to plant life in space. *Life, 4*(2), 205–216. https://doi.org/10.3390/life4020205.

Ibtisham, F., Wu, J., Xiao, M., An, L., et al. (2017). Progress and future prospect of in vitro spermatogenesis. *Oncotarget, 8*(39), 66709–66727.

Ijiri, A. (2003). Life-cycle experiments of medaka fish aboard the International Space Station. *Advances in Space Biology and Medicine, 9,* 201–216.

Ijiri, K. (1998). Development of space-fertilized eggs and formation of primordial germ cells in the embryos of medaka fish. *Advances in Space Research, 21*(8–9), 1144–1188.

Jakab, A., Schwartz, E., Kasprian, G., Gruber, G. M., Prayer, D., Schoft, V., & Langs, G. (2014). Fetal functional imaging portrays heterogeneous development of emerging human brain networks. *Frontiers in Human Neuroscience* online at https://doi.org/10.3389/fnhum.2014.00852.

Janmaleki, M., et al. (2016). Impact of simulated microgravity on cytoskeleton and viscoelastic properties of endothelial cell. *Scientific Reports, 6,* 32418. https://doi.org/10.1038/srep32418.

Jennings, R. T., Murphy, D. M. F., Ware, D. L., et al. (2006). Medical qualification of a commercial spaceflight participant: Not your average astronaut. *Aviation, Space and Environmental Medicine, 77*(5), 475–484.

Jonscher, K. R., et al. (2016). Spaceflight activates lipotoxic pathways in mouse liver. *PLOS One, 11*(5), e0155282. https://doi.org/10.1371/journal.pone.0155282.

Kapia, A., & Hsueh, A. J. (1997). Regularion of ovarian follicle atresia. *Annual Review of Physiology, 1997,* 349–363.

Kirschnick, U., Agricol, H.-J., & Horn, E. R. (2006). Effects of altered gravity on identified peptidergic neurons of the cricket *Acheta domesticus*. *Gravitational and Space Biology Bulletin, 19*(2), 135–136.

Lane, H. W., & Feeback, D. L. (2002). History of nutrition in space flight: Overview. *Nutrition, 18*(10), 797–804.

Latham, K. E., & Schultz, R. M. (2001). Embryonic genome activation. *Frontiers of Bioscience, 6,* d748–d759. https://www.bioscience.org/2001/v6/d/latham/fulltext.htm.

Leffler, E. M., Bullaughey, K., Matute, D., Meyer, W. K., Segruel, L., Venkat, A., et al. (2012). Revisiting and old riddle: What determines genetic diversity levels within species? *PLOS Biology, 10*(9), 1–9.

Loomer, P. M. (2001). The impact of microgravity on bone metabolism in vitro. *Critical Reviews in Oral Biology and Medicine, 12*(3), 252–261.

Louis, F., Deroanne, C., Nusgens, B., Vico, L., & Guignandon, A. (2015). RhoGTPases as key players in mammalian cell adaptation to microgravity. *Biomedical Research International, 2015,* 747693. https://doi.org/10.1155/2015/747693.

Lynch, M. (2010). Rate, molecular spectrum, and consequences of human mutation. *Proceedings of the National Academy of Sciences, 107*(3), 961–968.

Mark, S., Graham, B. I., Donoviel, D. B., et al. (2014). The impact of sex and gender on adaptation to space: Executive summary. *Journal of Women's Heath, 3*(11), 941–947.

Miao, Y.-L., Kikuchi, K., Sun, Q.-Y., & Schatten, H. (2009). Oocyte aging: Cellular and molecular changes, developmental potential and reversal possibility. *Human Reproduction Update, 15*(5), 573–585. https://doi.org/10.1093/humupd/dmp014.

Mistry, K. B., et al. (2012). A new framework for childhood health promotion: The role of policies and programs in building capacity and foundations of early childhood health. *American Journal of Public Health, 102*(9), 1688–1696.

Miyake, M., Hazama, A., Nielsen, S., & Shimizu, T. (2004). Effects of microgravity on organ development of the neonatal rat. *Biological Sciences in Space, 18,* 126–127.

National Academies of Science of the United States of America. (2006). *Space radiation hazards and the vision for space exploration*. Washington, D.C.: National Academies Press.

National Academies of Science of the United States of America. (2018). *A midterm assessment of implementation of the decadal survey on life and physical sciences research at NASA*. Washington, D.C.: National Academies Press.

National Academies of Science of the United States. (2013). *Reducing maternal and neonatal mortality in Indonesia: Saving lives, saving the future*. National Academies Press.

National Institutes of Health. (2019). *Preconception care and prenatal care*. See https://www.nichd.nih.gov/health/topics/preconceptioncare/conditioninfo/health-problems.

Neubert, J. B., et al. (1998). Effects of gravity on early development. *Advances in Space Research, 22*(2), 265–271.

Nikolaev, A., & Yang, E. S. (2017). The impact of DNA repair pathways in cancer biology and therapy. *Cancers, 9*(9). https://doi.org/10.3390/cancers9090126.

Ning, L.-N., Lei, X.-H., Cao, Y.-J., Zhang, Y.-F., Cao, Z.-H., Chen, Q., et al. (2015). Effect of short-term hypergravity treatment on mouse 2-cell embryo development. *Microgravity Science and Technology, 27*(6), 465–471.

Norbury, J. W., et al. (2016). Galactic cosmic ray simulation at the NASA Space Radiation Laboratory. *Life Sciences in Space Research, 8,* 38–51. https://doi.org/10.1016/j.lssr.2016.02.001.

Ogneva, I. V., Belyakin, S. N., & Sarantesva, S. V. (2016). The development of *Drosophila menalogaster* under different duration space flight and subsequent adaptations to Earth gravity. *PLOS One, 11,* e0166885. https://doi.org/10.1371/journal.pone.0166885.

Olson, W. M., Wiens, D. J., Gaul, T. L., Rodriguez, M., & Hauptmeier, C. L. (2010). *Xenopus* development from late gastrulation of feeding tadpole in simulated microgravity. *International Journal of Developmental Biology, 45,* 167–174.

Ombelet, W., Menkveld, R., Kruger, T. F., & Steeno, O. (1995). Sperm morphology assessment: Historical review in relation to fertility. *Human Reproduction Update, 1*(6), 543–557.

Page 2 of Delaney, L., & Smith, J. P. (2012). Childhood health: Trends and consequences over the life-course. *Future Child*, 22(2), 43–63.

Pan, D. (2007). Hippo signaling in organ size control. *Genes & Development, 21,* 886–897. https://doi.org/10.1101/gad.1536007.

Papaseit, C., Pochon, N., & Tabony, J. (2000). Microtubule self-organization is gravity-dependent. *Proceedings of the National Academies of Science of the United States of America, 97*(15), 8364–8368.

Pellegrini, M., Di Siena, S., Claps, G., Di Cesare, S., Dolci, S., Rossi, P., et al. (2010). Microgravity promotes differentiation and meiotic entry of postnatal mouse male germ cells. *PLoS One, 5*, e9064. https://doi.org/10.1371/journal.pone.0009064.

Ricci, G., Catizone, A., Esposito, R., & Galdieri, M. (2004). Microgravity effect on testicular functions. *Journal of Gravitational Physiology, 11*(2), 61–69.

Richard, S., Henggeler, D., Ille, F., Beck, S. V., Moeckli, M., Forster, I. C., et al. (2012). A semi-automated electrophysiology system for recording from *Xenopus* Oocytes under microgravity conditions. *Microgravity Science and Technology, 24*(4), 237–244.

Ross, M. D., & Tomko, D. L. (1998). Effect of gravity on vestibular neural development. *Brain Research Reviews, 28*(1–2), 44–51.

Russell, W. L., Russell, L. B., & Kelly, E. M. (1960). Dependence of mutation rate on radiation intensity. *International Journal of Radiation Biology, 1960*(Supplement), 311–320.

Schimmerling, W. (2010). Accepting space radiation risks. *Radiation Environments and Biophysics, 49*, 325–329.

Searby, N. D., Steele, C. R., & Globus, R. K. (2005). Influence of increased mechanical loading by hypergravity on the microtubule cytoskeleton and prostaglandin E2 release in primary osteoblasts. *American Journal of Physiology and Cellular Physiology, 289*(1), C148–C158.

Serova, L. V. (1989). Effect of weightlessness on the reproductive system of mammals. *Kosmicheskaia Biologiia i Aviakosmicheskaia Meditsina, 23*(2), 11–16.

Servick, K. (2015, April 27). Gene activation therapy prevents liver damage in mice. *Science*. https://doi.org/10.1126/science.aab2549.

Shimada, A., et al. (2005). Germ cell mutagenesis in medaka fish after exposures to high-energy cosmic ray nuclei: A human model. *Proceedings of the National Academies of Science of the United States of America, 102*(17), 6063–6067.

Simonsen, L. C. (2015). Mars mission and space radiation risks: Space radiation research and technologies for risk mitigation briefing to NAC HEO/SMD Joint Committee April 7, 2015. Online at https://ntrs.nasa.gov/archive/nasa/casi.ntrs.nasa.gov/20160006861.pdf.

Singh, K., Cubano, L., & Lewis, M. (2010). Effects of gravity perturbation on developing animal systems. *Advances in Space Research, 6*(12), 29–36.

Smith, C. M. (2011). *The fact of evolution*. New York: Penguin Random House.

Smith, I. M., & Forsman, A. D. (2012). Ovarian follicular and luteal development in the spaceflight mouse. *Gravitational and Space Biology, 26*(2), 30–37.

Souza, K. H., Black, S. D., & Wassersug, R. J. (1995). Amphibian development in the virtual absence of gravity. *Proceedings of the National Academies of Science of the United States of America, 9*, 1975–1978.

Tan, X. (2006). *The effects of microgravity on gametogenesis, fertilization, and early embryogenesis*. Paper delivered at 36th Committee on Space Research Scientific Assembly. Held 16–23 July 2006, in Beijing, China. Meeting abstract from the CDROM, #3631. Online at http://adsabs.harvard.edu/abs/2006cosp.meet.3631T.

Tash, J. S., & Bracho, G. E. (1999). Microgravity alters protein phosphorylation changes during initiation of sea urchin sperm motility. *FASEB Journal, 13*(9001), 843–854.

Tash, J. S., Johnson, D. C., & Enders, G. C. (2002). Long-term (6-wk) hindlimb suspension inhibits spermatogenesis in adult male rats. *Journal of Applied Physiology, 92*(3), 1191–1198.

Thibeault, S. A., Kang, J. H., Sauit, G., & Park, C. (2015). Nanomaterials for radiation shielding. *Materials Research Society Bulletin: Engineered Nanomaterials in Aerosopace, 40*(10), 836–841.

Tourmente, M., Gormendio, M., & Foldan, E. R. S. (2011). Sperm competition and the evolution of sperm design in mammals. *BMC Evolutionary Biology, 11*, 12. Online at http://www.biomedcentral.com/1471-2148/11/12.

US Social Security Administration Actuarial Tables. (2018). Online at https://www.ssa.gov/oact/STATS/table4c6.html.

Vaquer, A.-F., & Hadjantonakis, A.-K. (2013). Birth defects associated with perturbations in preimplantation, gastrulation and axis extension: From conjoined twinning to caudal dysgenesis. *Wiley Interdisciplinary Review of Biology, 2*(4), 427–442.

Vernós, I., González-Jurado, J.-, Calleja, M., & Marco, R. (1989). Microgravity effects on the oogenesis and development of embryos of *Drosophila melanogaster* laid in the Spaceshuttle during the Biorack experiment (ESA). *International Journal of Developmental Biology, 33*(2), 213–216.

Villanueva, M. M., Wong, M., Lu, T., Zhang, Y., & Wu, H. (2017). Interplay of space radiation and microgravity in DNA damage and DNA damage response. *Microgravity, 3,* 14. https://doi.org/10.1038/s41526-017-0019-7.

Wakayama, S., et al. (2017). Healthy offspring from freeze-dried mouse spermatozoa held on the International Space Station for 9 months. *Proceedings of the National Academies of Science of the United States of America, 114*(23), 5988–5993.

Wakayama, S., Kwahara, Y., Li, C., Yamagata, K., Yuge, L., & Wakayama, T. (2009). Detrimental effects of microgravity on mouse preimplantation development in vitro. *PLoS ONE, 4*(8), e6753. Online at pone.0006753 1..10.

Wang, Y., An, L., Jiang, Y., & Hang, H. (2011). Effects of simulated microgravity on embryonic stem cells. *PLoS ONE, 6,* e29214. https://doi.org/10.1371/journal.pone.0029214.

Woese, C. R. (2004). A new biology for a new century. *Microbiology and Molecular Biology Reviews, 68*(2), 173–186.

Wohlgemuth, D. J., & Murashov, A. K. (1995). Models and molecular approaches to assessing the effects of the microgravity environment and vertebrate development. *American Society for Gravitational and Space Biology Bulletin, 8*(2), 63–71.

Wolpert, L. (2008). *Triumph of the embryo*. Mineola, New York: Dover Publications.

Woodbury, R. A., Hamilton, W. F., & Torpin, R. (1938). The relationship between abdominal, uterine and arterial pressures during labor. *American Journal of Physiology, 121*(3), 640–649.

World Bank. (2018). *Life expectancy at birth.* See https://data.worldbank.org/indicator/SP.DYN.LE00.IN?end=2016&locations=US&start=1960&view=chart.

World Health Organization. (2006). *Promoting optimal fetal development: Report of a technical consultation*. Available at https://www.who.int/nutrition/publications/fetomaternal/9241594004/en/.

Wu, C., Guo, X., Wang, F., Li, X., Tian, C., Li, L., et al. (2011). Simulated microgravity compromises mouse oocyte maturation by disrupting meiotic spindle organization and inducing cytoplasmic blebbing. *PLoS ONE, 6*(7), e22214. https://doi.org/10.1371/journal.pone.0022214.

Xie, Y., Wang, F., Zhong, W., Puscheck, E., Shen, H., & Rapollee, D. A. (2006). Shear stress induces preimplantation embryo death that is delayed by the zona pellucida and associated with stress-activated protein kinase-mediated apoptosis. *Biology of Reproduction, 75*(1), 45–55.

Yamauchi, Y., Riel, J. M., Ruthig, V. A., Ortega, E. A., Mitchell, M. J., & Ward, M. A. (2016). Two genes substitute for the mouse Y chromosome for spermatogenesis and reproduction. *Science, 351*(6272), 514–516.

Yatagai, F., & Ishioka, N. (2014). Are biological effects of space radiation really altered under the microgravity environment? *Life Sciences in Space Research, 3,* 76–89.

Chapter 3
Population Genetics of Human Space Settlement

Abstract Just as we must see to the health of individual people when we venture permanently beyond Earth, we will have to see to the health of our populations—the gene pool, which in practical terms is composed of groups of families. In popular culture this often immediately brings up the spectre of 'genetic screening' for space settlement, as depicted in the 1990 film 'Gatacca'. In that film a draconian genetic-screening programme sought to establish a 'super race' of spacefarers. While the film had useful messages about genetic screening generally, it was entirely wrong about what will make for successful populations of people beyond Earth; it is genetic diversity, not uniformity, that is most likely to make for success beyond Earth, as new selective environments and pressures are encountered, and in fact themselves change through time in ways that we will only learn when we experience them. How to maintain genetic health, diversity, then, in the project of space settlement? This chapter tackles this fascinating question by evaluating the well-known drivers of population genetic change, namely mutation, migration, selection and drift. We will also explore important questions of humanity's Minimum Viable Population, a figure often of interest to space planners but one I suggest is a useful guide only for early populations which I recommend to grow rapidly as insurance against catastrophe, to which small populations are most vulnerable. I also present several formulae useful to estimation of populations of space settlers in various circumstances and mention other current models. Finally, I discuss genetic screening and testing in relation to the issue of the genetic constitution of space-settling populations, and make recommendations regarding extraterrestrial human population sizes based on a broad review of human and other population phenomena and the lessons of several billion years of Earth life evolution; these supply important lessons in the fossil and genetic records. This chapter updates and expands upon work I first published in *Acta Astronautica*, in 2014, regarding viable populations for interstellar voyaging. While that paper was interesting and useful, the fuller expression of its theory and application are found in this chapter.

> The Growth and Increase of Mankind is not so much stinted by anything in the Nature of the Species, as it is from the cautious difficulty most People make to adventure on the state of marriage, from the prospect of the Trouble and Charge of providing for a family—Edmund Halley, 1693

C. M. Smith, *Principles of Space Anthropology*, Space and Society,
https://doi.org/10.1007/978-3-030-25021-8_3

> Genetic…populations are defined by the extent to which individuals share in a common gene pool…they are determined in terms of degrees of biological relatedness…[from] parents and children within the nuclear family, as these show the maximal degree of relationship [to] the whole species' [gene pool]…Between the family on one hand and the whole species on the other, there are many intermediate categories [of populations] whose precise recognition often presents complex problems…however…[understanding these categories] is a vital prerequisite to understanding the nature of human variety. (Harrison and Boyce 1972: 3–4)

3.1 Introduction

While individual organisms require conditions conducive to healthy development and the life course, no life form is entirely individual; certainly not the highly-social higher primates, including humans. Humans typically live in groups of various populations and identities, from mated pairs to families, extended families, neighborhoods and villages, towns and other such communities, cities, states, nations and even multinational coalitions. Human life then requires many regulations of population concerns, at the least to keep populations somewhat tailored to the available resources, an issue of particular significance to the early stages of space settlement when agriculture is just being established and stocks to survive disaster are not yet fully established.

In the same way, genes do not exist in *vacuuo* but are components of reproductive populations or *demes*; and these are only subdivisions of entire species populations. Table 3.1 characterizes some common human population scales among people practicing some form of farming.

Though there is quite a bit of variation in human population size and structure, for rural farming settlements—which in Chap. 6 I argue will have significant

Table 3.1 Population sizes of various common human biological and social groupings

	Individual	Mated Pair	Nuclear Family	Extended Family	Local Community	Dispersed Community	Reproductive Population (deme)
Number	1	2	c.4	c.10	c.50	c.150	Multiple 00's to 000's

While there are variations worldwide, all cultures have some concept of a *cohabitating family*, whatever its actual makeup. Mated Pair refers to individuals in a substantial bond, and all cultures have some marriage institution, normally between just two individuals. Nuclear families are generally cohabitating people (living in the same domestic structure), often married or pair-bonded ('Mated Pair'). The extended family refers to the mated pair plus its offspring and sometimes associated relatives, like elders or visiting relatives. The local community refers to a social 'neighborhood' or other, often vaguely-defined, unit of face-to-face, sometimes daily interaction; it is typically not recognized institutionally. A more 'Dispersed Community' is an extension of the neighborhood and includes people known, but not necessarily interacted with daily or often; it could be a hamlet or small village, for instance. A reproductive population or *deme* refers to the people one encounters in life who may be considered potential mating partners, whether they are local or not. Table by the author

similarities with early off-Earth populations—settlement population in the several hundreds to about 1000 has been common in the human experience. While these populations sound small compared to modern urban populations, note that had reproductive links with other groups. For example, Fig. 3.1 indicates that in 19th century southern England individuals did mate outside their own village (though often not very far). The point for our topic is that just as we need to understand the biology of the individual (Chap. 2), humans live in linked populations and we must therefore understand human population genetics for the project of space settlement (we return to issues of settlement size and arrangement in Chap. 6 and Sect. 7.4.12 *Settlement Pattern*).

Another example is seen in the case of Polynesia. When contacted by Europeans in the 18th century AD, many island populations were less than 1000 (though the arithmetic mean of a sample of 35, tabulated in Table 3.2, is 5799, this is skewed by the few large populations, and Fig. 3.2 more accurately displays the real distribution), each of these subdivided into smaller, local villages (Younger 2008). But these were not reproductively isolated populations in a genetic sense; they were on average a day or three away from neighboring islands (mean distance 0.65 sailing days at a slow cruising speed for a Polynesian voyaging canoe of 13.5 kph). And inter-island marriages were common; one study has identified these linking villagers of the Society Islands with Mo'orea (c.200 km away), the Cook Islands (nearly 1000 km away), the Tuamotus (over 1300 km away), and the Austral Islands (770 km away) (Gunson 1997).

The genetic phenomena observed in populations of interbreeding individuals, as exemplified above, is studied in the field of *population genetics*. The focus of this chapter is to identify the population-genetics issues relevant to human space

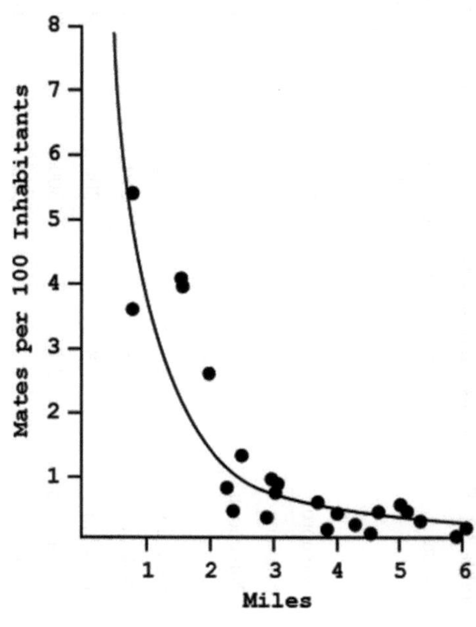

Fig. 3.1 Distance of marriage partner in 19th century Oxfordshire, England. Vertical scale is number of matings per 100 in a population, and horizontal scale is distance, in miles, of that mating from the villager's home. Diagram by the author after Boyce et al. (1967) as discussed in Fix (1999)

Table 3.2 Populations and distances to nearest neighboring population for 35 Polynesian Islands at European Contact

Island	Area (km^2)	Population at contact	Distance to nearest neighbor (km)	Distance to nearest neighbor (sailing days @ 27 kph)	Distance to nearest neighbor (hours @ 27 kph)	Distance to nearest neighbor (sailing days @ 13.5 kph)	Distance to nearest neighbor (hours @ 13.5 kph)
Nukuoro	1.7	124	177	0.3	6.6	0.5	13.1
Sikaiana	1.8	180	177	0.3	6.6	0.5	13.1
Nukunonu	4.7	275	60	0.1	2.2	0.2	4.4
Atafu	3.5	380	90	0.1	3.3	0.3	6.7
Fakaofo	4	500	60	0.1	2.2	0.2	4.4
Nukufetau	2.8	250	50	0.1	1.9	0.2	3.7
Funafuti	2.8	300	80	0.1	3.0	0.2	5.9
Nanumaga	3.1	300	50	0.1	1.9	0.2	3.7
Nui	1	300	120	0.2	4.4	0.4	8.9
Nukulaelae	1.8	300	120	0.2	4.4	0.4	8.9
Vaitupu	5.6	400	50	0.1	1.9	0.2	3.7
Niutao	2.8	450	100	0.2	3.7	0.3	7.4
Nanumea	2.7	500	60	0.1	2.2	0.2	4.4
Kapingamarangi	1.1	450	215	0.3	8.0	0.7	15.9
Pukapuka	7	600	530	0.8	19.6	1.6	39.3
Manihiki/Rakahanga	9.2	780	290	0.4	10.7	0.9	21.5
Tubuai	47.1	900	200	0.3	7.4	0.6	14.8
Tikopia	5	1200	135	0.2	5.0	0.4	10.0
Mangareva	13	1500	350	0.5	13.0	1.1	25.9
Tongareva	9.8	2000	180	0.3	6.7	0.6	13.3

(continued)

Table 3.2 (continued)

Island	Area (km^2)	Population at contact	Distance to nearest neighbor (km)	Distance to nearest neighbor (sailing days @ 27 kph)	Distance to nearest neighbor (hours @ 27 kph)	Distance to nearest neighbor (sailing days @ 13.5 kph)	Distance to nearest neighbor (hours @ 13.5 kph)
Mangaia	34.5	2500	177	0.3	6.6	0.5	13.1
Tahu Ata	70.8	1000	3	0.0	0.1	0.0	0.2
Ua Huku	82.4	1000	39	0.1	1.4	0.1	2.9
Ua Pou	112	1800	38	0.1	1.4	0.1	2.8
Fatu Hiva	78	3000	59	0.1	2.2	0.2	4.4
Nuku Hiva	345	6000	38	0.1	1.4	0.1	2.8
Hiva Oa	318	6500	3	0.0	0.1	0.0	0.2
Uvea	60	3000	180	0.3	6.7	0.6	13.3
Rotuma	44	3500	500	0.8	18.5	1.5	37.0
Futuna/Alofi	94	5000	180	0.3	6.7	0.6	13.3
Niue	260	5000	400	0.6	14.8	1.2	29.6
Tongatapu	259	8000	25	0.0	0.9	0.1	1.9
Rapa Nui	160	10,000	2600	4.0	96.3	8.0	192.6
Tahiti	1069	35,000	17	0.0	0.6	0.1	1.3
Hawa'ii	10,400	100,000	50	0.1	1.9	0.2	3.7

Data based on Younger (2008) and modified by the Author. The sailing speed estimations derive from Lewis (1972 [1994]). Table adapted from Gunson (1997) by the author

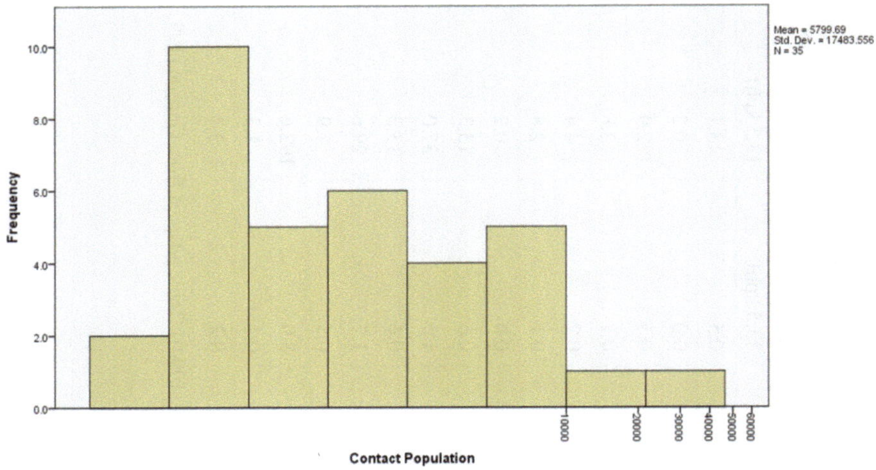

Fig. 3.2 Histogram of island populations tabulated in Table 3.2. The X axis is log scale and indicates skewness of the data, with many population figures below 5000. Data from Table 3.2, characterized by the Author in SPSS

settlement. To this end I define the main drivers of genetic change in populations, address how they may be considered in space settlement planning. I also address a few other population phenomena of interest including figures such as *Minimum Viable Population*, *Effective Population* and demographic structure, all of great interest (and discussed later in this chapter) when we consider the settlement of communities beyond Earth with the goal of becoming independent of Earth.

Throughout the chapter I focus on a timescale of roughly one to five generations, about 30–150 years. This figure has been selected for three main reasons. First, it is easily modelled by what we know of population genetics today, and does not strain any models of how populations change through time. Second, the timescale is one in which we may reasonably assume we are considering populations sharing some of our own basic cultural features; Mars families derived from Earth will likely be rather like modern people and are unlikely to radically differ bioculturally on this timescale; consider comparing modern people and those of Elizabethan England (c.500 years ago), in which comparison we see some differences but essential similarity culturally and biologically no significant difference. This is important because by the time such large cultural change has occurred, cultural traditions may well change how people make mate choices, the frequency of mating and so on, which *would* affect population genetics in an excellent example of biocultural evolution. Third, genetic and population-genetic technologies are likely to remain at least recognizable in this timescale, whereas a few centuries down the line we may well see such technologies becoming so distant from that of our own that it is unproductive to speculate on its effects.

It is important to note that on such a timescale we may or may not see significant genetic change on the population level; human behavior is complex and can accelerate or decelerate such change via technological and social mediation of behavior. It

Nature of Change	Mechanism of Change	Common Timescale of Change	Example
Speciation	reproductive isolation	thousands to millions of years	evolution of *Homo erectus* into *Homo heidelbergensis*
Substantial Gene Frequency Change	mutation, migration, selection, drift	centuries to many thousands of years	spread of genes allowing adult lactose tolerance among early cattle-rearing cultures
Culture Change	acceptance of innovation	highly variable, from *intergenerational* (e.g. a revolution) to *multigenerational* (e.g. church reform)	evolution of *Homo erectus* into *Homo heidelbergensis*
Acclimatization	physiological response	during the lifecourse of the individual	vasoconstriction in humans when exposed to cold

Fig. 3.3 Timescales for genetic change. Culture change is included here because cultural activity can influence genetics; for example, cultural mating taboos ensure that reproduction in human gene pools is not random, but can (a) increase close-kin marriages, in the case of socially-strategic marriages or (b) decrease close-kin marriages by social sanction against them. Diagram by the Author

is, though, likely that such populations will see selection acting against various genes and genetic *pathways* (complex consequences of genetic instructions) if they are not well-suited to conditions beyond Earth, resulting in individuals born with genetic conditions that are deleterious to the individual. The persistence of such genes or pathways depends in part on human behavior, which can, again, accelerate or decelerate the dispersal of such genes in a population. In the sections below I illustrate the main forces of change on the population level and comment on their relevance for space settlement at various timescales e.g. within-generation change and across-generation change (Fig. 3.3).

3.2 Essential Population Genetics Relating to Human Space Settlement

As outlined, evolution may be defined as any difference in the genetic properties of parent and offspring generations. Particularly in sexually-reproducing species, which contribute male and female DNA in sperm and egg production to the next generation—producing novelties (AKA 'mutations')—cloning is rare, and evolution occurs at each generation. Four essential factors condition the genetic composition of any population. They are discussed below, with notes on their implications in considering space settlement with the eventual goal of total independence from Earth.

3.2.1 Mutation

Mutation is the production of new genetic characteristics, also known as *diversity*. Mutagenesis occurs in many ways, including (a) mechanical deformation of DNA during replication, (b) alteration of DNA by mutagens such as radiation and certain chemicals, (c) the recombination of parental DNA in the production of gametes (sperm and egg cells), (d) the failure of mutation-repair processes, and (more so in prokaryotes than eukaryotes) (e) horizontal gene transfer, the acquisition of heritable DNA alteration during the course of life (Friedberg 2006). Whatever the cause, mutations generate differences between parent and offspring, and differences among members of any given generation. Currently, humans are estimated to have about 38 DNA base pair mutations per human gamete (sex cell, either egg or sperm) as a result of recombination during gamete formation, with each newborn carrying roughly 50–100 total mutations (genetic differences among the 3 billion base pairs of the human genome) from its parents and/or siblings (Lynch 2010: 966). Mutations may be *advantageous*, conferring an advantage to the carrier and making it more likely that the mutation will be passed on to the next generation, *disadvantageous*, conferring a disadvantage (e.g. poorer health) to the carrier, making it less likely that the mutation will be passed on to the next generation, or *neutral*, having no measureable effect on the likelihood that it will be passed on to the next generation; this does not mean, however, that neutral mutations never become more or less common in populations, as we will see below in Sect. 3.2.4, **Drift** (Fig. 3.4).

The *genetic diversity* of a population (D), an important quantity in long-term population planning, is a product of a mutation rate (μ), the population size (N_c or N_e, introduced below), and time (t), because mutations accumulate in a genome over time (Relethford 2001: 151). Note that t indicates *generations* (or *intergenerational interval*), not years. Human generation time (the number of years between birth and when an adult actually has offspring) has been estimated many times and within limits (ranging from c.25 to 30 among many sampled human populations) is "rather consistent regardless of the cultural and environmental differences among human societies" according to Matsamura and Forster (2008: 1504) and here I use 30 years, averaging their estimated 25–35 year generation times. We will return to these quantities later in this chapter.

Mutation rate is quite variable among species, subpopulations of species, families, individuals, sexes and even bodily tissues (Duret 2009) so no single mutation rate can be used in all calculations. Having said this, an average nuclear mutation rate per generation for modern humans has been estimated at 1.3×10^{-8} per generation (Leffler et al. 2012: 9); as mentioned, such a figure is used to help calculate figures that will be introduced below.

With regard to space settlement efforts, mutation rates across all tissues, individuals, sexes and so on would be increased by increased exposure to various forms of radiation, as explored in Chap. 2; however we must assume such exposure will be mitigated sufficient for long-term health and that it will not be a 'show-stopping' hurdle for space settlement (one of many recent studies suggests the radiation hazard

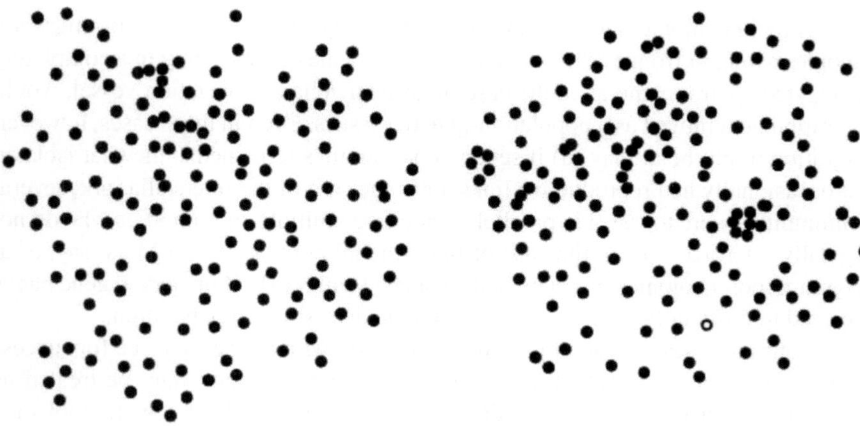

Fig. 3.4 Visual depiction of mutation as a force of change in population genetics. In generation 1 the individuals are all of the same life form (species), having the same essential characteristics of being circular and black. In generation 2 the life forms have had offspring, who are somewhat rearranged across the landscape. In the lower center a mutation, resulting from mating in generation 1, has appeared; it has the feature of a circular shape, but its color is white rather than black. This mutation may be beneficial, in which case its carrier might have more offspring than others, leading to the spread of this characteristic of white coloration. If the feature is deleterious, however, the individual is less likely to pass on the characteristic and it will likely disappear from the gene pool. If the feature is neither beneficial nor deleterious, it may persist in the population for generations and may, by chance alone, either become diluted or common. Diagram by the Author

of interstellar space is tolerable in some circumstances (Horneck et al. 2002). This is analogous to assuming, for the purposes of this book and study, that the various propulsion and navigation technologies will be available for human space settlement in a reasonably-expectable future. For the moment, increased mutation rate beyond Earth may be considered a concern more of the individual (addressed in Chap. 2) than the population, because it will be individuals themselves who are most guarded from mutation-inducing radiation such that the problem never appears on the population level.

3.2.2 *Migration*

Migration is a change in the genetic properties of a population as a result of individuals physically moving into or out of that population (Wijsman and Cavalti-Sforza 1984). Migration is driven by many motivators; two obvious kinds are "push" factors that compel migration due to local competition, and "pull" factors including the perception of new, nonlocal opportunities; note that human migration may be driven by cultural imperatives not found in other life forms (Gamble 2003). For the forseeable future, populations of humans on Mars or other near-Earth settlements (such as Lagrangian colonies) would have the opportunity for both in- and out-migration (people mating in from Earth and mating out from the colony); only in a completely segregated environment, as in the case of an interstellar colonization vessel, would migration be removed as a population-genetics issues. Even in these cases, however, migration might be at play (a) if several communities remained somewhat isolated but occasionally had reproductive links, or in the case of the interstellar ark, several communities were to travel in parallel, containing multiple communities who do not typically interbreed (or in the case of independent populations on Mars, say, or at Lagrange-point colonies or others in the solar system) or (b) if preserved gene banks are used to artificially 'in-migrate' new genetic diversity to a population.

For our purposes—those of providing humanity with the best chance for success in the project of space settlement—for the moment migration may be treated as a genetic health issue. As with individual health, the health of a gene pool may be compromised. Generally the greatest threat to a gene pool is *uniformity*. Broadly speaking, if all individuals are the same or nearly the same (low diversity), the chances of the population being unable to adapt to changes in the selective environment are reduced. For large populations of modern humans spread globally, this is a very unlikely case, but for the early generations of off-Earth settlers, populations will be low, and possibly dispersed, and so small populations may well be subject on occasion to strong selection against certain genetic characteristics. To guard against such a case we may recommend a general rule that the smaller the population, the more genetically diverse it should be. I would keep in mind that the people who eventually take up the project of space settlement are very likely to be the people who themselves form families and communities—they will not be 'assigned' such roles—and they will ensure such diversity with their own social mechanisms.

For the purposes of this book we may say that the most likely variety of migration significant to space settlement on the 30–150-year timescale would be the continual arrival of new genetic material from Earth (or other off-Earth populations) to any given off-Earth population; this would be genetic *in-migration* from the perspective of the off-Earth population. Nothing exotic would need to be done to avoid population genetics issues in such a scenario. Routine genetic screening and/or testing (see Sect. 3.6 below) will be an option as it is an option for many today. I argue that it is not our place to weigh heavily on such an issue of reproductive freedoms because how such tools will be utilized to ensure a desired level of diversity will be established by the settling populations themselves. I feel it is our place, in space anthropology, to advise what would be most healthy for a given space colony gene pool. The answer is simply, diversity. An example shows why it would be fruitless, at this time, to specify which gene forms (*alleles*) would be 'favorable' to carry beyond Earth.

A single base-pair mutation in the HBB gene (which has 1605 base pairs) on human chromosome 11 sometimes results in the carrier having malformed red blood cells which can impair blood oxygen transport leading to anemia and, often, death. On the other hand, some carriers do not have this malady *but* due to the unusual blood cell shape are more resistant to malaria. In this case, preventing all carriers of the more dangerous form of the HBB gene from being in a given population would prevent *both* the deleterious *and* advantageous gene form from being in a given population. This example is not about malaria but the fact that some genetic outcomes are advantageous and some are disadvantageous, and that it is very difficult to predict which characteristics will be advantageous because (a) we do not know what new selective pressures we will encounter and (b) because the universe changes, and is not static, we cannot know what characteristics will be advantageous or disadvantageous in the future. I believe the lesson is clearly made that as the environments beyond Earth are better explored and learned, we will have a better idea of what characteristics would be advantageous, but we must always recall that we cannot be sure how these environments will change and thus which gene forms will be advantageous over time. Furthermore, some alternate forms of a gene may persist in a population because they are functionally linked to critical developmental processes; eliminating the variant would eliminate the carrier. Finally, there are also some genetic conditions that are typically expressed only long after reproductive life (for example, Alzheimer's disease) and are not a detriment to a full and productive life. For these reasons the best policy will always be genetic diversity, giving a population a wide range of variation with which to survive environmental changes (Fig. 3.5).

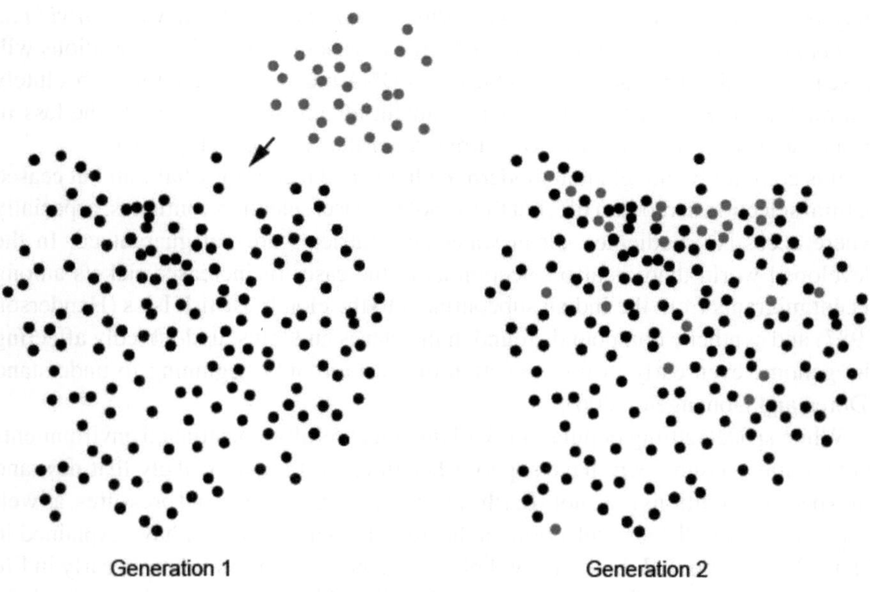

Fig. 3.5 Visual depiction of migration as a source of change in population genetics. In generation 1 the population on lower left is quite uniform, all being circular in shape and black in color; but arriving at the same time are individuals of the same species—they are circular—but with a variation, a gray color. Within a generation, interbreeding of these two slight variants of the previously-separated population results in new genes (gray) having 'migrated into' the original population, increasing diversity. Migration can go the other way as well, taking diversity out of a population

3.2.3 Selection

Selection is the favoring of certain individuals and characteristics in a population over others, often manifest in differential reproductive success of individuals such that over time, a population of a species or an entire species becomes tailored to its environment as poorer-fitted individuals become less numerous and more fit individuals become more numerous (Endler 1986; Hull et al. 2001). Selection can be *natural* (e.g. the effect of natural environment) or *cultural* [e.g. humans make careful mate choices, affecting genetics, and cultural information changes by an evolutionary process as well, a process beyond the scope of this paper; see Durham (1993) and Smith (2013)]. In nature the term 'selection' is unfortunate in that it implies an intent or directionality for which we have no evidence. Certainly with selective breeding human beings select for and against certain properties in our domesticated species, propagating those we prefer and eventually terminating those that we do not. But most selection is not the result of such choice; rather, factors such as temperature, precipitation, radiation, the presence of predators and so on determine which individuals (and thus which genes) will survive to form the next generation, and which will not. Selection, then, may be referred to as the *non-random differential survival of varying individuals in a population*. While variation has a degree of randomness (which variations will arise is determined by such random factors as DNA mutation), selection is absolutely not random; the better-fitted to an environment tend to propagate while the less-fit propagate less, leading to order over time rather than disorder (Fig. 3.6).

It is commonly thought that modern medicine and technology have all but ceased natural selection in modern life, but this is not the case; selection continues, especially where access to medicines, clean water and nutrients are not guaranteed. In the developed world there is also selection as in the cases of increased rickets among recent migrants from the Indian subcontinent to the cloudy British Isles (Henderson 1987) and synthetic compounds found in the uterus and fetus, undoubtedly affecting the genome even early in life, selection that we are only beginning to understand (Dorea and Donangelo 2006).

While space-settling populations will inhabit carefully-constructed environments and we must assume will have superior health care, it remains likely that new and unexpected conditions (e.g. non-Earth-normal gas atmospheres and pressures, as well as gravitation) will cause selection on the human population. As I have explained in Chap. 2 of this book, I think this will play out especially in gastrulation, early in life when things are quite fragile (Patrick et al. 2007). Also, no matter what technology is used to adapt, some new selection will surely shape the human genome before cultural and technological buffers can be invented by the earliest pioneers. Because we cannot profitably predict such selective pressures at present, selection is not discussed in this chapter, but rather in Chap. 2.

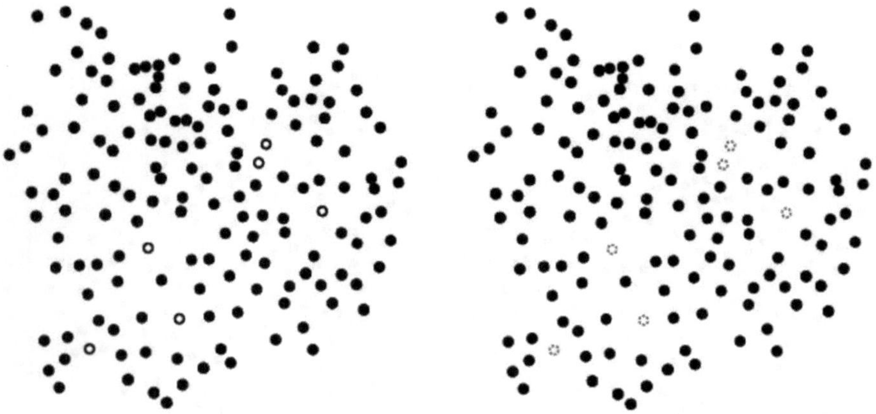

Fig. 3.6 Visual depiction of selection as a source of change in population genetics. In generation 1 the population is rather uniform, composed of round individuals mostly black in color with some white in color. Between generation 1 and generation 2, some environmental variable changes—let us say it becomes very cold—and that change is deadly to individuals of the white coloration. They die, or have so few offspring because of their poor health, that by generation 2 the characteristic of white coloration is extinct. This is selection 'against' a trait, making it less common, but there can also of course be selection *for* a given trait, making it *more* common. Central to evolutionary understanding today is that what changes will come along and reshape the population by such mechanisms is very unpredictable, especially over long time periods, such that evolution itself is largely unpredictable. What it is to be 'fit' today (white coloration in the Figure) may be a detriment tomorrow, when conditions have changed

3.2.4 Drift

Drift drives genetic change by a sampling phenomenon unrelated to selection; it reflects a degree of randomness affecting the likelihood that genes are propagated and is best understood as the fact that in a population of a given species, while on average an individual might have 1.3 offspring, in real life individuals not have 1.3 offspring but whole numbers of offspring, such as one, or two or three (Masel 2012: 704); over time, simply because of this fact, gene frequencies change even absent strong selection. Drift is most active in small, isolated populations with low population growth rates. This would not likely be the situation on early Mars or a Lagrangian colony, both of which would I think encourage early and rapid growth and in-migration of new genes. However it is a concern in modeling the exotic prospect of multigenerational voyaging with a rather fixed population over some generations. We will examine this case further below in this chapter (Fig. 3.7).

The principal effect of drift in closed populations can be estimated by modeling the loss of *heterozygosity*, which is a measure of the fact that in many genetic maladies, two copies of a deleterious gene (one from each parent and each variant of the single gene called an *allele*) must be present for the deleterious effect to be

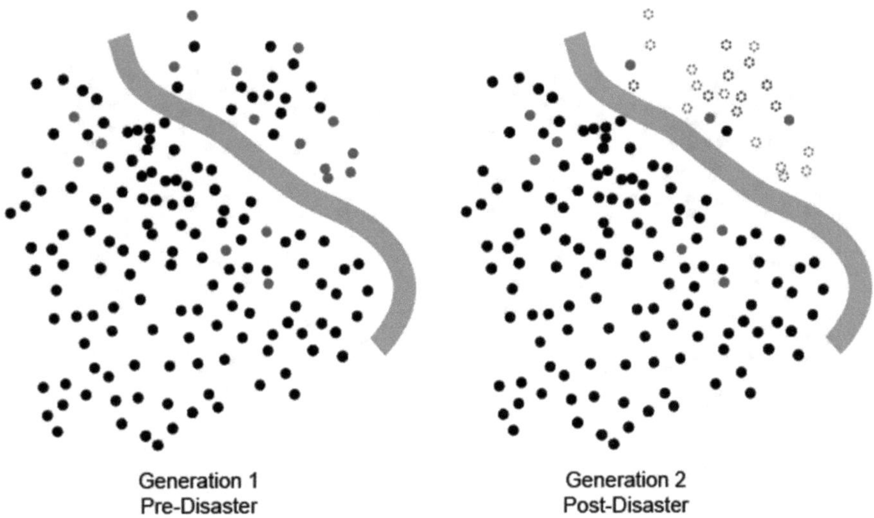

Fig. 3.7 Visual depiction of drift as a source of change in population genetics. In generation 1 the population of life forms is geographically spread both West and East of a river (gray meandering line), and each population has some individuals colored black and some colored gray; grays are much more common East of the river. Between generation 1 and generation 2 there is a natural disaster that ravages the population East of the river, killing many individuals (dotted lines indicate death). The survivors, a few gray-colored and a single black-colored individual, survived not because they were better-equipped to survive the disaster, they were simply lucky. As a simple consequence of this disaster, there will simply be fewer gray individuals mating into the larger population West of the river, and over time the gray coloration might become much less common

expressed in an individual (see the sickle cell example in 3.2.2). Heterozygosity refers to the amount of allele diversity in the gene pool; for example, if heterozygosity is 0.5, half of the alleles of a certain gene are deleterious and half not; this is an average for many deleterious genes in a human population. As a closed population that does not substantially increase in size continues to inbreed (as on a starship), the statistics of mate-selection dictate that the number of heterozygotes decreases, increasing homozygosity, or the chance that individuals will carry two copies of the deleterious allele (one from each parent), leading to expression of such genetic maladies. This inbreeding effect, often seen in cases of incest, results in the concentration of deleterious genes and reduced overall health of the population; close incest, for example, causes mental disorders (Adams and Neel 1967; Jancar and Johnston 2008) and one study has shown that first-cousin matings among Amish, Indian, French, Japanese, Pakistani, Swedish and Utah populations produced up to 22% infant mortality, roughly double the percentage seen in offspring of unrelated individuals (Jorde et al. 1999). Population size and density can of course affect the likelihood of close-relative mating. One example is seen in the case of the frequency of first-cousin marriage among rural populations, illustrated in Fig. 3.8. Here, mating between first cousins is more frequent in areas with lower population density and thus fewer nearby mate choices per person of reproductive age.

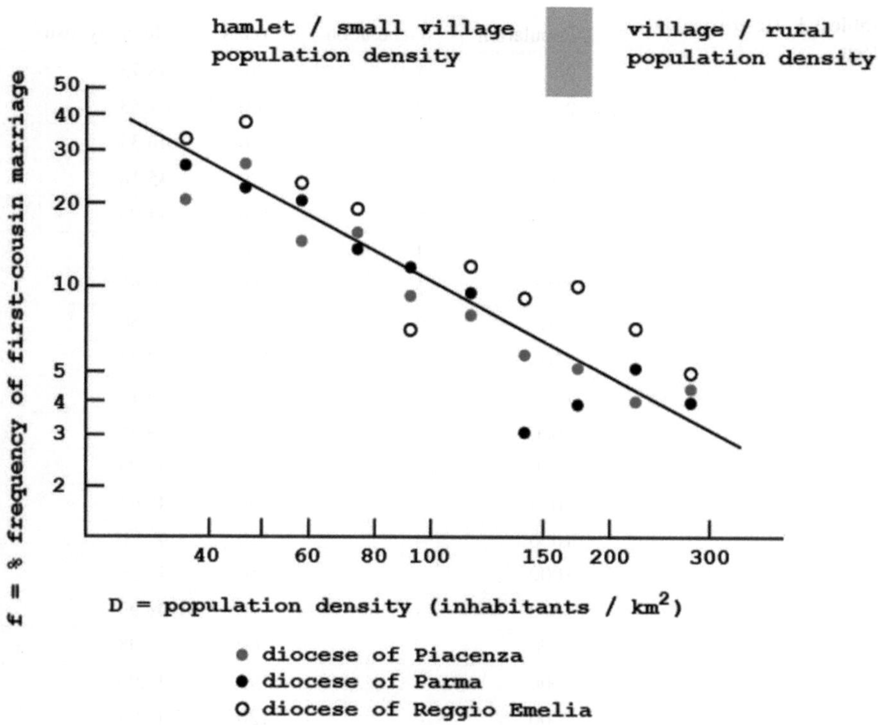

Fig. 3.8 Frequency of first-cousin marriage and settlement population density. Populations with a higher density (and, in actuality, a larger population) offer more potential mates per person, resulting in lower first-cousin marriages. This increases genetic diversity. Cross-cousin marriage has been shown (as mentioned in the text) to cause reduction in genetic diversity and increase in the frequency of genetic disorders in a population. The line indicates a mathematical relationship of f = 1.05/D. The diagram is adapted from Barigozzi et al. (1959), with modification by the author

The reduction of heterozygosity in a population, as we may see in the case of small, genetically isolated populations, is steady, and can be determined from the equation:

$$H = 1 - (1 - 1/2N_e + 1)^t$$

Where H = heterozygosity, N_e = effective population (discussed below) and t = elapsed generations. Table 3.3 shows that on the five-generation timescale considered in this book, heterozygosity (beginning at 50%) would drop to 44.05% in five generations in a population of 20 individuals; this six percent increase in homozygosity is roughly three times the amount allowed by animal breeders, who are familiar with the issues of inbreeding, and would be deleterious to the population. In populations even slightly larger, however, only one or less than one percent of heterozygosity is lost in five generations. This confirms geneticist M. W. Strickberger's estimation that (under realistic conditions) a population of 1000 sexually-reproducing life forms

Table 3.3 Heterozygous decay

Population	Generation	Years	Heterozygosity
20	1	30	48.75
20	2	60	47.53
20	3	90	46.34
20	4	120	45.18
20	5	150	44.05
150	1	30	49.83
150	2	60	49.67
150	3	90	49.50
150	4	120	49.34
150	5	150	49.17
1000	1	30	49.98
1000	2	60	49.95
1000	3	90	49.93
1000	4	120	49.90
1000	5	150	49.88
2000	1	30	49.99
2000	2	60	49.98
2000	3	90	49.96
2000	4	120	49.95
2000	5	150	49.94

Examples of heterozygous decay in a 5-generation timescale in various completely inbreeding populations with no migration, in or out. Heterozygozsity begins at 0.5 (50%)

would take several hundred generations to lose just 10% of their original heterozygosity (O'Rourke 1991: 711–713), so a Mars colony of such a founding size, for example, should be safe in this regard. In the interstellar ark conception mentioned above, these results suggest that strictly mathematically, from the perspective of heterozygotic decay and closed-population conditions on the order of 5–8 generations in duration, populations over 150 or so individuals would be sufficient as a closed, founding population beyond Earth. Figure 3.9 reiterates this point graphically. We return to such estimations later in this chapter.

3.2.5 *The Founder Effect*

The 'Founder Effect' (Fig. 3.10) is the phenomenon in which a subpopulation migrates away from the local origin, establishing a new population that no longer interbreeds with members of the original population. In this case, all descendant

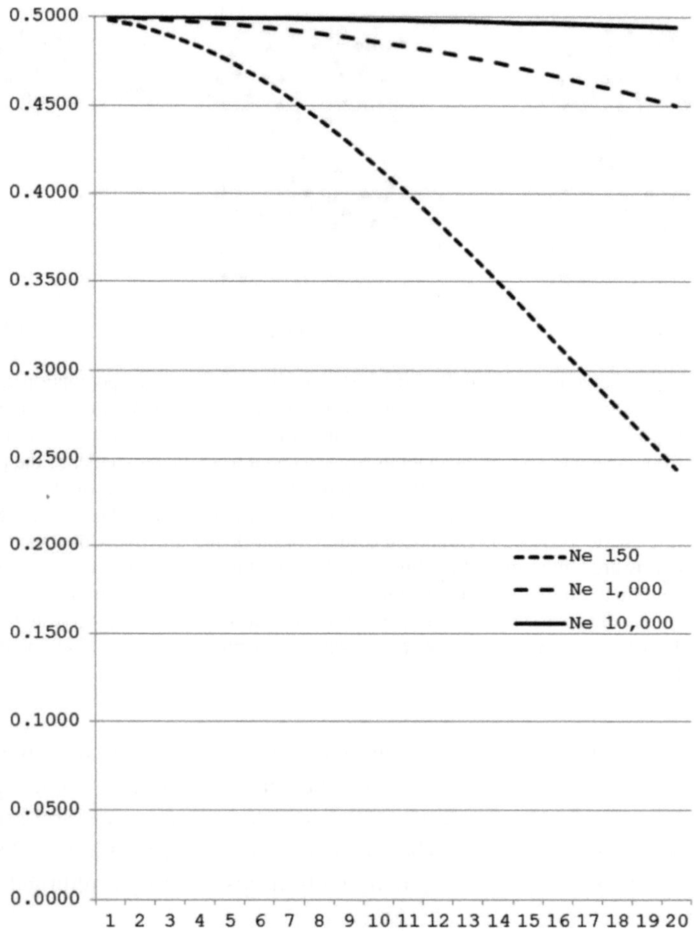

Fig. 3.9 Rate of heterozygotic decay in populations of 20, 150, 1000 and 10,000. Vertical axis is heterozygosity, starting at 0.50 (50%) for a given allele. Horizontal axis indicates generations. Model assumes a panmictic dioecious population under Hardy-Weinberg equilibrium and thus differs somewhat from real conditions, but indicates the essential point that populations approximating 10,000 are safe from heterozygous decay on the order of the 5–10 generation timescale envisioned by current beyond-Earth settlement planners

characteristics will be strongly conditioned by the *founding* population only, which might not be a statistically representative sample of the original population's diversity and will have its own unique characteristics, which will further diverge from the donor population over time (Barton and Charlesworth 1994).

Clearly, the Founder Effect is of critical importance to considering beyond-Earth settlement populations because they will be a subset of the entire human gene pool and (in this discussion) reproductively isolated from the human gene pool the moment they move away from Earth (we will explore the concept of sending preserved genetic

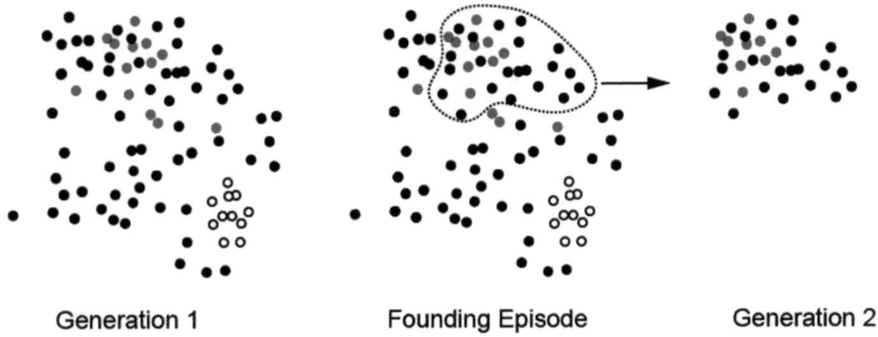

Fig. 3.10 Visual depiction of founder effect as a source of change in population genetics. In generation 1 the population of life forms is composed of black, white and gray shades, but they are all the same species of round animal. In a 'Founding Episode', a large group of the population's individuals from the upper area of the diagram are separated from the remainder of the population; perhaps they are cut off somehow, like a flock of birds blown by strong winds offshore, to an island where trade winds prevent them returning to their original home range. The second generation, then, due to the Founder Effect, is composed entirely of black and gray individuals, but none with the white coloring are represented. Note this is not because the white form was somehow 'selected against'; it was equally fit in the founding generation and would do just fine in the newly-occupied region. But simply due to a sampling issue, the white-coloration genes do not occur in the founding population. If that white coloration were advantageous somehow, generation 2 will be less healthy than it could otherwise be

material—sperm and egg banks—along with any settlers later in this chapter; here we focus on closed populations of living persons). The initial, founding genetic characteristics of colony population, then, particularly if it is small and expected to remain small for some time, would best sample as much of Earth's human diversity as possible; the number of people required to reflect this diversity is returned to below.

3.2.6 Inbreeding, Allee Effects and the Extinction Vortex

Several other population genetic issues must be addressed when considering an initial population of relatively isolated humans for multigenerational health (for Mars and Lagrangian colonies these are not so pressing an issue as for a closed, relatively small interstellar ark population). Two such population figures, N_e or '*Effective Population*' and MCP or '*Minimum Viable Population*', are discussed at length below. Breeding among small populations has the effect of increasing the frequency of deleterious genes in the population, popularly known as *inbreeding* (Harmon and Braude 2013; Mills and Smouse 1994). Thus, breeding population size and inbreeding concerns are directly related. Also, populations are most vulnerable to any catastrophe when they are small, not normally because the entire species could be killed during a catastrophe, but because only enough need to be killed to drop below the species' Minimum Viable Population and because such population 'bottlenecks' can have significant deleterious genetic effects (Masel 2012: 706). Small populations are also subject to *genetic fixity*, in which once-rare genes can rapidly become widespread, even if they are deleterious (they may be immune to selection if they are not immediately lethal and/or are linked to otherwise-indispensable functions); such drift-driven fixity occurs in about 4 N_e generations (Kimura and Ohta 1969: 766–770).

Small populations that suffer catastrophe can be subject to the *Allee Effect*, in which the demographic structure is so perturbed that "nongenetic reasons prevent population recovery" (Lande 1988: 1475), for example, if too many young are killed and breeding partners can no longer be found, leading to an *extinction vortex* phenomenon of obvious interest to the project of beyond-Earth settlement of humans and their domesticates.

3.2.7 Review

Mutation and migration essentially introduce variation into gene pools, while selection and drift erode variation. The trick to maintaining good gene pool health is to at least balance these processes. In nonhuman populations this is largely done by instinct (e.g. incest avoidance), whereas in humans, a small portion of instinct is elaborated by overriding cultural norms that carefully regulate reproductive activity. Regarding the first few generations of space settlement, the forces of selection, drift, mutation and migration are not as significant as population diversity and founder population issues. Multigenerational survival requires not just bare survival, but survival in profusion, such that catastrophes can be sustained without entering an extinction vortex. This is most clearly addressed by ensuring that that the founding population is large enough to persist over ecological time, e.g. the five-generation timescale noted in this paper.

3.3 Population Issues in Closed Settlements: Effective Population (N_e), Minimum Viable Population (MVP) and Census Population (N_c)

It is useful to estimate the minimum viable population size for humans in the case of space settlement for at least three reasons. First, such settlements might be cut off from Earth due to catastrophic long-term space weather, calamity on Earth, or even decisions made by Earth- or the beyond-Earth population. Second, such minimum populations may serve as starting points for studies of the population growth and other dynamics of such settlements. Finally, some scenarios, such as the closed-population 'Interstellar Ark' or 'Generation Ship' would be closed populations for some generations before arrival at an exoplanet system.

Since Sewall Wright first explored specific population issues in his 1931 publication of *Evolution in Mendelian Populations* (Wright 1931), two figures have been most discussed. Importantly, these figures were largely theoretically considered until the 1970s, when awareness of modern extinction rates in wild populations led to the rise of the field of conservation biology. Among this field's many concerns has been to identify the number of individuals, of a given species, required to maintain that population in good genetic health for some time (such figures help to determine the size of wildlife refuges, for instance).

One of these figures is *Effective Population*, or $N_{e.}$ Defined somewhat variably in the literature, N_e broadly estimates the minimum number of individuals of a species that can interbreed while producing viable offspring for some generations without suffering the effects of inbreeding (Strickberger 1985: 711). another way to think of it is as an ideal figure representing the number of individuals of a species that demonstrate the same amount of genetic variation as the entire population, or, alternately "…the number of individuals in an ideally behaving random-mating population that has the same magnitude of genetic drift as the actual population of interest." (Hawks et al. 2000: 8) (note that in all cases the sampled population is assumed to be genetically healthy on a multigenerational scale). N_e is normally lower than the actual number of organisms in a population because N_e only includes reproductively viable individuals (that is, sexually mature but not *senescent* or postreproductive). N_e varies by the reproductive biology of the species in consideration, the species' individual average longevity and other factors (see Wood 1987; Douglas 1966) for cultural reasons for variation in the human N_e). For many species, though, effective population ranges in the low thousands, and "…from bacteria to humans [N_e figures] are within an order of magnitude of each other." (Gillespie 2001: 2161; see also Flather et al. 2011: 314). In modern humans, who number into the billions, N_e is often calculated to be around 10,000, a figure we will return to below.

Though N_e is calculated in many subtly-varying ways, it can be largely thought of as a function dividing the amount of genetic variation in a sample of a species by an estimate of the mutation rate and population of that species. Also note that N_e is *always an estimate*, as it differs over time as the demographic history changes of a species changes. Also, different N_e figures can be calculated for different genes or other traits of a species. Effective population is real, but variable; informative, but has its explanatory limits. Though N_e is always an estimate it remains a genetically and realistically well-informed measure of the of the genetic diversity of a species, which itself broadly indicates *how many reproductive individuals are needed, at minimum, to maintain species genetic health (largely by avoiding the effects of inbreeding and loss of heterozygosity) over multigenerational time* (see Caballero 1994; Jorde et al. 2001: 2199; Ryman and Lairke 1991: 325), clearly a figure of interest in the topic of maintaining multigenerational, small-population gene pool health.

In non-human life N_e is often only 10–30% of the population (Frankham 1995), with nonreproductive young normally composing the bulk of the remainder, and only a small percentage of nonreproductive elderly composing the final fraction (this fraction is normally higher in highly social animals, such as primates, where elders may provide protection or alloparenting for the young). For a number of reasons,

N_e is normally 50% of the actual population in traditional human foraging cultures (where low populations are promoted) and only about 30% of the entire population in farming cultures (where high populations are promoted), issues returned to below (Relethford 2001: 147). Note that as humans are particularly active in making long-distance mating networks, and are highly mobile, the human N_e is well below the actual population of roughly six billion. Recent profiling of human genetic diversity has greatly improved our understanding of the human N_e (see e.g. Harpending et al. 1998; Voight et al. 2005).

The second figure, *Minimum Viable Population* (MVP) includes all members of the population, whether or not they are members of the reproductive age cohort (Reed and Bryant 2000; Shaffer 1991). While N_e is often calculated based on surveys of the genetic diversity of species, MVP is normally a result of surveys of the actual demographic structure (in terms of age and sex subdivisions or *cohorts*) of real-world populations of a species. Since among sexually-reproducing species populations include both reproductive and nonreproductive members at any given time, Minimum Viable Population is a more real-world estimation of the actual number of individuals required to maintain a population in good genetic health through time. The Minimum Viable Population is always larger than N_e and depends on many variables including the species' reproductive rate, gestation time, longevity per individual and other variables (Reed et al. 2003). MVP is particularly important in conservation biology's planning to maintain real-world populations in habitats scaled to support such a population. For most human populations, where breeding is maintained over large distances and among many genetically-variable sub-populations, MVP can be quite small (on Earth). However, in a closed system such as an isolated settlement on Mars or an interstellar generation ship where the population will be relatively fixed until a new planet is reached (where population can increase again) MVP should be relatively large compared to that on Earth. Because MVP literature emphasizes the *minimum* number of individuals to support endangered non-human population for some generations (often a century, a figure of importance to humans attempting to scale conservation efforts with human lifetimes and scales of political attention), I prefer to use the theoretically synonymous term N_c, or *census population*; this is the same as MVP but in its verbiage does not emphasize the *minimum* possible number, but rather a *reasonable* number, to sustain the population over a given time period. Census population, then (N_c, synonymous with MVP in conservation biology literature), is N_e (n of reproductive individuals) plus all nonreproductive members of the population (Avise et al. 1998). In human populations, N_c is often twice the size of N_e in small, foraging-hunting-gathering populations (where populations are deliberately kept low to avoid exhausting limited resources), and three times the size of N_e in agricultural populations (such as those of modern civilization supplied by agriculture, as will be the case in any currently-conceivable extraterrestrial settlement). However, because for theoretical purposes, early extraterrestrial settlements are being considered as relatively closed populations, in this study I am calculating MVP or N_c as (N_e(2)).

In summary: Effective Population (N_e) is the minimum number of reproductive individuals required to ensure genetic health over multiple generations (in our case, five at most), so if it is to be reproductively isolated for such a timespan its founding population should be large enough to sample a wide variety of the diversity of the human genome, because genetic diversity is key to the genetic health of a population over time. From this figure I will derive Census Population (N_c), the actual number of a real-world human population—distributed among a modern human sex- and age-cohort demographic structure—of the first generation of an off-Earth founding Population (FP), including reproductive and non-reproductive individuals, a figure selected to be sufficient to avoid inbreeding and genetic drift effects according to the duration of the projected population isolation and its unique properties (compared to Earth conditions) of (a) low-to-no population growth and (b) relatively small breeding pool with no opportunity for in-migration of new genes during its voyage or period of reproductive isolation. In this study I use five generations, each of 30 years (Fenner 2005: 420–422), for a total of 150 years. I adjust population figures to account for various likelihoods of large-scale disaster, which we can reasonably expect—or should at least prudently model for—in such a project.

3.3.1 Survey of Effective Population (N_e), Minimum Viable Population (MVP) and Census Population (N_c) Estimates for Various Life Forms

Before proceeding we must remember that N_e, N_c and MVP figures are calculated differently for different life forms, and there is active debate among scholars regarding methods to derive these figures. While none are calling to abandon calculation of these figures, it is clear that there are no 'magic numbers' that apply to all species (Lande 1995: 879; see also Flather et al. 2011; Wood 1987); at the same time, useful estimates can be made for conservation efforts (e.g. Rai 2003; Thomas 1990; Lynch and Lande 1998) significantly analogous to the interests of space settlement. One reason for the variation in calculating these figures is that sub-populations of a single species vary genetically, such that they may have a different N_e than another sub-population; also, one variable used to calculate N_e is the rate of mutation, that is the rate of new variation in the life form per generation; but this also varies both among and within populations (Leffler et al. 2012) and over time (Hey 2005); finally, the N_e might itself have changed through time, as is suspected of the genus *Homo* (Harpending et al. 1998), though it is still possible to make useful ballpark estimates.

To estimate a genetically-viable N_e and N_c (MVP) for a five-generation (c.150-year) timespan, I surveyed the research literature pertaining to these values published in the last 40 years, highlighting research (including meta-analyses) published in the last decade to account for new genomic knowledge that is reformulating evolutionary biology (see e.g. Smith and Rupell 2011). Material published earlier than about 1970 was normally much more theoretical and largely in need of revision. Since the 1970s, however, worldwide efforts to conserve species nearing endangerment has led to the establishment of the field of conservation biology, for which N_e and MVP are critically important figures. My search focused on animal life (rather than including plants or microbes) and, in particular, vertebrate life, and of the vertebrates, mammals, and of the mammals, primates, including earlier and modern *Homo*. Prehistorians have had an interest in ancient human demography for several generations, and have estimated N_e and MVP for various members of the genus *Homo* spanning the last two million or so years. Such estimates include recent studies using genomic data that have allowed high-resolution understanding of modern humanity's N_e and MVP. Such estimates also vary somewhat in their calculation of N_e for both methodological reasons and because humanity deviates in several ways from the Hardy-Weinberg equilibrium conditions (mentioned in Sect. 3.2.4) that are the theoretical foundation of N_e estimates; for example, humans exhibit (a) overlapping generations, (b) population subdivisions, (c) local population extinctions, recolonizations and migrations and (d) different reproductive patterns between the sexes and cultures. Nevertheless human N_e figures vary in a rather tight range, as will be seen below, when we consider the range of variation among all organisms sampled in this study.

Results of the survey are found in Table 3.4. They include low and high range estimates for N_e and N_c (representing MVP), averages of these figures, an indication of the life form studied ('Comp Sim' indicates a computer study only) and references. Many of these studies derived such figures from DNA diversity analyses, while PCA [Population Viability Analysis; see Reed et al. (2003)] was often used in conjunction with demographic and population structure data to arrive at N_c, synonymous here with MVP as described above.

Some results of this survey are tabulated for clarity in Table 3.5.

Eight summary results are immediately apparent from the summary:

1. There is no single universally-applicable N_e or N_c; this is expected because these figures are derived with slightly differing methods per study, for life forms with different life histories, reproductive biologies and lifespans and because these figures themselves change through time (Charlesworth 2009; Harmon and Braude 2013).
2. Almost no study sets N_e below 1500, often setting minimal N_c/MVP's of at least 5000 in nonhumans (Reed et al. (2003: 30) suggest populations "for long-term persistence may range from 2000…to > 13,000") and >3000 for modern humans; both, again, in the multiple thousands rather than hundreds, for example, and all below 100,000, setting the range between 10^3 and 10^4 in many cases. The exception (Wobst 1973) models Ice-Age hunter-gatherer populations, but is so abstracted that I dismiss it here.
3. N_e figures average just under 20,000 for all life forms and computer simulations combined, but significantly lower, just over 1500, for a few studies generalizing upon 'all animal life'. For hominoidea generally and hominins specifically, data derived largely from DNA diversity studies indicates an N_e on the order of 17,000–20,000, while for modern humans an N_e of just over 10,000 is indicated by many studies, with attendant N_c's of roughly 20,000 if modeling after historical foraging populations (who prefer smaller populations) or closer to 30,000 if modeling after historical preindustrial agricultural populations (who prefer larger populations).

Table 3.4 Survey of effective and census population data

Ne low	Ne high	MVP low	MVP high	Ne average	MVP average	Organism code	References
					5000	Animal	Lande (1995)
1000	5000	5000	25,000	3000	15,000	Animal	Lynch and Lande (1998)
					4500	Animal	Franklin (1980)
50	500			275	2475	Animal	Flather et al. (2011)
					2000	Comp Sim	Schultz and lynch (1997)
					5500	Comp Sim	Thomas (1990)
					>2500	Comp Sim	Whitlock (2000)
1000	5000				15,000	Comp Sim	Lynch and Lande (1998)
556	1250	5000	12,500		8750	Comp Sim	Franklin and Frankham (1998)
				12,300	24,600	Hominoidea	Yu et al. (2004)
				21,300	42,600	Hominoidea	Yu et al. (2004)
				25,200	50,400	Hominoidea	Yu et al. (2004)
					10,841	Insect	Traill et al. (2007)
					2200	Insect	Reed and Bryant (2000)
					5137	Large vertebrate	Traill et al. (2007)
9000	12,000	18,000	24,000	10,500	21,000	Later hominin	Takahata (1993)
8000	11,000	16,000	22,000	9500	19,000	Later hominin	Takahata (1993)
7000	9500	14,000	19,000	8250	16,500	Later hominin	Takahata (1993)
				18,000	36,000	Later hominin	Sherry et al. (1994)
				10,000	20,000	Later hominin	Harding et al. (1997)
		20,000	50,000	10,000	35,000	Later hominin	Relethford (2001)

(continued)

Table 3.4 (continued)

Ne low	Ne high	MVP low	MVP high	Ne average	MVP average	Organism code	References
10,000	100,000	20,000	200,000	55,000	110,000	Later hominin	Hawks et al. (2000)
				21,000	64,000	Later hominin	Hawks et al. (2000)
				8800	17,600	Later hominin	Hawks et al. (2000)
9000	17,500	18,000	35,000	13,250	26,500	Later hominin	Harpending et al. (1998)
				10,000	20,000	Later hominin	Takahata and Satta (1998)
				9640	19,280	Later hominin	Hey (2005)
6000	15,000	12,000	30,000	10,500	21,000	Later hominin	Zhao et al. (2001)
				10,659	21,318	Later hominin	Voight et al. (2005)
				10,000	20,000	Later hominin	Przeworski and Wall (2001)
					3876	Mammal	Traill et al. (2007)
					3500	Mammal	Newmark (Newmark 1987)
					>3200	Mammal	Soule (1987)
				10,400	20,800	Mod human	Nei and Grauer (1984)
				14,000	28,000	Mod human	Nei and Grauer (1984)
				10,000	20,000	Mod human	Hawks et al. (2000)
10,000	100,000	20,000	200,000	10,000	20,000	Mod human	Wang et al. (1998)
				15,000	30,000	Mod human	Dorit et al. (1995)
				10,000	20,000	Mod human	Hammer (1995)
3000	15,000	6000	30,000	9000	18,000	Mod human	Huang et al. (1998)
				10,000	30,000	Mod human	O'Rourke (1991)
				35,000	70,000	Mod human	Kaesmann et al. (1999)

(continued)

Table 3.4 (continued)

Ne low	Ne high	MVP low	MVP high	Ne average	MVP average	Organism code	References
		175	475		325	Mod human	Wobst (1973)
3100	7500	9300	22,500	5300	15,900	Mod human	Tenesa et al. (2007)
				5394	10,788	Mod human	Hey (2005)
				770	1540	Mod human	Hey (2005)
				10,000	20,000	Mod human	Chen and Li (2001)
				10,000	20,000	Mod human	Watterson (1975)
10,000	20,000			15,000	30,000	Mod human	Hey (2005)
				10,000	20,000	Mod human	Voight et al. (2005)
				10,000	20,000	Mod human	Voight et al. (2005)
					4196	Plant	Traill et al. (2007)
					3956	Small vertebrate	Traill et al. (2007)
				100,000	200,000	Various primates	Ayala (1995)
9000	68,000	18,000	136,000	38,500	77,000	Various primates	Bailey et al. (1992)
7000	57,000	14,000	114,000	32,000	64,000	Various primates	Bailey et al. (1992)
		3600	4500	500	4050	Various primates	Kinnaird and O'Brien (1991)
		50	20,000		10,025	Various primates	Harcourt (2002)
52,000	96,000			74,000	148,000	Various primates	Chen and Li (2001)
					4102	Vertebrate	Traill et al. (2007)
					4169	Vertebrate	Traill et al. (2007)
		2000	13,000		7000	Vertebrate	Reed et al. (2003)

See text for details

Table 3.5 Summary of population data surveyed

	Ne	MVP
Average	**18,187**	**31,810**
Animal	1638	–
Comp Sim	–	7813
Hominoidea	19,600	–
Insect	–	6521
Large vertebrate	–	5137
Later hominin	17,819	51,075
Mammal	–	3688
Modern human	**11,168**	**21,964**
Plant	–	4196
Small vertebrate	–	3956
Various primates	49,000	83,846
Vertebrate	–	5090
All hominins	14,494	21,888

See text for discussion

4. Only 13 of 61 studies yielded N_c/MVP figures for the sampled life forms under 5000.
5. N_e for large mammalian species only twice exceeds 100,000 (this is also noted in a smaller sampling in Hawks et al. 2000: 15).
6. N_e and N_c average (respectively) around 18,000 and 32,000 for the entire sample (n = 61 studies), 14,500–22,000 for all hominin estimates (n = 34 studies) and 11,000–22,000 for all modern human estimates (n = 18 studies, of which nearly half suggest an N_e approximating 10,000).
7. The metastudies (Reed et al. 2003; Traill et al. 2007; Lande 1995) all estimate N_c on the order of several thousand, ranging from around 4000 for vertebrates and mammals specifically to 5000–7000 for samples of hundreds animal species from multiple genera (Reed et al. 2003 studied 102 vertebrates while Traill, Bradshaw and Brook studied 212 varied species).
8. An important result not obvious on the table is that in these meta-analyses, an N_c on the order of several thousand (>1500, <8000) was not found to vary significantly by taxa (life form), trophic level (e.g. carnivore or herbivore), or latitude (geographical distribution).

While there is some significant variation in N_e and N_c estimates for modern humans at large (including purely theoretical modeling of ancient populations as well as empirically-based estimates using genomic diversity data), the average N_e of roughly 10,000 individuals closely approximates anthropologist J. H. Relethford's similar conclusion, based on his own survey of others' estimates and his own mathematical estimates (Relethford 2001). This figure is high if we consider space-colonization population estimates, such as anthropologist J. H. Moore's widely-discussed proposition of founding populations of 80–150 individuals (therefore an N_e of roughly 40–75) (Moore 1991a, b). Such small figures are based on anthropological observation that in many preindustrial human cultures, tribal or band populations of individuals are often of this scale—the 'Dunbar Number' of c.150 has also widely been publicized as a viable, self-contained human population. Such low-hundreds to low-thousands figures of 'natural human populations', when observed in modern non-industrialized cultures are often taken to reflect an ancient state of small, dispersed human populations, into the hundreds of thousands of years ago, considered a stable reproductive units over long time for members of the genus *Homo* (e.g. Wobst 1973).

There is a significant problem with these low figures, however, as estimates for viable founding populations of small and/or reproductively isolated communities. Even though human groups might have fallen into such low-hundreds figures for social (e.g. ethnic) and logistical reasons, they are (and in the past, were) never completely isolated populations. Human cultures have long-range mating networks and maintain large breeding populations even if they reside in smaller subunits observed by anthropologists (Kelly 2007); anthropologist R. B. Eckhard notes that even among historically-known foraging peoples, though they lived and traveled in such small groups, a larger collective gene pool (N_e) was maintained by "matings between members of immediately adjacent populations…[and modern foragers] have systematic (and often complex) arrangements for mate exchange among adjacent bands, either to make up for a shortage of mates of appropriate demographic characteristics or to ensure amicable relationships, or for other reasons." (Eckhardt 2000: 243). Prehistorian C. Gamble emphasizes such wide, integrating social and breeding networks as particularly distinctive of modern *Homo sapiens sapiens* (dating to after 100,000 years ago) as opposed to the Neanderthals (*Homo sapiens neanderthalensis*), who lived in apparently smaller territories and might well have had smaller breeding populations (Gamble 1999). While some human populations have been very low for some centuries (e.g. castaway island populations; see examples in Finney and Jones 1985; Kondo et al. 1991) I wish to emphasize that these are anomalous situations in which the population size and social structure was not the product of long cultural and natural selection, but simply an historical contingency…normally an accident. I discount these cases, opting for population sizes that have been the product of both cultural and biological evolution, not contingency, in our genus.

For all of these reasons, I feel the low-hundreds figures proposed as a founding MVP for a closed, multigenerational human space settlement would likely underestimate the actual size of a human gene pool required to maintain good multigenerational health. I feel the N_e of roughly 10,000 or 11,000—reflecting and concurring with high-resolution modern data on living populations as well as theoretical calculations—is a much more reasonable estimate. It is not necessary that that founding population be established all at once, but if its goal is independence from Earth, that number is what will be necessary and sufficient, I believe, for such independence and safety. We may proceed from that generalization towards more specifics.

3.4 Estimating and Modeling a Viable Population for a Five-Generation, Isolated, Beyond-Earth Human Colony

Table 3.6 can be used to evaluate the suitability of certain N_e and N_c figures for small and/or isolated human populations. The columns indicate various quantities described below. One factor built into the modeling is that of catastrophe. In a 5-generation period I expect some engineering or social catastrophe to occur. Just as we design safety factors into flight operations today, such as always flying with a fuel reserve, we should plan accordingly with regard to a human population. Aside from mechanical failure in beyond-Earth habitats themselves, the most likely large-scale catastrophe that could occur in the envisioned voyage would be the sweep of a plague through the population. Even the Earth's worst recorded human plagues have never killed more than 2% of the entire human population, but in smaller sub-units, such as Europe during the Black Plague of the 1300s, average mortality was as high as 30% (Galvani and Slatkin 2003). I use this mortality figure in planning for disaster in the calculations below.

The columns in Table 3.6 are arranged as follows, from left to right;

N_e = effective population = (n of reproductive individuals necessary to sustain genetic health while breeding among the closed population)
N_c = census population ($2N_e$) = total number of people in the population, including reproductive and non-reproductive members
P = probability of catastrophe during the 5-generation isolation period, from 1.0 (100%) to 0.03 (3%), dropping somewhat in each row to model different contingencies
L = lethality of catastrophe = percentage of N_c killed in a catastrophic disease sweep on the order of the European plague of the 14th century AD; here I use 0.30 (30%)
CF = Catastrophe Factor = $(N_c{*}L)P$ = n of individuals to supplement the isolated population, to offset catastrophe, based on the expected lethality of the disaster and one's comfort regarding modeling the likelihood of a disaster
FP, SF = Founding Population with Safety Factor = $(N_c + CF){*}[2 \text{ or } 3]$ = the original population (N_c) plus the number of replacements carried in event of a catastrophe (CF), multiplied by two or three, according to one's threshold for over-engineering. **These two columns represent the total number of people (N_c) as the founder population, adjusted for various probabilities of disaster; note that they always exceed N_e.**
N_c-Cat = for reference, the N_c minus 30% as in the case of catastrophe
N_e-Cat = for reference, the N_e minus 30% as in the case of catastrophe.

The remaining four columns (on the right side of the table) are most useful for evaluating whether, after catastrophe, certain critical population size limits are violated (discussed below).

Table 3.6 Effective and Cansus populations and adjustments for safety considering various catastrophe probabilities and 30% mortality in closed populations beyond Earth

Ne	Nc	P	L	CF	FP, SF = 2	FP, SF = 3	Nc-Cat	Ne-Cat	NcSF2-CAT	NcSF3-CAT	NeSF2-CAT	NeSF3-CAT
75	150	1.0	0.3	45	390	585	105	53	273	410	137	205
75	150	0.5	0.3	23	345	518	128	64	242	362	121	181
75	150	0.3	0.3	11	323	484	139	69	226	339	113	169
75	150	0.1	0.3	6	311	467	144	72	218	327	109	163
75	150	0.1	0.3	3	306	458	147	74	214	321	107	160
75	150	0.0	0.3	1	303	454	149	74	212	318	106	159
150	300	1.0	0.3	90	780	1170	210	105	546	819	273	410
150	300	0.5	0.3	45	690	1035	255	128	483	725	242	362
150	300	0.3	0.3	23	645	968	278	139	452	677	226	339
150	300	0.1	0.3	11	623	934	289	144	436	654	218	327
150	300	0.1	0.3	6	611	917	294	147	428	642	214	321
150	300	0.0	0.3	3	606	908	297	149	424	636	212	318
2000	4000	1.0	0.3	1200	10,400	15,600	2800	1400	7280	10,920	3640	5460
2000	4000	0.5	0.3	600	9200	13,800	3400	1700	6440	9660	3220	4830
2000	4000	0.3	0.3	300	8600	12,900	3700	1850	6020	9030	3010	4515
2000	4000	0.1	0.3	150	8300	12,450	3850	1925	5810	8715	2905	4358
2000	4000	0.1	0.3	75	8150	12,225	3925	1963	5705	8558	2853	4279
2000	4000	0.0	0.3	38	8075	12,113	3963	1981	5653	8479	2826	4239
3500	7000	1.0	0.3	2100	18,200	27,300	4900	2450	12,740	19,110	6370	9555
3500	7000	0.5	0.3	1050	16,100	24,150	5950	2975	11,270	16,905	5635	8453
3500	7000	0.3	0.3	525	15,050	22,575	6475	3238	10,535	15,803	5268	7901

(continued)

Table 3.6 (continued)

Ne	Nc	P	L	CF	FP, SF = 2	FP, SF = 3	Nc-Cat	Ne-Cat	NcSF2-CAT	NcSF3-CAT	NeSF2-CAT	NeSF3-CAT
3500	7000	0.1	0.3	263	14,525	21,788	6738	3369	10,168	15,251	5084	7626
3500	7000	0.1	0.3	131	14,263	21,394	6869	3434	9984	14,976	4992	7488
3500	7000	0.0	0.3	66	14,131	21,197	6934	3467	9892	14,838	4946	7419
11,000	22,000	1.0	0.3	6600	57,200	85,800	15,400	7700	40,040	60,060	20,020	30,030
11,000	22,000	0.5	0.3	3300	50,600	75,900	18,700	9350	35,420	53,130	17,710	26,565
11,000	22,000	0.3	0.3	1650	47,300	70,950	20,350	10,175	33,110	49,665	16,555	24,833
11,000	22,000	0.1	0.3	825	45,650	68,475	21,175	10,588	31,955	47,933	15,978	23,966
11,000	22,000	0.1	0.3	413	44,825	67,238	21,588	10,794	31,378	47,066	15,689	23,533
11,000	22,000	0.0	0.3	206	44,413	66,619	21,794	10,897	31,089	46,633	15,544	23,317

See text for discussion

N_cSF2-CAT = the N_c multiplied by safety factor of 2, minus 30%, the population lost in a catastrophe. **This is the n of people remaining after a 30% depopulation if the original population were doubled as a safety factor.**
N_cSF3-CAT = the N_c multiplied by safety factor of 3, minus 30%, the population lost in a catastrophe. **This is the n of people remaining after a 30% depopulation if the original population were tripled as a safety factor.**
N_eSF2-CAT = the N_e multiplied by safety factor of 2, minus 30%, the population lost in a catastrophe. **This is the n of reproductively active people remaining after a 30% depopulation if the original population were doubled as a safety factor**.
N_eSF3-CAT = the N_e multiplied by safety factor of 3, minus 30%, the population lost in a catastrophe. **This is the n of reproductively active people remaining after a 30% depopulation if the original population were tripled as a safety factor**.

3.4.1 A Note on Demographic Structure

The age and sex structure of a populations is its essential *demography*. As population cannot be allowed to grow beyond certain limits in the closed system of a somewhat isolated population, reproductive rights will not be as free as they are on Earth in many countries today. Certain age- and sex- structures of populations will have to be consciously maintained to give reproductive individuals plenty of non-relative mate choices. Table 3.7 sketches out sample demographic profiles for founding populations (N_c) of 18,000 and 40,000.

Three lessons learned from the analysis of Table 3.6 are particularly important to understanding Table 3.7:

1. **Viability over several generations is rarely considered possible in numbers** (N_e or N_c) **less than a few thousand**. While an N_e of 500 has been used for many species in conservation efforts, it is not a 'magic number'. Still, attendant N_c's normally measure in the low thousands at least, a figure acceptable to even scholars most critical of population viability estimates (Flather et al. 2011: 314).
2. **A cautious approach does not allow N_c to dip below about 7000 individuals**; this is particularly important in vertebrates (Reed et al. 2003).
3. **A cautious approach does not allow N_e to dip below about 10,000**; this is particularly important in the genus *Homo* (Relethford 2001; Takahata 1993).

In Table 3.6 we see that with N_e (reproductive population) at 75 persons for an N_c (total population) of 150 persons–the oft-cited 'Dunbar Number', also used by Moore (1991a, b)–even doubling or tripling the N_c with safety factors never brings populations into even the low thousands for either N_e or N_c. These populations would always be on the verge of disaster in the case of a Black-plague-like catastrophe. The same applies to the estimates with N_e at 150, giving an initial N_c of 300 persons, which is little better; and in both cases the N_e, even after doubling or tripling, is just at or even below 10% of the Ne of 10,000 that is apparently a good guide for human

Table 3.7 Sample demographic structure for closed populations beyond Earth at various founding populations. Top: founding population N_e 18,000. Bottom, founding population N_e 40,000

18k Population ($N_e = 9800$)			
N_e 6860 after catastrophe, N_c 13,720 after catastrophe			
Generation 1 (departing Earth)	M	F	Sum
Subadult	4000	4000	8000
Adult paired	4000	4000	8000
Adult unpaired	900	900	1800
Elder paired	50	50	100
Elder unpaired	50	50	100
Sum	9000	9000	18,000
40k Population ($N_e = 23{,}400$)			
N_e 16,380 after catastrophe, N_c 32,760 after catastrophe			
Generation 1 (departing Earth)	M	F	Sum
Subadult	6800	6800	13,600
Adult paired	6700	6700	13,400
Adult unpaired	5000	5000	10,000
Elder paired	1000	1000	2000
Elder unpaired	500	500	1000
Sum	20,000	20,000	40,000

health. Though these populations would not likely suffer from significant inbreeding effects (as discussed in Sect. 2.4) I consider them are demographically hazardous. For these reasons I discount both of these Founding Populations (75 and 150) as too risky to ensure multigenerational success if they are reproductively isolated for more than a generation or so.

With the founding N_e set at 2000 the recommendation that vertebrates over 1 kg body mass have an N_c of about 4000 (Traill et al. 2007) is met. In the first row we see that by adding 1200 individuals against the effects of a catastrophe, and then doubling the population, we achieve a population (N_c) of 10,400; if this population were to be struck with catastrophe of 30% lethality, 7820 individuals would survive, fulfilling the general guidance to keep the N_c at or above 10,400. Looking at the columns for N_e after catastrophe we see figures from a high of 5460 to a low of 2826; only one approaches even 50% of the recommended human N_e, and most are far below it. For this reason, founding populations on the order of anything under about 10,000, while they might preserve N_c, are risky in the event of disaster regarding maintenance of N_e. For these reasons I discount a Founding Population of N_e of 2000 as too risky for multigenerational success.

With the initial N_e set at 3500 the recommendation that vertebrates in general have an N_c of about 7000 is met (Reed et al. 2003). In the column IMP, SF = 2 we see that any figure, from 18,200 to 14,131 (N_c's doubled) allows the roughly 7000 individuals thought to be a viable multigenerational vertebrate population to flourish, even in the case of catastrophes, where we see 9800–19,110 individuals surviving. To preserve

N_e around 10,000 after a catastrophe (cell far right indicating 9555 survivors) can be done by tripling the population (column **IMP, SF = 3**) for a founding population of N_c 27,300. This is a reasonable Founding Population figure for what is planned to be, or might conceivably be for some multigenerational period, a closed population.

With the initial N_e set at 11,000 the average living human N_e (as reported in Table 3.2) is approximated. With N_c's adjusted to accommodate catastrophe, ranging from a low of 44,413 to a high of 57,200 when adjusted for catastrophe, recommendations for N_c's over c.5000 and N_e's around 11,000 after catastrophe are easily met by doubling the population (columns N_cSF2-CAT and N_eSF2-CAT). Tripling the population (column FP, SF = 3) invokes populations from nearly 70,000 to over 85,000, which appears to be overkill as just mentioned; doubling is sufficient to protect both N_c and N_e in a severe catastrophe.

In sum, I would consider roughly 10,000 an reasonable minimum N_e population, with an N_c at or exceeding 30,000 to be a safe Founding Population for closed, off-Earth settlements to be considered multigenerationally safe; this aligns well with the currently-known modern human N_e, which ranges from 9000 to 17,000, averaging somewhere in-between (Harpending et al. 1998: 1963), and would suffice to ensure enduring the modeled catastrophe. Such a figure would not have to be established immediately, but could well serve as a goal in population-growth strategy. Remember, larger populations are always more likely to survive any given disaster; and these populations should be both diverse and if not geographically widespread (unlikely in the case of a growing Mars colony or a generation ship) provision should be made for viable sub-populations of the Founding Population to survive independently through reasonably-expectable catastrophes.

3.4.2 Recommendations Regarding Founding Populations Beyond Earth

Considering what is known about vertebrate, mammalian, nonhuman primate and hominin genetic variation and population dynamics, closed extraterrestrial populations that would be isolated and could not grow significantly for some time should not number less than several thousand colonists at the outset. A total Founding Population (N_c) of c.10,000 would just barely preserve the recommended 7000 individuals in a vertebrate population in the event of a reasonably-expectable catastrophe. Somewhat safer prospects are found with populations beginning around 14,000 individuals, but until the founding N_c is on the order of 18,000, the critical human N_e of roughly 10,000 could not be maintained in the event of the modeled catastrophe. Any founding population with an N_c above 40,000 would be very likely to survive in good health over five generations, even in the event of a Black Death-style plague with a mortality figure of 30%. Figure 3.10 depicts the earliest stages of such a settlement on Mars.

So far I have employed various safety factors, catastrophe lethality and probability of disaster per generation, and so on. These variable quantities should be altered, and different risks taken in other models, exhaustively, until we have a better understanding of these issues. I discuss some of this modelling below, in the specific case of the generation ship or 'Worldship'. First, however, I address one of the most common questions I am asked about the prospect of human settlement beyond Earth, 'will we become a new species?'.

3.4.3 The Question of Speciation

Earth life forms are classified in many ways for various analytical purposes. Naturally-occurring groupings of life forms that breed among themselves are recognized as *species*. A literature review indicates that while the boundaries of some species can be blurry, they are identified scientifically on the grounds of "morphological, ecological, behavioral, and molecular information" (Mayr 1996; de Queiroz 2005) and species:

- live in reproductive isolation from other life forms
- share a mate-recognition system
- occupy the same ecological niche
- share a common ancestor.

Earth life forms change, of course, through time by the evolutionary process, so it is recognized that while species are real, and do exist (e.g. whales do not mate with porpoises), "[s]pecies is a stage in a process, not a static unit" (T. Dobjhansky as quoted ion Gould and Eldredge 1982: 312). Over time, this evolution leads to *speciation*, "the formation of two or more new species from one ancestral population" (Wood and Reiseberg 2002: 415). Speciation occurs as a result of *reproductive isolation* in which two subpopulations (let us say A and B) of a previously single population, become separated and then accumulate genetic variations that prevent members of populations A and B from having healthy offspring. For example, over time, subpopulations of the North American threespine stickleback (*Gasterosteus aculeatus*) diverged in the water column, some spending most of their time feeding nearer the surface (*limnetics*), while others stayed nearer the lake floor (*benthics*). Over time, the females of the deeper-dwelling and shallower-dwelling groups have come to prefer to mate with males from their own deeper- or shallower- populations, leading to reproductive isolation (Boughman 2001). Other causes of reproductive isolation are listed in Table 3.8.

Clearly if human populations beyond Earth cease to mate with Earth-folk, it is possible that the two populations (Earth and off-Earth) would accumulate enough variations that if they did attempt to mate, they could either not do so, or the offspring would be so unhealthy that they themselves could not have offspring. This would be speciation.

In wild populations of animal life, most speciation takes place over a long period; many thousands or even millions of generations. There is no fixed rate of evolution

Table 3.8 Mechanisms of reproductive isolation that can lead to speciation

Kind of isolation	Explanation	Example
Temporal isolation	Different mating schedules emerge	American field crickets reach reproductive age in different seasons, preventing them from meeting and mating
Ecological isolation	Different habitats are occupied	Some Japanese ladybugs will mate only on the plant they prefer to eat; however, members of the same species eat different plants
Behavioral isolation	Different behaviors become established	Moths will attempt to mate only with females releasing a specific pheromone; fiddler crab females prefer males with larger claws
Mechanical isolation	Male and female sexual organs are physically incompatible	Closely related fruit fly species' male and female genitals simply do not fit as "lock and key"
Hybrid inviability	Hybrids of diverging populations do not develop properly	Crosses of sheep and goats do not survive to reproductive age
Hybrid sterility	Hybrids of diverging populations cannot have their own offspring	Crosses of horses and donkey makes sterile offspring

Table by the Author

for the many millions of Earth species [reflected in the title of one book addressing speciation, *Rates of Evolution* (Campbell and Day 1987)]. This is largely because both mutation rates and the rate of environmental change vary. One model describing this variable evolution rate is called *punctuated equilibrium*, referring to rapid pulses of change (punctuations) occurring between periods of more common equilibrium or stability. A recent review (*The Million Year Wait for Macroevolutionary Bursts*) indicated that among mammals, significant evolutionary divergences (speciations) normally occur over a time period of hundreds of thousands to millions of years (Udeya et al. 2011). Note that his makes speciation much easier to observe in the fossil record (which represents vast time periods) than in nature, where gradual change is constant but normally so subtle to observe directly. Occasionally, much more rapid speciation is seen, occurring in tens or hundreds of generations (Hadid et al. 2014), but these appear to be exceptions. Theoretically, speciation is expected to be more rapid if the founding population of a new habitat is small and does not represent that original population's full genetic variation, and this has been occasionally observed in laboratory settings with life forms that have a short generation time (Weinberg et al. 1992).

With regard to space settlement, then, we can see that there is no reason that human populations reproductively isolated from Earth would not over time (a) accumulate their own local variations and (b) be unable to have healthy offspring if they were to mate with Earth populations. This would be speciation. It would likely take many generations and is unlikely in the timeframe considered in this book. First, populations beyond Earth will probably continue to have mating relationships with Earth folk, both by returning to Earth on occasion and by the arrival of new Earth-folk to whatever colonies exist beyond Earth. This would constitute gene flow, though it may be very small in volume, statistically speaking. Second, even if some beyond-Earth population became entirely reproductively isolated from Earth populations, what we know of mammalian genetics suggests it would still be many generations before enough variations accumulated in either/or the Earth or beyond-Earth populations to preclude healthy reproduction. It *could* happen, though, rapidly, in such extraordinary circumstances, if specific mutations related to the reproductive system were involved. This could cause 'reproductive mismatch' that might prevent fertile mating. We do not know the likelihood of this possibility as it would depend on a currently entirely hypothetical case. If populations beyond Earth wish to remain biologically viable with the Earth genome, they would carefully monitor the reproductive and, I think, developmental, genetics early in the lifecourse, examined thoroughly in Chap. 2.

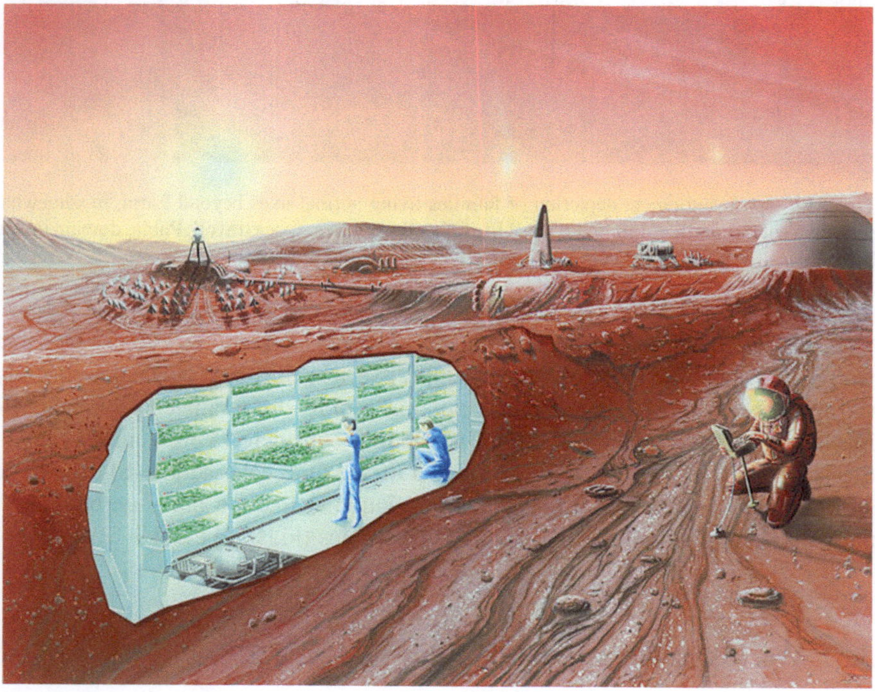

Fig. 3.11 Early-stage Mars settlement. Here the concept is a self-sustaining research center that would eventually grow in population, shifting from being a special-use site (science support) to a lifelong habitation. The concept image shows use of Mars regolith as shielding for inhabited architecture. Images depicting families and communities 'in space', rather than the more Spartan 'science station' appearance (see Fig. 3.11), will be important motivators unless the space-settlers simply 'self-select' and choose to go beyond Earth without any such motivation. Public domain image from NASA-Ames Research Center

Fig. 3.12 Space settlement depiction of families living normal lives beyond Earth, in somewhat suburban conditions, highlighting normal life rather than 'exotic space travel'. Public domain image by D. Davis for NASA publication "Space Settlements: A Design Study", 1975

Finally I will note that if speciation were to occur in populations beyond Earth, that would likely be, from an evolutionary perspective, a healthy sign of adaptation. Only if it were the result of a dramatic mutation that rapidly differentiated Earth and off-Earth populations in a deleterious way would it be considered a problem. But again, this is an hypothetical case. We may say that for the timescales addressed in realistic space-settlement planning, the issue of speciation is likely a nonissue because it normally takes hundreds to thousands of generations (Figs. 3.11 and 3.12).

3.5 Worldship Population Studies

The concept of the *worldship*, a vast vessel carrying thousands of people to another star, has figured in space-settlement thought for decades (for a recent review see Hein et al. 2012). Since 2010, for various reasons, there has been a resurgence of interest in this concept. I have become involved with the Icarus Interstellar research group, based in Europe, who have been modelling Worldships or 'Generation Ships' for nearly a decade. Their actual focus is on propulsion, specifically providing humanity with near-light-speed propulsion by the end of the present century. I was invited in 2011 to review the issue of the appropriate population for such voyage, which I published in 2014. Below I present a revisitation of the topic as written at the invitation of the European Space Agency's *Advanced Concepts Team,* with worldship researcher Andreas Hein of MIT and Centrale Supélec, and Strasbourg University astrophysicist Frédéric Marin. While the worldship concept is exotic, it is a fascinating topic and goal, and its modeling drives much interesting and worthwile research.

3.5.1 Time and Space Boundaries of the Thoughtscape

The project of interstellar voyaging is ultimately meant to preserve and spread human life. Therefore it 'builds out' from the central concerns of the human body. For the exploratory period of short-term spaceflight, the concerns of the individual body were measured in days and months, or up to a year. These are the scales of biology and flight physiology, as noted in Chap. 1. As we move towards consideration of permanent space settlement and even interstellar voyaging to exoplanets, concern must expand to include issues of individual bodies arranged as families, families arranged as communities, communities as a population (a 'deme'), and populations as cultures. These are the special domain of demography, population genetics and the scientific study of humanity, anthropology. In particular, anthropology studies human *biocultural* evolution as humanity adapts both by gene and (moreso in the last 100,000 years) culture.

Determination of a worldship population depends on the objective. Our objective is to allow a genetically- and culturally-healthy population arrive at an exoplanet, where they may land and begin a new world for humanity. Figure 3.13 indicates the 14 stars within 10LY of Earth; propulsion engineering and other issues related to the feasibility of reaching each are explored elsewhere (Long 2016). Exoplanet discoveries are burgeoning and the current paradigm is that there are "…2 ± 1 planets in the habitable zone of each [Milky Way galaxy] star" (Boivard et al. 2015: 3608). The closest habitable planet may well be within 10LY of Earth. We will likely know within a decade as new exoplanet characterization tools are developed; these are scheduled to include the *James Webb Space Telescope* (Beichman et al. 2014), the *Extremely Large Telescope* (Rodler 2018), and the *ExoLife Finder* (Berdyugina et al. 2018).

The 10LY distance is selected as a boundary here for reasons of time and space. Ten light years represents a distance just reasonably possible to reach, with reasonably-expectable worldship engineering, in several centuries, if reasonably-expectable propulsion speeds are achieved (Heller 2017; see also *Journal of the British Interplanetary Society Special Volume 65* on 'Worldships'). If the objective is to land a healthy population of humans (and their many domesticates) on an exoplanet after some centuries, we must know how many humans are required to establish a new population that itself will be multigenerationally viable. A number of such estimates have been made since the 1980s and are discussed below.

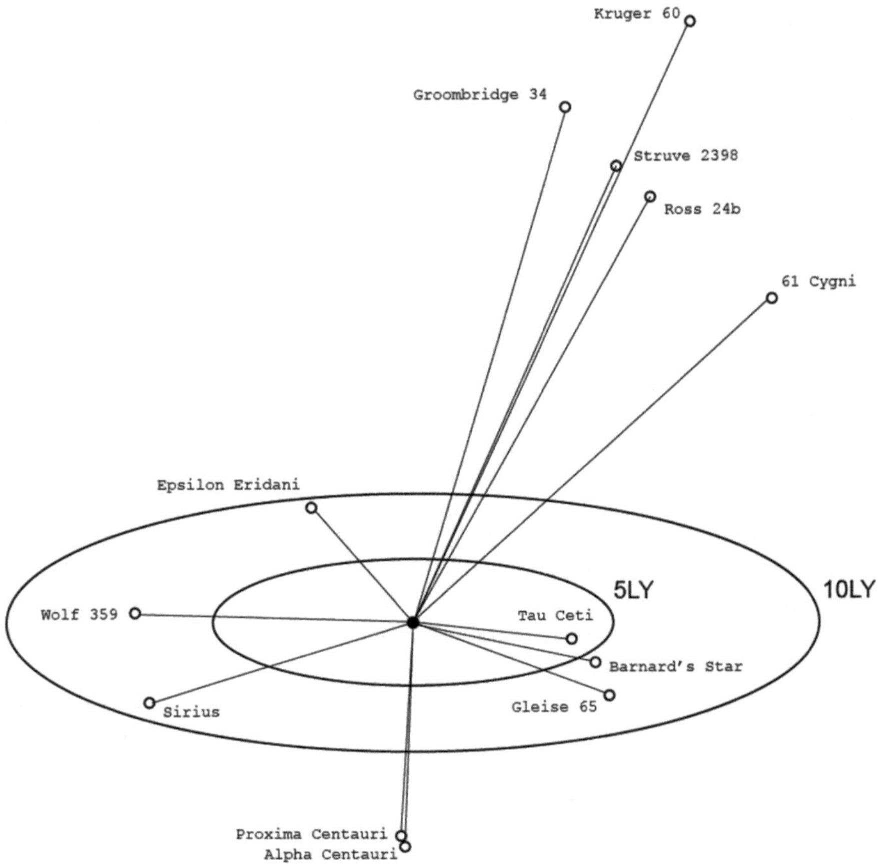

Fig. 3.13 Schematic diagram of stars within 10LY of Earth. Note that Luhman 16A, Luhman 16B and WISE 0855-0714 are brown dwarfs stars and Sirius B is a white dwarf star, and that no planets orbiting these stars would likely be habitable. This reduces the n of candidate targets, but also the n of systems to search and characterize. Rings at 5 and 10 Light Years from the sun (center) are on the galactic plane. Figure by the author

3.5.2 Biological Health: Estimates of Worldship Populations to Date

How many humans are required, as a founding population, to ensure that future generations live in good multigenerational health? The question has been addressed mainly by (a) population geneticists, for theoretical interest, (b) conservation biologists, to help conserve species at risk of extinction, (c) anthropologists, with an interest in human mating patterns and prehistoric dispersals, (d) space-settlement planners envisioning *open populations* that may be expected to continue to bring in new members over time and (d), space-settlement planners envisioning *closed pop-*

ulations that cannot be expected to acquire new members over the course of some journey. Our worldship interest is in the latter category.

Current worldship plans do not suggest the worldship to return to Earth or to voyage indefinitely. Rather, this population is envisioned as *departing* Earth, *traveling* to an exoplanet, and then *establishing* the population on the exoplanet so that it may grow. Such a voyage then includes two *founding* populations, the founders departing Earth and the founders serving as the original stock on the exoplanet from which subsequent generations will derive.

Establishing the first founding population is envisioned to be relatively easy, as humanity will have the Earth's diversity and population of people from which to choose a genetically healthy gene pool, composing a population. The constitution of the second founding population, however, is of most interest as it must itself be intact and healthy after the period of travel. The question then becomes, what do we need to do to ensure a bioculturally healthy population after some centuries as a closed population? Note that this is a more involved question than "what is the minimum number of people who can survive a multicentury voyage to an exoplanet". I return to this issue below after a review of the main estimates so far made to inform the worldship-planning community.

Seven carefully-researched estimates of interstellar voyage founding populations have been published (Table 3.9). For reference here I will distinguish between the founding population departing Earth (D1 for deme 1) and founding population arriving at an exoplanet (D2 for deme 2). Table 3.1 indicates these five estimations and also SIMOC, an agent-based model currently in development.

Anthropologist John Moore (1935–2016) referenced ETHNOPOP, a program written in C++ with his colleagues Dan Yu and Wenqiu Zhang in the 1990s, in his 2003 book chapter *Kin-based Crews for Interstellar Multi-Generational Space Travel* (Moore 1991a, b) His computer program was based on his extensive knowledge of hunter-gatherer band and tribe mating practices that linked small bands of travelers genetically over large landscapes. Essentially Moore's model was concerned with demographics, the age- and sex-structure of a given population, because of his original interest in using it to model prehistoric human dispersals across the globe. The program began with a population of a given size, and age- and sex-structure, and at each computing cycle (representing a generation, or c.30 years), evaluated each population member's likelihood of mating, death, and a few other variables. Thus it was an agent-based model, though of limited power compared to today's models. Still, his application of the model to spacefaring suggested to him that a founding population (D1) of 80–150 individuals would be sufficient to avoid inbreeding over a multi-century voyage (Fig. 3.14). The second founding population (D2) mentioned above was not a concern for Moore as he actually proposed that his voyagers would return to Earth, where they could again have the luxury of mating into a large and diverse population.

Subsequently in 2012 I took on the task of identifying an "Interstellar Migrant Population" (D1) for Icarus Interstellar's *Project Hyperion*, to assist in the reference design for a worldship. My approach, published in 2014 as a research paper titled *Estimation of a Genetically Viable Population for Multigenerational Interstellar Voyaging: Review and Data for Project Hyperion* (Smith 2014) was statistical and

Table 3.9 Seven published and/or current worldship population estimates

Model and references	Developed	Model type	Spacefaring simulations?	Recommended founding population (D1)	Current regard
ETHNOPOP Moore (1991a, b)	1990s	Demographic	Few	<300	Likely low
SMITH Smith (2014)	After 2013	Statistical	Several	>7500, ideally 14,000–44,000	Lower and middle figures reasonable; higher figures too high
GARDNER-O'KEARNY Smith (2014)	After 2013	Statistical agent-based	Several	>2000	Possibly reasonable
HERITAGE Marin (2017)	After 2015	Monte Carlo agent-based	Many, ongoing	>5000	Possibly reasonable
HERITAGE Marin and Beluffi (2018)	After 2015	Monte Carlo agent-based	Many, ongoing	Mathematically minimum 98, likely ideally larger	Possibly reasonable
HERITAGE Marin et al. (2018)	After 2015	Monte Carlo agent-based	Many, ongoing	c.500	Possibly reasonable
HEIN/SMITH/MARIN (in press)	**2019**	Monte Carlo agent-based and anthropological	Yes, ongoing	Some multiples of 500–1000 person "village modules"	Biologically and culturally realistic and reasonable
SIMOC Studies underway at this writing	After 2015	Monte Carlo agent-based	Many proposed via distributed parallel computing	–	–

Results of SIMOC, noted in the final row, are imminent and not yet published. D2, mentioned in the text, is not noted in this table

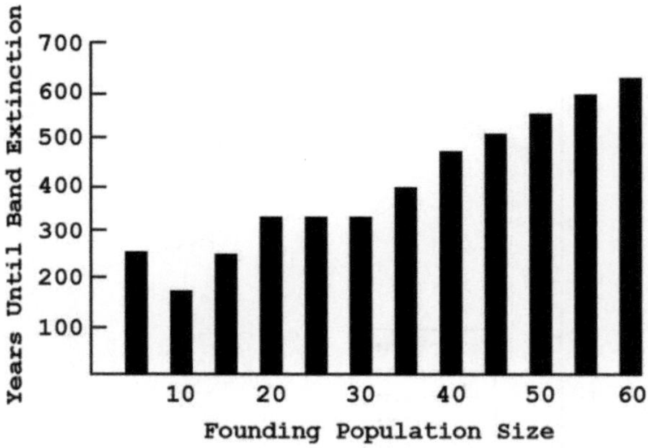

Fig. 3.14 Moore ETHNOPOP simulation showing years to demographic extinction for closed human populations. 'Bands' of people survive longer with larger starting sizes, but these closed populations all eventually become extinct due to demographic (age and sex structure) deficiencies. Figure adapted by the author after Moore (2003)

largely based on population genetics. I surveyed the research literature for various animal species' Minimum Viable Populations, figures below which natural animal populations did not drop in nature; this was discussed earlier in this chapter. With these as a context, I considered humanity's bioculturally-evolved natural populations, which, in the c.10,000 range, were not so different from the average for vertebrates. I considered also the effects of the primary population-dynamics processes of genetic mutation, migration, selection and drift, as well as the likelihood of catastrophe *en route* to a given exoplanet. From these figures I derived a simple formula to estimate the ideal D1 population given the desired D2 population, suggesting "anywhere from roughly 14,000 to 44,000" individuals as entirely safe D1 populations sufficient to ensure a D2 population equal to or greater than the common human breeding population of c.10,000. Lower figures for D1, in the 7500 range, I mentioned, were on the edge of reasonable, but I strongly suggested that 10,000 should be the absolute minimum D1 population setting off from Earth. My paper's final approach was to run agent-based computer simulations, written in MATLAB by my colleague William Gardner-O'Kearny. These were agent-based, establishing a D1 population and observing its change over each computing cycle, during which—as in Moore's model—each agent's age, probability of death, probability of finding a mate (based on its own age and sex), and other variables, was computed, such that the population changed over time. This model revealed that, strictly demographically, populations in the c.2000 range could survive for some centuries as closed systems (Fig. 3.15). While noting this, I did not suggest it as a viable D1 population, for reasons I will return to below.

In 2017 astrophysicist Frédéric Marin published the first in a series of papers (ongoing) revealing results of his HERITAGE agent-based program titled *Heritage: A Monte Carlo Code to Evaluate the Viability of Interstellar Travels Using a Multigenerational Crew* (Marin 2017). The approach used Monte Carlo methods (repeated random sampling of repeated simulations) to identify system properties emergent over time due to the properties of many individual agents. HERITAGE was written in C/C++ and is characterized by a large number of variables being evaluated

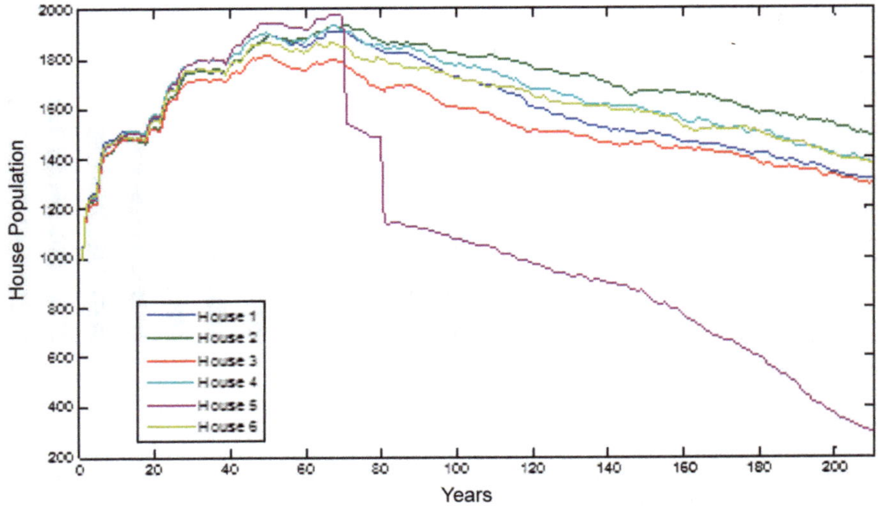

Fig. 3.15 MATLAB results simulating independent population fates over 210 Years. Each 'House' is a separate population, for example, a village or largely-isolated or isolable (for quarantine purposes) interstellar voyaging 'worldship' travelling, perhaps, in parallel. Each population begins at 1000 individuals and is allowed to double in about three generations to grow to just under 2000, the pre-set carrying capacity. The collapse of House 4 came in the form of an 'extinction vortex', discussed in the text, in which a random catastrophe so depopulated the breeding population that age- and sex- structures were disturbed, making it increasingly difficult to find mates. The general decline in populations resulted from aging populations and concomitant low replacement levels

at each computing cycle per individual agent. Access to superior computing power allows Marin to run many hundreds or thousands of these simulations in the Monte Carlo approach (in the US a related method is referred to as 'bootstrapping') to identify general system properties based on many specific cases (simulated voyages). Figure 3.16 indicates the HERITAGE flowchart and input parameters. Marin and his colleagues ran multiple simulated multicentury voyages that include the following results. Simulations of Moore-like populations (D1 population set at 150) over 200 years resulted in unhealthy inbreeding (due to the small population) and population reduction at the end of the voyage to about 33% of the original population, so that D2 (exoplanet founder population) would be about 50 individuals. Simulations of populations based on my figures of D1 set at 14,000 people departing Earth were shown to "be more efficient at mixing the genetic pool in order to ensure a safe sixth generation [D2 founder population]…and even with a severe catastrophe the mission is not compromised…[this scenario] is the only one to achieve the goal of…bringing a genetically healthy crew to another distant planet" (Marin 2017: 193).

Marin's second paper HERITAGE paper (*Computing the Minimal Crew for a Multi-Generational Space Travel Towards Proxima Centauri b*), coauthored in 2018 with C. Beluffi, applied an improved version of HERITAGE which included more complex mating and other reproductive rules, again evaluated at each computing cycle for each simulated member of the population. In this simulation, populations

Parameters	Values	Units
Number of space voyages to simulate	1000	–
Duration of the interstellar travel	2000	years
Colony ship capacity	500	humans
Overpopulation threshold	0.9	fraction
Inclusion of Adaptive Social Engineering Principles (0 = no, 1 = yes)	1	–
Number of initial women	49	humans
Number of initial men	49	humans
Age of the initial women	20	years
Standard deviation for the age of the initial women	1	years
Age of the initial men	20	years
Standard deviation for the age of the initial men	1	years
Number of child per woman	2	humans
Standard deviation for the number of child per woman	0.5	humans
Twinning rate	0.015	fraction
Life expectancy for women	85	years
Standard deviation for women life expectancy	15	years
Life expectancy for men	79	years
Standard deviation for men life expectancy	15	years
Mean age of menopause	45	years
Start of permitted procreation	30	years
End of permitted procreation	40	years
Initial consanguinity	0	fraction
Allowed consanguinity	0	fraction
Life reduction due to consanguinity	0.5	fraction
Possibility of a catastrophic event (0 = no, 1 = yes)	1	–
Fraction of the crew affected by the catastrophe	0.3	fraction
Year at which the disaster will happen (year; 0 – random)	750	years
Chaotic element of any human expedition	0.001	fraction

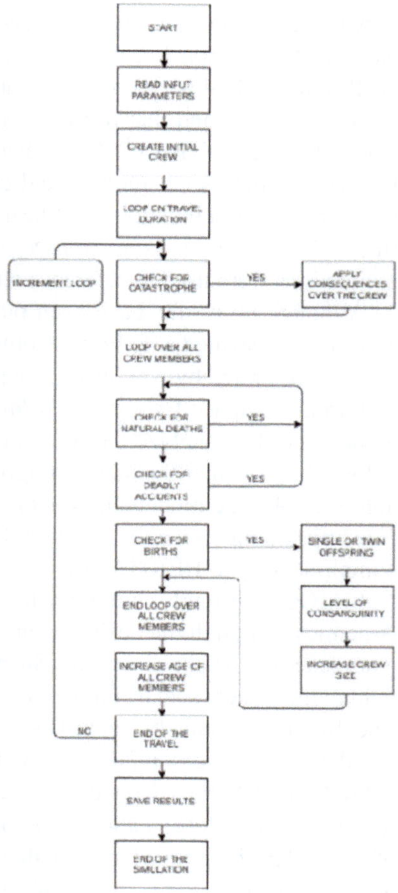

Fig. 3.16 HERITAGE input parameters and flow diagram Courtesy of F. Marin. See text for discussion

starting with D1 numbering 150 individuals (and a worldship capacity of 500) survived not only centuries, but over six millennia, in good genetic and demographic health; an even smaller D1 Fig. (98 individuals) was also identified as viable on this timescale. This was attributed to "adaptive social engineering principles" that change the mating rules *en route*, rather than applying a single rule throughout the voyage. This is an entirely reasonable adjustment of HERITAGE and is encoded in IF-THEN constructs such as the following: "*If the amount of people inside the vessel is lower than the [worldship's predetermined capacity], the code allows for a smooth increase of the population by allowing each woman to have an average of 3 children (with a standard deviation of 1). When the threshold is reached, HERITAGE impedes the couples' ability to procreate but allows women that were already pregnant to give birth even if the total number of crew members becomes marginally higher than the threshold.*" (Marin and Beluffi 2018: 433). While the study proposed very low D1 figures (compared to any other estimates) the authors did caution in the paper that "the impact of mutation, migration, selection and drift is not included in HERITAGE…[so] we emphasize that the minimum crew of 98 settlers we found is a lower limit…" and that further work might well suggest a larger D1 figure. The main advantage of HERITAGE is that it more accurately models real human mating behavior, which is not random and can thus, by consciously avoiding inbreeding, support smaller populations. In Marin's third paper reporting results of HERITAGE (the 2018 *Numerical Constraints of the Size of Generation Ships*), 500-person D1 populations were used as reference study (Marin et al. 2018).

Although no results have been published at this writing, the SIMOC (Scalable Model of an Isolated, Off-World Community) multiagent simulation is near completion. Orchestrated by Kai Staats as a project of the Arizona State University School of Earth and Space Exploration's *Interplanetary Initiative*, SIMOC is designed to model, and then analyze the results of, the physical characteristics of an off-Earth colony. In particular the habitat's agricultural, life-support, recyclable and consumable variables are modelled, as is the health of each 'colonist' placed in the system (see https://simoc.space/). I discuss SIMOC more in Chap. 7, but mention it here as complimentary to HERITAGE.

SIMOC is currently configured to model, as mentioned, physical rather than social dynamics, although the designers have expressed an interest in the interactions and emergence of social phenomena. Such social phenomena have been addressed in the field of multiagent social simulations, capably defined as "…the intersection of three scientific fields, namely, agent-based computing, the social sciences, and computer simulation" (Davidsson 2002). It will be very interesting to compare the results of SIMOC and HERITAGE. Current plans include comparison of SIMOC simulations with real-world closed habitat experimentation at the University of Arizona, a form of 'ground truthing' in which a mathematical model may be improved by comparison with, and then better modelling of, real-world systems. Social phenomena that may emerge in SIMOC and other multiagent simulations could be of great interest. At this writing we are aware of, but have not been able to review, W. S. Bainbridge's 2019 book, *Computer Simulations of Space Societies* (Bainbridge 2018).

3.5.3 Reasons for Estimate Variations for D1, Earth-Departing Interstellar Population

It is natural that a variety of population sizes have been proposed for D1, the Earth-departing founding population, as researchers from different backgrounds have brought various approaches to this question. I believe that some of the variation derives from different conceptions of human populations and human behavior over time.

Moore's gravitation towards low figures come from a long-term anthropological perspective recognizing that hunting and gathering cultures have survived for many thousands of years in low population densities, so that just few centuries should be relatively easy for a D1 population less than several hundred. However, Moore's figures appear somewhat too low as he did not really account for the fact that while humans may live together in breeding populations (demes) numbering in the hundreds [the famous 'Dunbar Number' of c.150 individuals is often quoted regarding hunter-gatherer group size (Dunbar 1993)], such populations always have reproductive links with other groups as mentioned throughout this chapter. Also, Dunbar's figures largely reflect populations of hunting-and-gathering folk who move seasonally over large landscapes, whereas in all conceivable worldship designs the subsistence mode would be agriculture, which is characterized by residential sedentism. Residential sedentism, worldwide and throughout prehistory, always leads to higher populations, as discussed later in this book.

My own anthropological biases led me to highlight larger population figures because human populations are always linked to others, normally in the thousands of individuals, figures approaching the c.7500 population range for naturally-evolved populations of naturally-occurring vertebrates. I am is also conditioned by an emphasis on catastrophe: for Moore, human populations have generally survived quite well even small populations in particular because they have often had large landscapes and many resources available; a local catastrophe could be averted by moving to new resource territories, and if one group actually became extinct, humanity was so widespread that others always continued. But considering the prospect of a closed population carrying all their resources with them, I have the expectation that eventually some catastrophe will strike, and for this single, isolated population there is nowhere to go, no 'geographical reserve'. To be sure to arrive in relatively safe populations (D2), I have gravitated towards particularly large Founding Populations (D1).

For Marin's first 2018 approach with C. Beluffi, there is an attempt to reduce D1 to an absolute minimum as revealed in the paper title, *Computing the Minimum Crew for a Multigenerational Space Journey Towards Proxima Centari b*, with the paper's function stated as "to quantify the minimal initial crew necessary for a multi-generational space journey to reach Proxima Centauri b with genetically healthy settlers". Here the focus is on propulsion at speeds achieved today with the Parker Solar Probe, resulting in the need to voyage for an estimated 6300 years to *Proxima Centauri b*. The philosophy driving the search for the minimum viable population here is that of the "scarcity paradigm" of crewed spaceflight. In this paradigm, we must identify the minimum mass to transport to reduce cost. Marin and Beluffi's comparatively small figure of less than a hundred individuals is identified as a viable D1 figure under certain adaptive mating rules that may change over the course of a journey—back, then, to figures closer to the Moore thoughtscape. This addresses human behavior more subtly than my own initial, statistical approach, by taking into account that the statistical approach assumes random mating among individuals of a population. However, human cultural guidelines assure that mating is not random, but highly structured, for example, to avoid incest. Future population modelling will best be most productive if carried out with both approaches.

3.5.4 *Biological Health: Where Are We Today on Estimates of Worldship Populations?*

From strictly mathematical, statistical and genetic perspectives we may say that Earth-departing D1 founder populations of humans, numbering in just the low hundreds of people, could survive for centuries or even some millennia in health sufficient to serve as D2 (exoplanet founder) populations when mating is cleverly devised to avoid inbreeding. I mentioned this in my 2014 paper, for instance, stating that "any population over 100 or so" would avoid some of the chief problems of small populations on such timescales. Marin and Beluffi demonstrate this with the high-fidelity HERITAGE program that capably simulates human social engineering to manage population health.

I feel the swing of the pendulum in worldship population estimates is somewhat too wide. While the smallest figures may work biologically, they are rather precarious for some generations before the population has been allowed to grow. Figures on the higher end of the my 2014 estimate, multiple tens of thousands, may be offputting 'nonstarters' for those just being introduced to the conversation. In a paper currently in press, F. Marin and I collaborated to suggest figures somewhere between these extremes, with Earth-departing (D1) figures on the order of 1000 persons. Because useful modelling is still underway, there is a practical way to use such an estimate even at this early date. We have therefore proposed that for habitat design and modelling at present, 1000-person 'modules' or 'villages' be designed, that could at a later date simply be multiplied as elements of a worldship 'cluster'. This way, the Earth-departing population could be set to any figure one wants, for example 3 villages composing 3000 people, or 10 for 10,000 or just one for a departing population of 1000 (Fig. 3.17). While this modularity does increase mass (as compared to a single-vessel design using the most efficient enclosure of space by material) my colleagues and I feel the modularity is worth the tradeoff. For instance, multiple, independent 'villages' travelling in parallel, each with a population of c.1000, would reduce the possibility of a catastrophe wiping out the entire population. Travelling in parallel would allow people to visit other 'towns' for pleasure, cultural exchange and marriage and reproduction, but also to be 'quarantined' (culturally and/or biologically) if desired. The population on the order of 1000 per 'village module' is also viable culturally, as I discuss fully in Chap. 6.

3.6 Conclusions and Recommendations

Human populations will be rearranged beyond Earth into new sizes and distributions in order to fit the given habitats we choose to settle. However populations are maintained with social regulation of reproduction, they will have to be more carefully monitored, for some early generations at least, than we are accustomed to on Earth today. If we presume, however, that such regulation will likely be devised by the

Fig. 3.17 A multigenerational vessel schematic designed by Steve Summerford. Revolving for 1-g conditions around a central hub are eight habitations; in the conception invoked in this chapter. Each village would be composed of c.1000 people, as in many small towns on Earth today, for a total population of 8000 in this concept. Propulsion and other systems are kept at some distance to the rear. Figure copyright and courtesy of Steve Summerford (see Summerford 2012)

space settlers themselves, to keep their populations healthy and within the limits supportable by a given habitat, there is no reason to invoke unfair or unjust directives dictated from 'on high' by the Population Police. These regulations will develop internally, among the settling culture(s). So long as populations in the multiple thousands are maintained for relatively isolated communities, they should be quite safe as Founding Populations, but I would suggest that the population plan be to grow quite rapidly for safety.

3.6.1 Genetic Testing and Genetic Screening

Clearly, even smallish (compared to Earth) populations of humans can, from a population-genetics perspective, survive beyond Earth should the life-supporting technologies be provided. The question of the constitution of these populations naturally arises and is one of the most common questions I am asked when discussing the prospect of space settlement: 'who gets to go, and who would be left behind?' The question derives more from common conceptions based on science fiction stories than from science itself; indeed I find most commonly people cite the 1990 film

Gatacca when imagining a draconian 'top-down' selection process where a corporation or state determines who is able to participate in space exploration or settlement. The determination is made genetically, and the film's premise is a worthy caution against such projects as *eugenics*, the genetic manipulation (by artificial selection) of a genome in order to create a *master race* (in this case, a race of spacefarers). But while eugenics projects did exist early in this century they have been abandoned, in part as geneticists have shown that the race concept is of no biological value. Old textbooks of genetics and physical anthropology, for example, had entire chapters on the characteristics of different races, but today what were once called 'races' are known to be populations exhibiting some segment of a spectrum of continuous variation in such traits as skin or hair color, all within the same genus, *Homo*, and the same species, *sapiens*. The essential problem with the *master race* conception—generally and specific to this discussion—is that population biologists measure population health by genetic diversity, not uniformity. Genetic diversity gives a population multiple ways to survive, for example in the case of sudden, unpredictable selection against certain gene carriers. Genetic uniformity means all individuals are subject to the same potential disaster, no matter how 'fit' they are at a given moment. Considering this, but also the issues of inbreeding and small populations discussed earlier in this chapter, what may we say about the genetic constitution of populations of prospective space-settlers?

There are really only two main issues to consider in the genetic makeup of founding populations of space settlers. First, the relatively small populations would best be more diverse than homogenous. How diverse, and what sort of diversity is established or ensured will vary case by case. Note again I propose this will not be achieved by draconian 'selection boards' but by spontaneous assembly and decision-making of space-settling people themselves. Second, the populations and their descendants would be better off if they did not carry abnormally-high frequencies of lethal or debilitating genetic disorders; but even a little math would show that they could even support some carriers of lethal conditions…and in some cases, the conversation regarding what gene carriers to support will be very similar to the conversation on Earth. For example, the great astronomer Stephen Hawking had motor neuron disease, an ultimately lethal genetic disorder. But Hawking lived a full and productive life; would spacefaring populations 'select against' individuals predicted to have this condition, perhaps though genetic screening or testing? The discussions will be no different than they are on Earth unless the founding populations are extremely small, on the order of tens of people, which have not been proposed in any case.

With these common-sense guidelines in place it is actually rather difficult to imagine what genetic malady would be culturally selected against—carriers being prevented from space settlement—in ways that differ from such reproductive decisions made daily on Earth. We proceed with birth and caretaking for people with many genetic disorders, people who have full and productive lives, including artists and scientists. I see no reason why such people would not participate in space settlement, even in the frequencies that they are encountered in Earth populations. Still, genetic testing and screening, discussed below, will likely be used to best understand the genetic character of populations of space settlers, just as they are used today on Earth for people's general knowledge.

3.6.2 Genetic Testing

Genetic testing "uses a variety of techniques to determine if someone has a certain genotype [is carrying certain gene forms]....[identifying] those who have or may carry a genetic disease, those at risk of producing a genetically defective child, and those who may have a genetic susceptibility to environmental agents" (Cummings 2003: 472). Currently over 6000 genetic disorders are known, though in many cases they are known only crudely, in the *association* of a gene variant with a set of phenotypic expressions or symptoms, though lacking detailed understanding of the causal mechanisms. In practical terms, only a small percentage of these are currently monitored; the *Genetic Disease Foundation*, for example, based in New York, lists 281 such genetic disorders for which patients are regularly prenatally tested. Table 3.10 indicates some of these conditions.

3.6.3 Genetic Screening

While genetic testing is voluntary, *genetic screening* is normally policy, "carried out on populations...as a matter of law" (Cummings 2003: 472). What screening laws would be implemented in the selection of space settlers will be, I think, determined by those settlers themselves. As they are on Earth today, those decisions will be informed in part by the space-settler's cultural conceptions regarding "four important ethical and legal principles: autonomy, confidentiality, privacy, and equity, (Andrews et al. 1994: 247). Currently in the United States, GINA, the *Genetic Information Nondiscrimination Act* provides federal legal protection for individuals "from genetic discrimination in health insurance and employment" (GINA 2019). A similar protection in Europe is provided by the 2018 *General Data Protection Regulation*. The ethics of genetic screening is an active field and will certainly be examined as individuals and groups forming a space-settling culture make decisions about their own genetic information.

At this point I would only propose, from the anthropological perspective, that unless very small founding populations are proposed, numbering in the dozens or so, with no plans for the population to grow or increase its variety over time—a proposition I have not seen seriously proposed in any space-settlement plans—the ethics of genetic testing and screening for space settlers need not be significantly different from these ethics on Earth. If the population proposed is so small that the Earth-population incidence of certain genetic maladies is a significant problem, then I suggest the population is already too low and a larger population should be proposed. In the case of small settler populations that plan to grow rapidly, as I

Table 3.10 Selected metabolic disease and birth defects diagnosable by prenatal testing

Disease or Birth Defect	Presentation	Frequency	Lethality	Treatable?	Prevalence	Prognosis
Adrenogenital syndrome	Sex organ abnormalities	1:15,000	Low	Symptomatic, systemic	Somewhat higher in Central Europeans	Survivable, with possible stature reduction
Citrullinemia (types I and II)	Blood pollution by ammonia buildup/nervous system abnormalities	1:60,000/1:150,000	Moderate	Symptomatic, systemic	Type II more common in Japan	Survivable, but with cognitive impairment
Cystic fibrosis	Lung and other organ disorders	1:3000	Moderate	Symptomatic	More common in EuroAmerican Whites	Often lethal before 40
Fabry disease	Somatic cell fat accumulation	1:50,000 (males)	Low	Symptomatic, systemic	Panethnic	Variable, but often survivable into 60s
Gaucher disease	Somatic cell fat accumulation	1:30,000	Highly variable	Symptomatic	More common in Eastern European Jews	Variable, highly debilitating through life
I-Cell disease	Skeletal abnormalities and mental retardation	Very low	Low	Symptomatic	Panethnic; slightly higher rate in Japan	Variable but often lethal before 10
Lesch-Nyhan syndrome	Physical and cognitive abnormalities	1:400,000	High	symptomatic	Panethnic	Often lethal before 30

(continued)

Table 3.10 (continued)

Disease or Birth Defect	Presentation	Frequency	Lethality	Treatable?	Prevalence	Prognosis
Mannosidosis	Skeletal, facial and intellectual abnormalities and retardation	1:500,000	Variable	Symptomatic	Panethnic	Variable but often lethal before 10
Marfan syndrome	Connective tissue disorders affecting heart, eye and skeleton	1:15,000	Low	Symptomatic	Panethnic	Often survivable well into old age
Muscular dystrophy (x-linked)	Muscle wasting	1:4000	Variable	Symptomatic	More common in EuroAmerican Whites	Variable but may be survived into old age
Progeria	Pre-pubescent accelerated aging	1:4,000,000	High	Symptomatic	Panethnic	Commonly lethal before 15
Sandhoff disease	Nervous system degeneration, ataxia, bone abnormalities	1:100,000	High	Symptomatic	Panethnic	Commonly lethal before 5
Spina bifida	Spinal and associated disorders	1:1000	Low	Symptomatic	Panethnic	May be survivable into old age
Tay-Sachs disease	Nervous system disorders	1:300,000	High	Symptomatic	More common in Eastern European Jews	Commonly lethal before 5

(continued)

Table 3.10 (continued)

Disease or Birth Defect	Presentation	Frequency	Lethality	Treatable?	Prevalence	Prognosis
Thalassemia	Blood oxygen starvation	1:45,000	Variable but often low	Symptomatic, systemic	More common in mediterranean populations	Variable but often survived into old age
Werner syndrome	Post-pubescent accelerated ageing	1:200,000	High	Symptomatic	Panethnic; slightly higher rate in Japan	Commonly lethal before 50
Xeroderma pigmentosum	Impaired DNA repair mechanism, severe ultraviolet radiation sensitivity	1:200,000	High	Symptomatic	Panethnic; slightly higher rates in Japan, North Africa and Middle East	Variable but often lethal before 30

Presentation indicates symptoms; Frequency is an average figure; Lethality is a generalization; Treatability regards whether the symptom and/or the genetic disorder itself is addressed; Prevalence is a generalization, note that it is based towards First-World populations whose genetics are better-known than in other regions; Prognosis is a generalization due to often highly variable severity of the disorder. Data from Cummings (2003) and surveys in databases of the National Institutes of Health Genetic Home Reference, the World Health Organization and the Genetic Disease Foundation

suggest elsewhere in this book for, for example, Mars settlement, only the first few generations might have somewhat more close attention to genetic screening and testing than we do on Earth today. Once the population is into a thousand or so, I see no real need for extraordinary genetic screening or testing, beyond what is already carried out on Earth. I also see no reason that people with genetic maladies, anywhere on a spectrum of severity, could not be cared for and have lives just as meaningful beyond Earth just as they do on Earth.

3.6.4 Practical Considerations for Beyond-Earth Population Diversity and Population Size

A review of the population genetics of space settlement indicates that on the timescale considered in this book (several generations) and the degree of reproductive isolation considered in this book (not very isolated as in the case of early Mars and Lagrangian settlements, completely isolated in the case of Worldships, though those may use gene banks or gene editing to artificially introduce diversity on some kind of schedule) the population genetics issues differ.

For the relatively closed population, I suggest a thoughtscape in the multiple thousands as founding populations, with the main issue over time being the delivery of a large and diverse population—even despite *en route* catastrophe—to an exoplanet. Such a population may be sustained, it seems, with adaptive population level planning *en route*. Theoretical work in this field continues, as mentioned, in my own studies and those of Frederic Marin and others.

For the significantly-less closed population, which I expect to have more reproductive links with Earth for some time—such as early Mars or Larangian-point settlements—I suggest a thoughtscape highlighting genetic diversity and, as rapidly as possible, population increase to levels that would rebound without significant population assistance from outside, such that the population becomes genetically diverse and numerically large as rapidly as is practical. These features, along with being geographically widespread (either actually or practically, by means of the ability to quarantine or otherwise insulate subpopulations from catastrophes), are as mentioned the population-level features of very long-lived species, and what, I believe, space settlement should be shaped to emulate.

3.7 Summary Points

- Human populations normally have reproductive links with other populations, such that small 'tribe' populations (in the dozens) are misleading regarding humanity's Minimum Viable Population

- A practical Minimum Viable Population (MVP) for humanity is in the range of 1000–10,000 reproductive individuals
- Reproductive subpopulations of humans (demes) need not be of the MVP size, but should be in social and biological contact with other populations to form a total MVP on the order of 1000–10,000 reproductive individuals
- Establishing beyond-Earth space settlement plans using the Minimum Viable Population is a dangerous approach; larger populations are safer
- Maintaining a genetically healthy human population beyond Earth would be a matter of monitoring the universal factors of mutation, migration, selection and drift
- Human reproductive freedoms currently enjoyed by some Earth populations will be somewhat curtailed by cultures settling habitats beyond Earth in somewhat small populations, for some generations
- The smaller the founding populations beyond Earth, the greater degree of social engineering will be required to ensure genetic and demographic health over time
- After establishment of early, small founder populations, human settlements beyond Earth should be encouraged to (a) establish stored agricultural stocks sufficient to sustain them though reasonably-expectable catastrophes and then (b) promote rapid population growth
- Space settlement should not be planned with a paradigm of scarcity, but of abundance
- Space settlement plans should be shaped to support human populations that are large, spatially widespread and genetically diverse; supporting these features should be built into space settlement plans from the beginning, rather than haphazardly, and later.

References

Adams, M. S., & Neel, J. V. (1967). Children of incest. *Pediatrics, 40*(1), 55–62.

Andrews, L. B., Fullarton, J. E., & Holtzman, N. A. (1994). Assessing genetic risks: Implications for health and social policy. *Social, Legal, and Ethical Implications of Genetic Testing*, (8), 247–289. https://www.ncbi.nlm.nih.gov/books/NBK236044/. Washington (DC): National Academies Press (US).

Avise, J. C., Ball, R. M., & Arnold, J. (1998). Current versus historical population sizes in vertebrate species with high gene flow: A comparison based on mitochondrial DNA lineages and inbreeding theory for neutral mutations. *Molecular Biology and Evolution, 5,* 331–344.

Ayala, F. (1995). The myth of eve: Molecular biology and human origins. *Science, 270,* 1930–1936.

Bailey, W. J. K., Hayasaka, K., Skinner, C. G., Kehoe, S., Cieu, L. C., Slightom, J., & Goodman, M. (1992). Reexamination of the African hominoid trichotomy with additional sequences from the primate beta-globulin gene cluster. *Molecular Phylogenetics and Evolution, 1,* 97–135.

Bainbridge, W. S. (2018). *Computer simulations of space societies*. New York: Springer.

Barigozzi, G., et al. (1959). Schema della distribuzione geografica dei gruppi sanguigni in Italia. In L. P. Hollander (Ed.), *7th Congress of the Proceedings International Society of Blood Transfusion, Rome* (pp. 266–269). Basel: Karger.

Barton, N. H., & Charlesworth, B. (1994). Genetic revolutions, founder effects and speciation. *Annual Review of Ecology, Evolution, and Systematics, 15,* 133–164.

Beichman, C., et al. (2014). Observations of transiting exoplanets with the James Webb space telescope. *Publications of the Astronomical Society of the Pacific, 126,* 1134–1173.

Berdyugina, S. V., et al. (2018). *ExoLife finder (ELF): A hybrid optical telescope for imaging exo-earths.* Astrobiology Science Strategy for the Search for Life in the Universe National Academies of Sciences, Engineering, and Medicine (White Paper). Online at https://www.planets.life/wp-content/uploads/2018/03/ExoLife_Finder_2018_white_paper.pdf.

Boivard, T., Lineweaver, C. H., & Jacobsen, S. K. (2015). Using the inclinations of Kepler systems to prioritize new Titius–Bode-based exoplanet predictions. *Monthly Notices of the Royal Astronomical Society, 448,* 3608–3627.

Boughman, J. W. (2001). Divergent sexual selection enhances reproductive isolation in sticklebacks. *Nature, 411,* 944–948.

Boyce, A. J., Kucheman, C. F., & Harrison, G. A. (1967). Neighbourhood knowledge and the distribution of marriage distances. *Annals of Human Genetics, 30,* 335–338.

Caballero, A. (1994). Development of the prediction of effective population size. *Heredity, 73,* 657–679.

Campbell, K. S. W., & Day, M. F. (Eds.). (1987). *Rates of evolution.* Allen and Unwyn: London.

Charlesworth, B. (2009). Effective population size and patterns of molecular evolution and variation. *Nature Reviews Genetics, 10,* 195–205.

Chen, F. G., & Li, W.-H. (2001). Genomic divergences between humans and other hominoids and the effective population size of the common ancestor of humans and chimpanzees. *American Journal of Human Genetics, 68*(2), 444–456.

Cummings, M. R. (2003). *Human heredity* (6th ed.). Pacific Grove, California: Thomson.

Davidsson, P. (2002). Agent based social simulation: A computer science view. *Journal of Artificial Societies and Social Simulation, 5*(1). See http://jasss.soc.surrey.ac.uk/5/1/7.html.

de Queiroz, K. (2005). Ernst Mayr and the modern concept of species. In J. Hey, W. M. Fitch, & F. Ayala (Eds.), *Systematics and the origin of species: On Ernst Mayr's 100th Anniversary* (pp. 243–266). Washington, DC: National Academies Press.

Dorea, J. G., & Donangelo, C. M. (2006). Early (in uterus and infant) exposure to mercury and lead. *Clinical Nutrition, 25,* 369–376.

Dorit, R. L., Akashi, H., & Gilbert, W. (1995). Absence of polymorphism at the AFY locus on the human Y chromosome. *Science, 268,* 1183–1185.

Douglas, M. (1966). Population control in primitive groups. *British Journal of Sociology, 17,* 263–273.

Dunbar, R. I. M. (1993). Coevolution of neocortical size, group size and language in humans. *Behavioral and Brain Sciences, 16*(4), 681–735.

Duret, L. (2009). Mutation patterns in the human genome: More variable than expected. *PLOS Biology, 7*(2), 217–219.

Durham, W. H. (1993). *Coevolution: Genes, culture and human diversity* (1st ed.). Stanford: Stanford University Press.

Eckhardt, R. B. (2000). *Human palaeobiology*. Cambridge: Cambridge University Press.

Endler, J. (1986). *Natural selection in the wild.* Princeton, New Jersey: Princeton University Press.

Fenner, J. N. (2005). Cross-cultural estimation of the human generation interval for use in genetics-based population divergence studies. *American Journal of Physical Anthropology, 128*(2), 415–423.

Finney, B., & Jones, E. M. (1985). *Interstellar migration and the human experience*. Berkeley: University of California Press.

Fix, A. G. (1999). *Migration and colonization in human microevolution*. Cambridge: Cambridge University Press.

Flather, F. H., Hayward, G. D., Beissinger, S. R., & Stephens, P. A. (2011). Minimum viable populations: Is there a 'Magic Number' for conservation practicioners? *Trends in Ecology and Evolution, 26*(6), 307–316.

Frankham, R. (1995). Effective population size/Adult population size ratios in wildlife: A review. *Genetical Research, 66*, 95–107.
Franklin, J. R. (1980). Evolutionary change in small populations. In M. E. Soule & B. A. Wilcox (Eds.), *Conservation biology: An evolutionary-ecological perspective* (pp. 135–150). Sinauer: Massachussetts.
Franklin, J. R., & Frankham, R. (1998). How large must populations be to retain evolutionary potential? *Animal Conservation, 1*, 69–70.
Friedberg, E. C. (2006). Mutation as a phenotype. In L. H. Caproale (Ed.), *The implicit genome* (pp. 39–56). Oxford: Oxford University Press.
Horneck, G., Mileikowski, C., Melosh, H. J., Wilson, J. W. Cucinotta, F. A., & Gladman, B. (2002). Viable transfer of microorganisms in the solar system and beyond. In G. Hornbeck & C. Baumstarck-Khan (Eds.), *Astrobiology: The quest for the conditions of life* (pp. 57–76). New York: Springer.
Galvani, A. P., & Slatkin, M. (2003). Evaluating plague and smallpox as historical selective pressures for the CCR35-Delta32 HIV-Resistance allele. *Proceedings of the National Academy of Sciences, 100*(25), 15276–15279.
Gamble, C. (2003). *Timewalkers*. Cambridge: Harvard University Press.
Gamble, C. (1999). *The palaeolithic societies of Europe*. Cambridge: Cambridge University Press.
Gillespie, J. H. (2001). Is the population size of a species relevant to its evolution? *Evolution, 55*, 2161–2169.
GINA. (2019). See http://www.ginahelp.org/.
Gould, S. J., & Eldredge, N. (Eds.). (1982). *Genetics and the origin of species by Theodosius Dobzhansky. Columbia classics in evolution series*. New York: Columbia University Press.
Gunson, N. (1997). Great families of Polynesia: Inter-island links and marriage patterns. *Journal of Pacific History, 32*(2), 139–179.
Hadid, Y., et al. (2014). Sympatric incipient speciation of spiny mice Acomys at "Evolution Canyon", Israel. *Proceedings of the National Academies of Science of the United States of America, 111*(3), 1043–1048.
Hammer, M. F. (1995). A recent common ancestry for human Y chromosome. *Nature, 378*, 376–378.
Harcourt, A. H. (2002). Emprical estimates of minimum viable population sizes for primates: Tens to tens of thousands? *Animal Conservation, 5*, 237–244.
Harding, R. M., Fullerton, S. M., Griffiths, R. C., Bond, J., Cox, M. J., Schneider, J. A., et al. (1997). Archaic African and Asian lineages in the genetic ancestry of modern humans. *The American Society of Human Genetics, 60*, 772–789.
Harmon, L. J., & Braude, S. (2013). *Conservation of small populations: Effective population sizes, inbreeding, and the 50/500 Rule*. In S. Braude & B. S. Low (Eds.), *An introduction to methods and models in ecology, evolution, and conservation biology* (pp. 125–138). Princeton: Princeton University Press.
Harpending, S., Batzer, M. A., Gurven, M., Jorde, L. B., Rogers, A. R., & Sherry, S. T. (1998). Genetic traces of ancient demography. *Proceedings of the National Academy of Sciences of the United States of America, 1995*, 1961–1967.
Harrison, G. A., & Boyce, A. J. (1972). The framework of population studies. In G. A. Harrison & A. J. Boyce (Eds.), *The structure of human populations* (pp. 1–16). Oxford: Clarendon Press.
Hawks, J., Hunley, K., Lee, S.-H., & Wolpoff, M. (2000). Population bottlenecks and pleistocene human evolution. *Molecular Biology and Evolution, 17*, 2–22.
Hein, A., Pak, M., Putz, D., Buhler, C., & Reiss, P. (2012). World ships—Architectures and feasibility revisited. *Journal British Interplanetary Society, 65*, 119–133.
Henderson, J. B. (1987). The importance of limited exposure to ultraviolet radiation and dietary factors in the etiology of Asian rickets: A risk-factor model. *QJM: An International Journal of Medicine, 63*, 413–442.
Hey, J. (2005). On the number of new world founders: A population genetic portrait of the peopling of the Americas. *PLoS Biology, 3*(6), 0965–0975.

Huang, H., Fu, Y. X., Chang, B. H., Gu, X., Jorde, L. B., & Li, W. H. (1998). Sequence variation in ZFX introns in human populations. *Molecular Biology and Evolution, 15*(2), 138–142.

Hull, D. L., Langman, R., & Glenn, S. (2001). A general account of selection: Biology, immunology and behavior. *Behavioral and Brain Sciences, 24,* 511–528.

Jancar, J., & Johnston, S. J. (2008). Incest and mental handicap. *Journal of Intellectual Disability Research, 34*(6), 483–490.

Jorde, J. B., Carey, J. C., Bamshad, M. J., & White, R. L. (1999). *Medical Genetics*. St. Louis: Mosby.

Jorde, L. B., Watkins, W. S., & Bamshad, M. J. (2001). Population genomics: A bridge from evolutionary history to genetic medicine. *Human Molecular Genetics, 10*(20), 2199–2207.

Kaesmann, H., Weibe, V., & Paabo, S. (1999). Extensive nuclear DNA sequence diversity among Chimpanzees. *Science, 286,* 1159–1162.

Kelly, R. J. (2007). *The foraging spectrum: Diversity in hunter-gatherer lifeways*. New York: Eliot Warner Reprints.

Kimura, M., & Ohta, T. (1969). The average number of generations until fixation of a mutant gene in a finite population. *Genetics, 61,* 763–771.

Kinnaird, M. F., & O'Brien, T. G. (1991). Viable populations for an endangered forest primate, the tana river crested Mangabey (*Cercocebus galeritus galeritus*). *Conservation Biology, 5*(2), 203–213.

Kondo, Y., Bruhweiler, F., Moore, J., & Sheffield, C. (Eds.), (1991). *Interstellar travel and multi-generational space ships*. Wheaton, Illinois: Apogee Books.

Lande, R. (1988). Genetics and demography in biological conservation. *Science, 241,* 1455–1460.

Lande, R. (1995). Mutation and conservation. *Conservation Biology, 9*(4), 782–791.

Leffler, E. M., Bullaughey, K., Matute, D., Meyer, W. K., Segruel, L., Venkat, A., et al. (2012). Revisting and old riddle: What determines genetic diversity levels within species? *PLoS Biology, 10*(9), 1–9.

Lewis, D. (1972 [1994]). *We, the navigators*. Honolulu: University of Hawaii Press.

Long, K. F. (2016). Project Icarus: Development of fusion based space propulsion for interstellar missions. *Journal of the British Interplanetary Society, 69,* 289–294.

Lynch, M. (2010). Rate, molecular spectrum, and consequences of human mutation. *Proceedings of the National Academy of Sciences of the United States of America, 107*(3), 961–968.

Lynch, M., & Lande, R. (1998). The critical effective size for a genetically secure population. *Animal Conservation, 1*(1), 70–72.

Marin, F. (2017). Heritage: A Monte Carlo code to evaluate the viability of interstellar travels using a multigenerational crew. *Journal of the British Interplanetary Society, 70,* 184–195.

Marin, F., & Beluffi, C. (2018). Computing the minimal crew for a multi-generational space journey towards proxima centarui b. *Journal of the British Interplanetary Society, 71,* 431–438.

Marin, F., Beluffi, C., Taylor, R., & Grau, L. (2018). Numerical constraints on the size of generation ships: from total energy expenditure on board, annual food production and space. *Journal of the British Interplanetary Society, 71,* 382–393.

Masel, J. (2012). Rethinking Hardy-Weinberg and genetic drift in undergraduate biology. *BioEssays, 34*(2012), 701–710.

Matsamura, S., & Forster, P. (2008). Generation time and effective population size in polar eskimos. *Proceedings of the Royal Society B: Biological Sciences, 275,* 1501–1508.

Mayr, E. (1996). What is a species and what is not? *Philosophy of Science, 63,* 262–277.

Mills, L. S., & Smouse, P. E. (1994). Demographic consequences of inbreeding in remnant populations. *American Naturalist, 144*(3), 412–431.

Moore, J. H. (1991a). Kin-based crews for interstellar multi-generational space travel. In Y. Kondo, F. Bruhweiler, J. Moore & C. Sheffield (Eds.), *Interstellar travel and multi-generational space ships* (pp. 81–88). Wheaton, Illinois: Apogee Books.

Moore, J. H. (1991b). Kin-based crews for interstellar multi-generational space travel. In Y. Kondo, F. Bruhweiler, J. Moore, & C. Sheffield (Eds.), *Interstellar travel and multi-generational space ships* (pp. 81–88). Wheaton, Illinois: Apogee Books.

Moore, J. H. (2003). Evaluating five models of human colonization. *American Anthropologist, 103*(2), 395–408.
Nei, M., & Grauer, D. (1984). Extent of protein polymorphism and the neutral mutation theory. *Evolutionary Biology, 27*, 73–118.
Newmark, W. D. (1987). A land-bridge island perspective on mammalian extinctions in Western North American Parks. *Nature, 325*, 430–432.
O'Rourke, D. (1991). Genetic considerations in multi-generational space travel. In Y. Kondo, F. Bruhweiler, J. Moore & C. Sheffield (Eds.), *Interstellar travel and multi-generational space ships* (pp. 89–99). Wheaton, Illinois: Apogee Books.
Patrick, P. L., Tam, L., & Loebel, D. A. F. (2007). Gene function in mouse embryogenesis: Get set for gastrulation. *Nature Review Genetics, 8*, L368–L381.
Przeworski, M., & Wall, J. D. (2001). Why is there so little intragenic linkage disequilibrium in humans? *Genetical Research, 77*, 143–151.
Rai, U. K. (2003). Minimum sizes for viable population and conservation biology. *Our Nature, 1*, 3–9.
Reed, D. H., & Bryant, E. H. (2000). Experimental tests of minimum viable population size. *Animal Conservation, 3*(2000), 7–14.
Reed, D. H., O'Grady, J. J., Brook, B. W., Ballou, J. D., & Frankham, R. (2003). Estimates of minimum viable population sizes for vertebrates and factors influencing those estimates. *Biological Conservation, 113*(1), 23–34.
Relethford, J. H. (2001). *Genetics and the search for modern human origins*. New York: Wiley-Liss.
Rodler, F. (2018). Exoplanet research in the era of the extremely large telescope. In H. J. Deeg & J. A. Belmonte (Eds.), *Handbook of exoplanets* (pp. 1105–1120). New York: Springer.
Ryman, N., & Lairke, L. (1991). Effects of supportive breeding on the genetically effective population size. *Conservation Biology, 5*(4), 325–329.
Schultz, S. T., & Lynch, M. (1997). Mutation and extinction: The role of variable mutational effects, synergistic epistasis beneficial mutations, and degree of outcrossing. *Evolution, 51*(1997), 1363–1371.
Shaffer, M. L. (1991). Minimum population sizes for species conservation. *BioScience, 31*, 131–134.
Sherry, S. T., Rogers, A. R., Harpending, H., Soodyall, H., Jenkins, T., & Stoneking, M. (1994). *Alu* evolution in human populations: Using the coalescent to estimate effective population size. *Genetics, 147*(1994), 1977–1982.
Smith, C. M. (2013). Comment on 'An Evolutionary framework for cultural change: Selectionism versus communal exchange. *Physical Review Letters, 10*, 156–167.
Smith, C. M. (2014). Estimation of a genetically viable population for multigenerational interstellar voyaging: Review and data for project Hyperion. *Acta Astronautica, 97*, 16–29.
Smith, C. M., & Ruppell, J. (2011). What anthropologists should know about the new evolutionary synthesis. *Structure and Dynamics, 5*(2), 1–13.
Soule, M. E. (1987). *Viable populations for conservation*. Cambridge: Cambridge University Press.
Strickberger, M. W. (1985). *Genetics*. New York: Prentice Hall.
Summerford, S. (2012). Colonized interstellar vessel: Conceptual master planning. Internally-distributed paper available at http://www.icarusinterstellar.org/colonized-interstellar-vessel-conceptual-master-planning.
Takahata, N. (1993). Allelic geneology and human evolution. *Molecular Biology and Evolution, 10*, 2–22.
Takahata, N., & Satta, Y. (1998). Footprints of intragenic recombination at HAL locus. *Immunogenetics, 47*, 430–441.
Tenesa, A., Navaro, P., Hayes, B. J., Duffy, D. L., Clarke, G. M., Goddard, M. E., & Cisscher, P. M. (2007). Recent human effective population size estimated from linkage disequilibrium. *Genome Research, 17*, 520–526.
Thomas, C. D. (1990). What do real population dynamics tell us about minimum viable population sizes? *Conservation Biology, 4*(3), 324–327.

Traill, L. W., Bradshaw, C. J. A., & Brook, B. W. (2007). Minimum viable population size: A meta-analysis of 30 years of published estimates. *Biological Conservation, 139*(2007), 159–166.
Udeya, J. C., et al. (2011). The million-year wait for macroevolutionary bursts. *Proceedings of the National Academies of Science of the United States of America, 108*(38), 15909–15913.
Voight, B. F., Adams, A. M., Frisse, L. A., Qian, Y., Hudson, R. R., & Di Rienzo, A. (2005). Interrogating multiple aspects of variation in a full resequencing data set to infer human population size changes. *Proceedings of the National Academy of Sciences, 102*(51), 18508–18513.
Wang, D. G., Fan, J.-B., Siao, C.-J., Young, P., Sapolsky, R., Ghandour, G., et al. (1998). Large-scale identification, mapping, and genotyping of single-nucleotide polymorphisms in the human genome. *Science, 280*(5366), 1077–1082.
Watterson, G. A. (1975). On the number of segregating sites in genetical models without recombination. *Theoretical Population Biology, 7,* 256–276.
Weinberg, J. R., Starczak, V. R., & Jorg, D. (1992). Evidence for rapid speciation following a founder event in the laboratory. *Evolution, 46*(4), 1214–1220.
Whitlock, M. C. (2000). Fixation of new alleles and the extinction of small populations: Drift load. *Beneficial Alleles, and Sexual Selection, Evolution, 54*(2000), 1855–1861.
Wijsman, E. M., & Cavalti-Sforza, L. L. (1984). Migration and genetic population structure with special reference to human. *Annual Review of Ecology and Systematics, 15,* 279–301.
Wobst, H. M. (1973). Boundary conditions for palaeolithic social systems: A simulation approach. *American Antiquity, 39,* 303–309.
Wood, T. E., & Rieseberg, L. H. (2002). *Speciation: Introduction, encyclopedia of the life sciences* (Vol. 17, p. 2002). London: Nature Publishing Group.
Wood, J. W. (1987). The genetic demography of the Gainj of Papua New Guinea: Determinants of effective population size. *American Naturalist, 129*(2), 165–187.
Wright, S. (1931). Evolution in Mendelian populations. *Genetics, 16,* 97–159.
Younger, S. M. (2008). Conditions and mechanisms for peace in precontact Polynesia. *Current Anthropology, 49*(5), 927–934.
Yu, N., Jensen-Seaman, M. I., Chemnick, L., Ryder, O., & Li, W.-H. (2004). Nucleotide diversity in gorillas. *Genetics, 166,* 1375–1383.
Zhao, J., Li, W. J., & Xiong, M. M. (2001). Population-based linkage disequilibrium mapping of QTL: An application to simulated data in an isolated population. *Genetic Epidemiology, 21*(Suppl 1), S655–S659.

Chapter 4
Cultural Adaptations in Human Space Settlement

Abstract Just as the human genome will be changed as it adapts to conditions beyond Earth, our second channel of adaptive capacity—culture—will also change. Indeed it has been more than a million years since biology was our main mode of adaptation; rather, since the evolution of complex cultural behaviors including complex tool use, symbolism, and language, we as a species have survived and adapted less *because* of our biology than *despite* our biology. It is behavioral flexibility that led Inuit people of the high Canadian Arctic to invent effective cold-weather clothing and to mark their broad landscapes with cairns and statuary as aids to navigation, and it is behavioral flexibility that led Australian aborigines to carve hunting territory maps directly on their hunting tools. In the same way, adapting our "ways of living", our cultures, to conditions new to the human experience will allow us to survive and ultimately thrive, long-term, beyond Earth. And we need not simply guess about how our cultures will change, or what parts of them will change, we can once again use the tools and insight of evolutionary processes to help us predict, or at least anticipate, cultural adaptation beyond Earth. In this chapter we use these tools to describe and evaluate the adaptive potential of some 'Cultural Universals', complex, nongenetically-transmitted behaviors or domains of behavior found globally in all cultures, but somewhat differing by both the cultures' material and cultural selective environments. We may expect these—at least—to change as new environments are encountered beyond Earth, and we may productively predict some of this change so that it does not come as a shock, and even productively make recommendations about how to tailor some of our behavior in these domains to the new conditions of space settlements. At the least, this allows a structured, evolutionarily-informed approach to the obvious and fascinating—but very fuzzy and open-ended—question, 'what will change when we go to space?'.

> …is not just because human cultural evolution is cumulative that it is distinctive. It is distinctive because of *what* accumulates, and what accumulates is change that is complex, adaptive, and open-ended because it is born in a conceptual network powered by both associative thought (which detects relationships of correlation) and strategic thought (which detects relationships of causation)—Liane Gabora (From *The Cultural Evolution of Socially Situated Cognition*, Gabora 2008)

C. M. Smith, *Principles of Space Anthropology*, Space and Society,
https://doi.org/10.1007/978-3-030-25021-8_4

4.1 Introduction

Any change in the heritable properties of a population over generations may be termed *evolution*, and from what we have seen in Chaps. 2 and 3 we can be sure that the human genome will evolve on beyond Earth just as it continues to evolve here on Earth today (Mostafavi et al. 2017).

But genetic information is, in the human lineage, only one channel or stream of adaptive, evolving information; another is culture (Whiten et al. 2011). Cultural information evolves as surely as biological information, and the scientific recognition of this fact even in the last few decades (see for example the entire volume of *Proceedings of the Royal Society B* titled "Culture Evolves") reshapes how we conceive of the change of culture, our main adaptive mechanism.

The term culture was introduced in Chap. 1 as socially-transmitted guides to behavior. Here it is important to explore the term a bit further; this will also be important to understanding Chap. 6.

The term *culture* has been defined many ways in the last 150 years and it remains the subject of active debate among archaeologists. Despite details, however, archaeologists and their colleagues in other, closely-related disciplines, refer to culture as a learned and shared set of ideas that guide some human behavior. By referring to culture as learned, we differentiate it from instinct. Instinctual behavior is that directed strictly by neural anatomy. Some human behavior is shaped by instinct but many of our guides to our behavior are learned. We learn by observing and, often, imitating our parents and peers; sometimes we are formally taught. Instinct, then, in a human infant, includes crying out when hungry. Learned behavior, culture, is typically much more complex. For example, as guides to behavior consider the set of navigation instructions that a young Polynesian seafarer learns from their mentor. Those navigation instructions are very specific cultural information that is moved from mentor to student according to certain customs; they are not passed on in neural architecture but they are important to adaptation and survival.

Note also that culture is also referred to as being shared. Humans live in groups—from families to communities to entire civilizations—which share certain learned behaviors. A person born and raised in urban Japan, for example, learns a set of ideas about personal space, gender roles, appropriate foods for certain occasions (and so on) that differ from those learned by a person born and raised in, say, rural Australia. The origins of those Japanese and Australian ideas, and even their specific content (what is appropriate to eat at a wedding?), however fascinating, are for the moment irrelevant; the point is that groups of humans share sets of ideas that guide behavior, and groups that differ in their basic behavioral range are often referred to as different cultures.

Culture is, then, at root, information, organized we may say as ideas. Those ideas (reducible to sets of interconnected neurons) are archived in the minds of individual people. These ideas do not become encoded in DNA, rather they are passed to the next generation by the learning mentioned above, technically referred to as enculturation.

Enculturation might be informal, as when a youngster observes and imitates its parents or siblings, or it could be highly formal and structured, as when an Arctic hunter learns about the properties of different kinds of sea ice. In either case, cultural information—sets of ideas—are moved from one mind to another in the phenomenon of cultural transmission.

The significance of all this is that one of humanity's most distinctive characteristics is the complexity, richness, and diversity of its cultures worldwide and through time. Some other animals, arguably, transmit social information from parent to offspring, as in chimpanzee societies where mothers seem to very carefully instruct their young in the making and use of tools such as termite-mound probes made from branches, or wooden clubs used to crack open nuts on the surface of a flat rock, used as an anvil. Belgian anthropologists Christophe Boesch and Hedwige Boesch-Achermann claim even to have observed a female chimpanzee correct an error in her offspring's use of a nut-cracking club, which they call an instance of *active teaching*. But while some other life forms do have cultures, those cultures are far less complex than in humanity. For instance, human culture has specific rules for an infinitude of behaviors, such what flag to fly when bringing a boat into a harbor, to how to greet the Queen of England, what gestures are appropriate in a marriage ceremony, and so on. One reason for this complexity is that humans absolutely rely on cultural information to survive; we need very specific instructions to navigate not only the physical realities of life—finding and acquiring water, food and nutrients, and keeping ourselves warm, or cool—but to navigate equally-critical social relationships. For a number of reasons, humans are very social animals, and exclusion from socialization can be intolerable; it's often used as punishment, as in ancient Iceland where outlaws were often punished by banishing them to the wilderness, totally deprived of social interaction.

Two caveats; first, while culture is meant as a guide to behavior, there is normally a spectrum of adherence to that guidance. Some people adhere very tightly to cultural guides, and some actively disdain them, and many people exist somewhere in-between. Second, it can be difficult to 'draw a line around' even modern cultures, whose members can vary internally from one end of an economic spectrum to the other, for instance. And cultures can contain subcultures, such as 'Deadheads' (enthusiasts of the Grateful Dead rock band) in modern America; are Deadheads 'American'? A visiting German tourist might say yes, certainly, whereas Deadheads themselves might say, no, absolutely not, they are something else entirely.

Archaeologists extend the culture concept backwards in time, referring to archaeological cultures, which are assumed ancient human groups known by a distinctive set of artifacts or artifact attributes (this will be of importance in Chap. 6 of this book). For example, archaeologists will use the term 'Lapita culture' to refer to the ancient Polynesian people who made and used distinctive pottery, fish hooks, seagoing voyaging canoes and other artifacts dated to about 3000 years ago. Lapita culture, we might say, spread through the Pacific, discovering islands by ship, and we can trace their spread through the islands by dating these distinctive sets of artifacts they appear earliest in Western Polynesia and later in Eastern Polynesia.

Archaeologists also commonly use the term material culture, referring to objects made or modified by humans. This term recognizes the fact that when people make or modify things, they often do so in a way that is informed by their culture, and this is very useful for archaeologists. For example, when indigenous Saami people (reindeer-herders of the European Arctic) build a shelter, they build a structure that shelters them from the wind, insulates them from outside temperatures, and keeps them from laying directly on the frozen ground. Native Inuit people of the Canadian Arctic also build shelters, but because of many differences between Saami and Inuit lives and cultures, Saami and Inuit shelters differ; their material culture (in this case, shelters) could be studied to learn about their lives. Material culture adaptations will be particularly important in this chapter of this book.

Finally, culture is often used in the term *culture change*. This refers, simply enough, to the change, over time, of cultures, both modern and ancient. Since culture is a shared set of ideas, culture change is essentially the change of ideas shared by a group of people. Culture change is a deceptively simple term; models of culture change—anthropological ideas about the exact mechanisms of cultural change, about what drives it, and how it plays out over time—are many and complex, but are not particularly pertinent at this point.

Table 4.1 summarizes the main uses of the word culture in this book.

Table 4.2 summarizes the historical background of conceptions of culture change, which in some ways parallel the historical conceptions of biological change. In this book I use the Modern Darwinian approach to culture dynamics.

As culture is considered in this book an information set that is changed by evolutionary dynamics, Table 4.3 summarizes these essential evolutionary processes of Replication, Variation and Selection (necessary conditions for evolutionary change) in both genetic and cultural information. Like genes, information is *replicated* (culturally, in the form of ideas), *varies* in its expression, and those variable expressions of ideas do not all survive to be transmitted to the next generation, but some are more-often copied than the rest (the nonrandom differential propagation of information termed *selection*). Culture, then, does not merely change by a process analogous to biological evolution, it does indeed evolve, albeit in its own way (details of culture evolution processes informed by the Modern Evolutionary Synthesis are beyond the scope of this book: see however (Gabora 2013; Smith, Gabora and Smith 2018). The fact that culture changes by an evolutionary process is not trivial; while we may be at a loss to predict the historical contingencies that shape the *cultural selective environment* (e.g. could we have predicted the counter-culture movement of the 1960s, or the mass suicides at Jonestown, or China's 14th Century retreat from an exploring nation to one focused internally, or the fine details of Shaker religious reforms of the 19th Century?) it at least allows us to better structure our investigations of and even predictions about cultural evolution. In the same way, the questions that biologists ask about living things radically changed with the recognition that evolution occurs (following wide and rapid acceptance of Darwin's concepts in 1859–1860).

The most important lesson for this book is that some culture change should be understandable in evolutionary terms, and that culture change need not be seen as a black box or entirely contingent, but sometimes demonstrating adaptive patterning. In the same way that understanding the 60-million year evolutionary history of the primate order helps us better understand the adaptive diversity of the c.200 living

Table 4.1 Uses of the word culture in this book

	Definition	Example	Notes
Culture as information	Learned, shared guides to behavior	Guides indicating the appropriateness of potential mating partners; beliefs about the supernatural; specific languages	At root, culture is information that resides in the minds of individual people; it is passed from person to person by many systems of communication, including speech, writing, dance and gesture
Culture as identity	A group of people sharing certain learned guides to behavior	English culture, Japanese culture, Maori culture, Aztec culture	While it can be hard to draw lines around a culture, and cultures can contain subcultures, on other scales cultures as different as Alaskan native and urban German (for example) can be discerned
Archaeological culture	A presumed group of ancient people represented by specific characteristics evident in sets of artifacts (see 'Material Culture', below)	'Thule' culture of the Alaskan and Canadian Arctic, dated to over 1000 years ago, represented by distinctive artifacts including bone snow-goggles, certain harpooning equipment for hunting sea mammals, and dog sleds	Archaeologists are careful not to 'reify' archaeological cultures; reification is thinking of an abstract concept as real. Some archaeological cultures are easier to discern from others, just as in living cultures
Material culture	Objects made or used by humans (artifacts)	The material culture of ancient Mediterranean seafaring includes the use of boats made of wooden planks, whereas the material culture of Southeast Asian seafaring includes the use of boats made of bamboo lashed together	Cultural information—shared by groups that have a common identity—guides differences in how groups produce material culture
Culture change	Change of ideas (guides to behavior) shared by groups	Early Icelandic (Viking) culture was polytheistic, with many Scandinavian gods, but in 999 AD the Icelanders converted to Christianity	Culture change can be fast, slow, or anywhere in-between, and it might be easy, difficult or impossible to detect in material culture

Table 4.2 Historical conceptions of cultural and biological change through time

Period of western civilization	Conception of biological change	Conception of culture change
Greek	There is little change through time; life forms are 'types', material expressions of ideals	There is little change through time; cultures are 'types' most akin to tribes or families, with little historical dimension or depth
Medieval	There is little change through time; life forms were created for specific purposes	There is little change through time; cultures are simply remnants of original tribes described in the Old Testament; they are 'types' that can change only in trivial ways
Enlightenment Predarwinan	Life forms are 'types' that change little and are best understood descriptively; they *may* change by effects of the environment on the individual life form	Cultures are rather well-defined 'types', though they may change by effects of their own local, historical conditions
Early Darwinian	Life forms are not 'types' but current manifestation of an historical process; they change due to both random processes and an internal drive towards 'perfection', a 'climb' or 'progression' that is perhaps divinely inspired	Cultures are 'types' representing discrete stages of cultural ascent from 'Savagery' to 'Barbarism' to 'Civilization'. The change from one to another is driven by an internal 'striving' or other (rather mysterious) governing power
Modern Darwinian	Life forms are historical documents manifesting traces of the past and present; they change according to ultimately simple but proximally-complex laws of replication, variation and selection of biological information (genes) in selective environments. Although equilibrium and order do emerge as ecosystem properties, there is no central controller or decision-maker in the biological evolutionary process	Cultures are groups of individuals sharing an array of guides to behavior, guides that have been shaped by multiple features including the material environment (studied by human behavioral ecology), historical contingency (studied by historians), the cultural environment (studied by cultural anthropologists with an historical awareness) and even individual behavior (studied by anthropologists highlighting agency). Cultures are adaptive mechanisms at least partially tailored for survival in the physical and social senses; they may be well- or poorly-adapted, and adaptations may or may not benefit all members of the culture

Table 4.3 The evolution of cultural and biological information

Process	Biology	Culture	Example
Replication	Chemical replication of *genes*	Cultural replication of *memes*	**Biology**: Meiosis and mitosis **Culture**: Enculturation and teaching children or peers
Variation	Mutagenesis occurring on DNA	Cultural novelty arising in cognition	**Biology**: Base pair substitution; DNA strand break with imperfect repair **Culture**: Creative thought; productive, new associations of information
Selection	Differential propagation of *genes* among a deme	Differential propagation of *memes* among a population	**Biology**: Reproductive success of bison better able to survive cold winters than their peers due to genetic features **Culture**: Propagating cultural information via propaganda, or preventing its propagation via censorship
Adaptation	Fixation of once-novel *genes* promoting biological survival among new selective pressures	Fixation of once-novel *memes* promoting biological and/or social survival among new selective pressures	**Biology**: Evolution of a variety of bird beak forms in Galapagos Island finches from single ancestral, mainland beak form **Culture**: Development of Polynesian inheritance rules that promoted chief's offspring to explore for new islands to settle rather than remain on one island
Evolution of new forms	Reproductive isolation leading to mating inviability among previously mating-viable demes; *cladogenesis*	Conceptual fixation and incompatibility with e.g. Ultimate Sacred Postulates of previously-compatible culture; *ethnogenesis*	**Biology**: Evolution of early *Homo* from late members of the genus *Australopithecus* **Culture**: Reformation of Icelandic immigrant culture concepts in USA Great Lakes region to accommodate some American norms
Evolutionary transition	Substantial reorganization of genetic information	Substantial reorganization of cultural information	**Biology**: Appearance of sexual reproduction, involving recombination **Culture**: Abandonment of medievalism and rise of Renaissance thought in the West

primate species, understanding this allows us to better understand how culture has allowed humans to succeed and may continue to succeed beyond Earth.

Tables 4.1, 4.2 and 4.3 should considerably enrich the reader's understanding of culture as an evolvable, adaptive mechanism.

We need only address one more issue, that of *intentionality*. While evolution in the rest of the animal kingdom is undirected by the evolving entities themselves, among humanity there is the potential to proactively drive significant culture evolution. Two examples on opposite ends of a spectrum of selection are *propaganda*, the deliberate spread of certain information in a population, and *censorship*, deliberate prevention of the spread of such information. In these cases, humans with particular intents become the selective agent shaping which information is propagated (selected *for*) and which is not propagated (selected *against*). In contrast to such *cultural selection*, *natural* selection, has no intent or purpose, so we may be excused for thinking that cultural selection will be inexplicable except with knowledge of the intent of cultural selective agents or 'selectors'. But it is not so clear; for example, whatever the cultural system, it must actually work to some degree to provide people with basic requirements such as food, water, nutrients and even essential group-dynamic functions including the administration of justice and rules of inheritance. In this way, cultural information must be at least minimally adaptive, or it will not persist long; we know of many cases in which such a divergence of cultural guides to behavior, and the genuine survival needs of the population, result in *cultural maladaptation* (see Edgerton 1992). This leads to the study of power relationships in human cultures, for example we may as to whom is a certain cultural behavior adaptive? Slavery, for example, does not benefit the slave, but does benefit the slave-owner (recognizing this does not mean that I advocate slavery, only that it is an example of inequal benefits in a given set of human relationships). Rather than being drawn into a deep discussion of such issues and what cultural behaviors are ethically sound from the perspective of space settlement, here I will simply programmatically refer readers to essential principles of the United Nation's *Universal Declaration of Human Rights* (UDHR; see), a body of statements (drafted in 1948) that capably outlines most of the Enlightenment values adopted by most democratic societies in the last few centuries. Such principles will be reviewed and remade as humanity settles beyond Earth, and it certainly is not my goal in this book to say how such specific decisions are carried out, only the DHR is likely a good starting point for such consideration.

How does the understanding of culture as an adaptive, evolving information system assist with the aims of this book, namely to make success more likely in the human project of space settlement? Let us begin with the essential lessons of this chapter so far:

- Culture is socially (not sexually) transmitted guides to behavior, and it is a significant tool of human adaptation and survival
- Culture can be adaptive, shaping behavior to fit the conditions of a given environment
- While culture is highly variable, it must minimally address basic survival requirements—both biological and cultural—of a given group of people

The first point above establishes that human survival to date has been biocultural, involving not *just* nature (biological evolution) *or* culture (cultural evolution) but both information systems as adaptive mechanisms. My paradigm for space settlement is that for it to succeed and persist it must be genuinely adaptive and evolutionary to ensure multigenerational success rather than short-term, goal-driven success. Recognition of the reality and nature of biocultural evolution, then, is important in understanding the tools, the adaptive mechanisms, that may be deployed to help humanity succeed beyond Earth.

The second point above establishes that culture may be transformed for adaptive purposes. Rather than thinking of changing ancient traditions, for example, to better suit conditions on Mars or a Langrangian colony as 'social engineering' (with a negative connotation), we may think of such, genuinely, as adaptation (with a positive connotation).

The third point above is programmatically the most important to the aims of this book. It indicates that however variable human culture is worldwide today and through the past. It has resulted in some patterning useful to our understanding of humanity's *adaptive potential*—the *cultural reaction norm* we might say—certainly of interest to space settlement planning. This is evident in light of recurring patterns of cultural evolution and adaptation see in the study of *cultural universals*, discussed in the remainder of this section.

As mentioned, in all human populations a number of basic issues must be solved for sheer survival; an individual needs certain quantities of calories, fresh water and nutrients daily, and the way these are obtained can be strongly conditioned by the immediate resource environment. This has led to humanity, worldwide, developing or evolving cultural traits—"human universals"—that reflect basic problem-solving for the kinds of life form that we are, biologically and socially (see full discussion in Brown 1991). Table 4.4 lists some of these cultural universals; for the programmatic purposes of this book I consider title these *Domains of Adaptation*.

While this differs widely in detail worldwide, all humans have solved basic subsistence needs in one of four main modes of subsistence; *foraging* (hunting and gathering), *pastoralism* (raising domesticated animals), *horticulture* (low-intensity farming and animal domestication) or *agriculture* (intense farming with complex irrigation and animal domestication). These work for humanity, as opposed to, for example, photosynthesis (not available to [directly] process energy in most animal life). And, the mode of subsistence strongly conditions many cultural variables; for instance, foraging people are normally rather *egalitarian*, with few instances of social ranking, while among horticulturalists and agriculturalists, social ranking (inequal access to resources by population members) is common. In this case, elementary issues such as social ranking ethos are strongly conditioned by subsistence mode. There are plenty of other factors that play into the shape of a culture, and responsible anthropologists are careful not to be 'ecological determinists'. But we can learn from such cases.

Where to begin? It's easy to be overwhelmed when we begin to think about how human culture will change on multigenerational voyages. To order my thoughts, I have begun to tackle this issue by researching the above-mentioned

Table 4.4 Common cultural universals or domains of adaptation

Domain of adaptation	Concept	Examples
Language	Specific spoken and gestural (bodily) systems of communication, including vocabularies and grammars	Some languages assign gender to nouns, while others do not
Ethics	Concepts of right and wrong, justice, and fairness	Some cultures execute murderers, while others do not
Social roles	Rights and responsibilities differ by categories such as age (child, adult), gender (man, woman), and status (peasant, King)	Cultures differ in the ages at which people take on certain rights and responsibilities, and specifically what those rights and responsibilities are
The supernatural	Concepts regarding a universe considered fundamentally different from daily experience	Different cultures worship different gods, goddesses, and other supernatural entities
Styles of bodily decoration	Human identity is often communicated by bodily decoration, either directly on the body or with clothing	Some cultures heavily tattoo the body while others communicate identity more with clothing styles
Family structure	Concepts of kinship or relations between kin, and associated ideas such as inheritance	Some cultures are polygynous, where males have several wives, and some are polyandrous, where females have several husbands
Course of life stages	Lives of individuals are segmented across the lifespan	Some cultures have formal rituals commemorating leaving behind a stage of life and entering another, while others are less formal and may have different stages demarcated
Reproductive behavior	Regulation of reproductive behavior, including incest rules	Cultures differ in the age at which sexual activity is permitted
Food preferences	Concepts of what are appropriate food and drink in certain situations	Some cultures eat certain animals while others consider them unfit to eat
Aesthetics	Concepts of ideals, beauty, and their opposites	Some cultures value visual arts more than song, and vice versa
Ultimate sacred postulates	Central, unquestionable concepts about the nature of reality	Some cultures consider time to be cyclic while others consider it linear

"human universals". Again, these are domains of behavioral regulation common to all human cultures. For example, all cultures, in one way or another, regulate sexual behavior, and incest in particular. All human cultures also have basic moral systems for dealing with truth-telling, property theft, and killing. And all human cultures have distinctive languages, styles of bodily decoration, family structure and so on. Remember, the ways that these issues are solved differs by culture (leading to tremendous cultural diversity worldwide), but a set of universal cultural facets is a good place to start when we consider what we might expect to evolve beyond Earth.

Rather than generally cast about, then, grasping for what might change over generations of human cultural evolution during future voyages to distant worlds, each of us guided by our own specific ideas and interests, we can systematically use the tools of anthropology and the principles of evolutionary theory to make useful predictions and recommendations to care for the cultural health of extraterrestrial settler populations. Moreover, since these domains of action are universal, it is likely that they are adaptively rather plastic and have been useful in the success of ancient cultures, and therefore we may consider them likely to change when communities meet extraterrestrial selective environments, both material and socio-cultural.

4.2 Cultural Universals and Human Space Settlement

In short, for space settlement planning addressing the very fuzzy question of 'what will change when we settle space?' the domains of cultural adaptation addressed in Table 4.4 are a good start. We may explore their adaptive range of expression so far on Earth, and implications for space settlement, in the following chapter subsections.

Note that of the subsistence modes mentioned above as structuring factors of culture, it is intensive agriculture that is most likely to be practiced by humanity in any conceivable beyond-Earth settlement for some time, rather than hunting and gathering, or low-intensity horticulture, for instance.

It is important to note that any specific cultural change, from Earthly lifeways, in many aspects of human life we explore below might be considered 'trivial' in an adaptive sense, such as a culture's varieties of artistic expression or subtleties of their religious traditions. But I argue that these are not trivial at all. In fact the sum of these slight or large differences *constitute the differences between cultures.* If we have an interest in how cultures will adapt to conditions beyond Earth, then, we must consider many such small features of the cultural guides to behavior that shape the myriad decisions that are made every day by individual minds, though minds equipped with somewhat-shared guides to decision-making that we delineate and describe as distinctive 'cultures'.

How will human cultures differ on Mars, or in orbital colonies or Lagrangian point colonies or even in interstellar colony ship, from human cultures on Earth? Not, I think, conceivably in their essential features of organizing the population and its activities, many of which promote biological survival *through* cultural survival. It is these essential features that are outlined below. Rather, it is in the specific ways that beyond-Earth cultures address these features that they will differ from those of Earth. I argue that these are entirely expectable and point out to critics of the concept of space settlement that each human culture has changed before, and in fact cultural change, just like biological change is both the norm and is often a sign of healthy adaptation to chanced cultural circumstances. As the world has changed, cultures have adjusted their behaviors, many times, to change. Knowing that the universe changes, that it is not static, in fact, is a basic argument against a fundamentalism of any kind which proposes that a particular culture or religion currently has all answers to all questions—because as humanity matures and explores, encountering new phenomena, even new questions will be posed, and might not be answerable to ancient systems of thought. The overall message regarding human culture change beyond Earth is that we should expect it, and that such change is likely to be a healthy sign of adaptation. Below we may examine some of the main features of human adaptive cultural tools, and how they are expressed in humanity past and present, as guidelines to how they may be adjusted to conditions encountered in the new living conditions beyond Earth. We shall examine, then, the cultural 'toolkit' for adaptation, just as we have examined the human biological toolkit of adaptation in prior chapters on genetics.

4.2.1 Language

Language	Specific spoken and gestural (bodily) systems of communication, including vocabularies and grammars	Some languages assign gender to nouns, while others do not

The bulk of cultural guides to behavior are communicated with some form of language, a form of animal communication with many features unique to humanity. Many life forms communicate (technically: *emitting and receiving information*) but human communication is uniquely rich in its complexity and, particularly, in its use of metaphor and symbolism, and its open-ended nature, allowing and promoting the unique association of previously-unassociated ideas which result in *creativity*. In this last way language not only is used to convey cultural survival information, but serves to promote the variation of ideas that is the very source of unique, inventive, proactive human innovations allowing survival in a myriad environments worldwide and, we propose, beyond Earth.

Note that here I refer to *language* as the system of communication unique to a given culture (it may include gesture and other nonverbal communication) as contrasted with speech which simply refers to the production of vocalizations. Speech is not necessary for linguistic communication as attested by people who cannot produce vocalized speech but communicate entirely well with sign language, for example. Language does require however a capacity for understanding and using *grammar*, which here refers to the rules of language construction.

The essential unit of language vocalization is the phoneme, loosely referring to a distinctive sound, such as the difference in English between 'r' as in the word 'run' ands 's' as in the word 'sun'. A survey of the phoneme number diversity among roughly 2000 modern languages recently indicated that (a) most languages have from 20 to about 60 phonemes and (b) "geographic axes of greatest phonemic differentiation correspond to axes of genetic differentiation, suggesting that there is a relationship between human dispersal and linguistic variation…" (Creanza et al. 2015). Currently several thousand languages are spoken globally, however most human speak one of a handful of languages, namely English, Spanish, some variety of Hindi or some variety of Chinese. Premodern linguistic diversity has been mapped to environmental diversity (Maffi 2005) and languages have at least had one important function in the naming and description of the many of humanity's inhabited environments.

All cultures, then, have languages that at least—but importantly—in part are used to name and organize thought about the local environment, both physical and cultural. Languages are not necessarily static but change through time, again indicating their use as adaptive tools. Rates of linguistic change may be rapid, slow, or each of these for some period in the life of a given culture. Words that are infrequently used can be lost over time, while more commonly-used words are retained longer; several surveys have found that language 'splitting', in which a new language requiring special training is derived from an ancestor language, typically occur on the order of a century or so (Lieberman et al. 2007). Of particular interest to off-Earth settlement planners is the finding that language changes may be faster in smaller than larger communities (Nettle 1999), however this pattern has been challenged and it has been suggested that the rate of language change is unlikely to be generalized and depends on many unpredictable contingencies (Wichman and Holman 2009).

Reasons for language change are many; new words (*vocabulary*), for example, might be 'imported' to a language as new phenomena are encountered, as in the case of *cacao*, referring to the domesticated bean of central America, where it was first encountered by Europeans in the first quarter of the 16th Century AD and—with physical transport of beans back to Europe—common in the English language by the 1550s AD. Language change may of course occur internally as well, as in the case of invention of new words for new inventions, such as *byte*, referring to a unit of computer information, dating to 1956 AD. Language change may be resisted as well as promoted within a culture; promotion of new terms and words may be done consciously, as in favoring them over a synonym for whatever reason, while resisting change may be done by social sanction against the use of novelites. No general rule has been identified regarding such changes, and they must also be considered highly contingent on events in the life of a given culture.

Members of a single culture may speak multiple languages and/or multiple varieties or *dialects* (regional variants) of a single language. The *language of work* refers to "the language used to converse with work colleagues and supervisors …" (Palmer

1997: 265) and of course subtle but meaningful variations on a given language are used on formal occasions, in informal settings, for artistic purposes, and so on. Still, languages are not simply social constructs, when communication between speakers of exclusively one or the other of two languages cannot functionally communicate we may indeed say that they speak different languages.

Language includes nonvocal communications including *gesture* and body language that can be critical to rich communication. Figure 4.1 illustrates some gestures and body language as depicted in a 17th-century book of manners, indicating subtle differences in use of the body to communicate different messages while eating. Note that slight variations can communicate significantly different messages. Such variations are particularly likely to emerge in the 0.38 g gravity field of Mars, leading to slight variations in body language in Mars cultures.

Considering these generalizations, what may we say about the likelihood and nature of language change as it applies to settlement beyond Earth? We may begin with an examination of Table 4.5, which indicates some of the units of human vocalized language and some prospective likelihoods and general rates of change in these units; it may be used to consider the rate and nature of language change as it plays out in the adaptation of languages to environments settled beyond Earth. From this table I surmise that we can expect vocabularies to change, both by dropping use of older words and phrases that no longer 'map to a useful reality' and by the invention of new words and phrases to reflect and order new realities. For example, on Earth we have in English the expression 'as big as a whale', as we are familiar with oceans and whales. Oh Mars, for example, or other places without liquid seas and large swimming animals, such a phrase might be kept for some time as a relic or used as a quaint expression, but is likely to be dropped from common use rather quickly

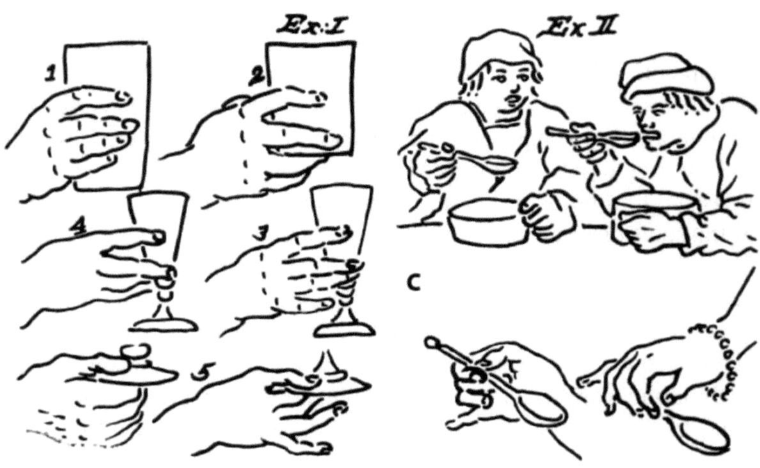

Fig. 4.1 Gestural Communication in Eating. Retrieve source from my files. Perhaps replace with diagram of standing postures and their 'body language messages'

Table 4.5 Human language elements and their likelihoods and rates of change in space settlement scenarios

	Brief definition	Likelihood of change on the order of a century	Likelihood of change on the order of a generation (30 years)
Phonemes	Smallest sound unit of a language	Medium to Low	Medium to High
Grammar	Rules of language use	Low	Low
Syntax	Rules of phrase, clause and sentence formation	Low	Moderate
Vocabulary	Unique identifying words	High	High

as new generations are born on Mars, never visit Earth, and know whales only as images in books or other media. It is just by such omissions and inventions of words, phrases and ideas that languages, worldviews, entire bodies of literature, and cultures themselves change over time.

In sum we may say that it is difficult to imagine language *not* changing in the new circumstances of new environments beyond Earth. New words and phrases are very likely to be innovated to describe new physical phenomena: it is true that natives of the high Arctic have many words for what people unaccustomed to snow or sea ice simply call snow, or sea ice, and in the same way the new character of light, or gravity, on Mars (for example) will demand new linguistic description. The invention of American English is treated comprehensively in Mencken's (1937) classic *The American Language*, in which he notes that (Mencken 1937: 174):

> New words, of course, are no more produced by 'the folk' than are new ballads: they are the inventions of concrete individuals…[which must be transmitted to a populous, often by those with the ability to reach many through various media, such as] the elder Roosevelt [who] was responsible, either as coiner or propagator, for…*nature-faker*, *big-stick* and *embalmed-beef.*

This reminds us that the like other innovations, linguistic innovations may be more rapidly accepted if they are associated with higher-regarded people, or people who have the means to widely distribute their innovations. This is not always the case, however; subversive linguistic innovations can also spread when there is discontent with establishment language. The point being that there are many and complex factors involved in language change, making it rather difficult to predict. But, again, the point in this book is not to proscribe what language change will or should occur beyond Earth, but to point out that language change may be investigated as a means of culturally adapting to habitats beyond Earth.

It is also entirely possible that new dialects will emerge from the founding language 'stocks'; this may in part be due to different gravity fields (e.g. on Mars), different ambient atmospheric pressure in new habitats (such as orbital colonies that may wish to use lower pressure to reduce construction costs) and even conscious choice, to differentiate the settling culture from its origins on Earth. New rules for language construction I think are far less likely to be innovated in the first generations of an off-Earth colony as they are more structural and less less-often changed historically.

The lesson here is one that we will find throughout much of this chapter; namely that we may well expect aspects of a given adaptive domain (in this case, language) to be altered as part of the *adaptation to new circumstances beyond Earth*; however precisely how is unlikely to be predictable. As mentioned, natives of the North American Arctic have many terms for different varieties of sea ice and land snow, all of which aid in their survival—some snow types are suitable for building snow houses, and some sea ice types are less stable for travel than others. How such adaptation will play out in language beyond Earth is unclear, and it is not clear to me why we would try to predict it in detail. Our most constructive use of the information we have reviewed is to state the following:

- Some language change will be adaptive and make life better for off-Earth settlers
- We can expect vocabularies to include new words to describe new experiences beyond Earth, and to occasionally 'retire' words and concepts that are no longer relevant to the human experience
- Regional or settlement-specific dialects are likely to emerge, in part as markers of independence from Earth
- Some language items may be retained as connections to Earth, however these are likely to be used mainly in special rather than everyday situations
- Body language is particularly likely to change where the gravity field is not exactly that of the Earth, affecting nonverbal communication and other such behaviors as performance art and sports.

4.2.2 Ethics

Ethics	Concepts of right and wrong, justice, and fairness; the practical application of broader moral values	Some cultures execute murderers, while others do not

Ethics are the *practical application of moral values*. All cultures have moral and ethical guidelines to deal with problems of truth-telling, the management of property, and the management of unjust killing; of course, cultures also have ethical guidelines to other domains of behavior. In many cultures, essential moral codes are enshrined in religious systems, and are so central to a culture's identity that they are included in the Ultimate Sacred Postulates. Both moral systems, however, and ethics, may be determined and applied, respectively, secularly. This has been some of the work of the field of *moral philosophy*. As an example of a set of secularly-determined moral values that could guide more practical ethical decisions, below I list Gert's 'ten principles of humanistic ethics' (Gert 1988):

- Do not kill.
- Do not cause pain.
- Do not disable.
- Do not deprive of freedom.
- Do not deprive of pleasure.
- Do not deceive.
- Do not cheat.
- Do keep your promises.
- Do obey the law.
- Do your duty.

Of course here I do not suggest that these should be the 'rules' for people settling beyond Earth, only that the field of moral philosophy supplies many such proposals for secular ethics (others may be found in the 1948 Universal Declaration of Human Rights). Naturally, many venturing beyond Earth will wish to carry their religious traditions' moral and ethical codes, and will continue to use them to shape cultural behavior. I do not think it is productive to make suggestions about what sort of moral and ethical systems 'would' or 'would not' work beyond Earth. For example, one might suggest that a very flexible, liberally-minded ethical system would be best to accommodate the many new circumstances we will find beyond Earth. At the same time, though, one might propose that only a highly conservative culture, with rigid guidelines to behavior, would properly protect members of the culture from just these new circumstances. I think both are somewhat correct; 'settlement culture' will at first be rather limited in resources and inherently dangerous (compared to suburban life on Earth today), and in such conditions, strict guides to behavior in the form of core moral values and ethical practicalities could well protect, in particular, children from dangers. On the other hand, an overly-rigid moral structure might indeed not be suitable to the new conditions of life beyond Earth, thus fomenting the contradictions and tensions of behavioral guides that have been outmoded by conditions changed from when they were established.

May we suggest anything productive about ethical system change in settlements beyond Earth? A cross-cultural review of the moral systems of many human cultures reveals some commonly-recurring specific domains of behavior addressed by moral systems and their derived ethics (Newman 1976; O'Niell and Petrinovich 1998; Schwartz 1992; Thornhill 1991; Westermarck 1906); they are listed below in no particular order:

- truth-telling
- unjustified killing
- mutual aid
- punishment for moral /ethical transgressions
- incest
- management of property

Considering the likely cultural conditions of life beyond Earth, we may suggest that several of these ethical domains are likely to require adjustment. Regulation of sexual behavior and *incest* will likely be amplified in smaller and isolated populations because of the potential for genetic maladies as noted in Chap. 3. The *management of property* will likely also be affected as settlement resources might well, early on at least, be limited, such that any item of property is valued more highly than in an environment of plenty. How *justice* is administered (e.g. *punishment*) will depend on how a given extraterrestrial culture establishes its system of justice. On the other hand, it is hard to imagine that prohibitions on lying and truth-telling, unjustified killing or mutual aid (benevolence and giving aid) would be radically altered in any circumstances; people must share a reality for a culture to function at all, and both the taking of human life and offering of assistance to those in need are highly regulated in all cultures.

In short we cannot say precisely how a given culture's ethical codes will change beyond Earth, and it would I think be counterproductive to try to predict this in great detail. Rather we may suggest that cultures be ready to make alterations to such moral systems and their ethical derivatives.

4.2.3 Social Roles

Social Roles	Rights and responsibilities differ by categories such as age (child, adult), gender (man, woman), and status (peasant, King)	Cultures differ in the ages at which people take on certain rights and responsibilities, and specifically what those rights and responsibilities are

The activities carried out by members of a culture are often assigned to people given various social roles. Many cultures divide certain activities by *gender*, for example, which is a social assignment that in most cases maps to *biological sex* (male or female). In the native cultures of the North American Arctic, women are tasked with the building and maintenance of certain watercraft, whereas males have more hunting duties. Among the Amish, traditional male roles include farming tasks while women took on the bulk of childrearing. Such identities and roles are somewhat more fluid in many modern Western cultures, where they have been changing, largely, since the 'counterculture revolutions' of the late 1960s and early 1970s, which saw a broadening of women's social roles. Even before this, necessity drove some significant social role adjustments in the United States, as in World War II women were employed for work previously largely carried out by men entirely because of cultural conventions of the pre-WWII era, illustrated in Fig. 4.2.

Can we predict what social roles will emerge in human settlements beyond Earth? Probably not, and it is not our business, but will be the business of the colonists

Fig. 4.2 'Rosie the Riveter', Illustrating a Shift in Social Role Norms in the United States During World War II. Public domain image

themselves. We may generalize that as in many small-population, agricultural communities, an emphasis is put on an individual's ability to do useful work rather more than on the person's various life-stage or other identities.

4.2.4 The Supernatural

The Supernatural	Concepts regarding a universe considered fundamentally different from daily experience	Different cultures worship different gods, goddesses, and other supernatural entities

Every culture has concepts of some existence or realm of action beyond the immediately-obvious world of daily, material, natural existence, a realm of the *supernatural.* Not all members of a culture necessarily agree that such supernatural realms or powers exist, but each culture has some concepts structuring such beliefs; and we should recall that only in the last two centuries or so has it been possible to make a conscious choice not to believe ion such realms or venerate their inhabitants and/or powers, due to the rise of scientific inquiry and the Euro-American abandonment (in part) of the church as the ultimate authority. Whatever one's personal beliefs, it is likely that at least some of the people settling space will carry to their new settlements concepts of the supernatural. How such conceptions will change is, again, unpredictable, but we may again confidently say that there exists potential for significant change and for significant fortification of Earth-based supernatural conceptions.

For example, encounters with new phenomena have had both effects. For the Catholic church, the European discovery of the New World in the 15th and 16th centuries initially brought on a radical ideological entrenchment, to the extent that newly-encountered people would be forced into conversion to Christianity at the risk of death. In the long run, however, Western culture at large had some of its religious ideologies profoundly shaken by the discoveries in the New World, including tens of millions of human beings, and thousands of kinds of plants and animals, never mentioned in the Old Testament. While the Church tried mightily to accommodate these (Native Americans, for example, were said to be simply remnants of the Lost Tribes of Israel), within a century such ancient questions as the age of the Earth (once simply calculated using the Old Testament) were under scientific, secular investigation; the Old Testament had not stood up as a guide to the natural world in its encounter with the New World. On this case, encounters with the new led to both entrenchment and change, as the West finally took up science and secular knowledge as a guide to Nature, leaving the bible only as a moral authority document. The Catholic church in Europe has largely followed suit, with Catholic schools teaching evolution as a factual process—though one put in motion by a supernatural Creator—and maintaining its own staff of scientists and astronomers at the Pontifical Academy of Sciences, established in 1603 (Vatican 2018).

Will such radical change occur during humanity's exposure to new phenomena beyond Earth? It is perhaps most likely when the essential foundations of a given supernatural tradition are challenged by new phenomena. These, described in Sect. 4.2.11 as *Ultimate Sacred Postulates*, are unquestionable foundational explanations about why and how the universe is as it is. But they may be challenged. Finding life beyond Earth may challenge Christianity's USP (just one of many) that '*In the beginning, God created the heaven and the Earth*' by some, though others may interpret this phrase so broadly that it can accommodate such a finding. It will be to others to identify the Ultimate Sacred Postulates of the supernatural traditions of different cultures and to identify which of them are most likely to be challenged by the new phenomena most likely to be encountered beyond Earth. Findings that could challenge modern human cultural USP's are difficult to predict but may include the following; note that most address the issue of human specialness or centrality in the solar system and perhaps the universe:

- there is no life on Mars, and never has been = *we are a great exception in the solar system*
- there is or has been life on Mars = *we might not be so unusual in the universe*
- evidence of an extraterrestrial civilization = *we are certainly not an exception in the universe*
- we are unable to adapt to conditions beyond Earth = *possibly reinforcing the concept that humanity was 'made' for Earth alone*
- we are able to adapt to conditions beyond Earth = *eroding the concept that humanity was 'made' for Earth alone*

Generally speaking, world religious traditions (past and present) may be classified as falling into one of four main varieties; these have been ably classified by anthropologist Wallace (1966), who highlighted access to the supernatural world as a significant ordering variable in the expression and organization of religions worldwide and through time. In *shamanic* religions, individuals have direct, unscheduled contact with members of their supernatural world. *Shamans* and lifelong guardian spirits may assist in this contact as they have special knowledge. Such religion is rare today but traditionally has been practiced mainly by the most mobile of human societies including native peoples of the Arctic and Australia. The shamanic world is *animistic*, populated by supernatural beings animating all (not just living) things. Although some deities are more powerful than others, among shamanistic religions there is typically no supreme god.

Regular rituals carried out in special places and at special times, giving members access to a supernatural world populated by many supernatural beings are key features of *communal religions*. Such were common among native people of the Americas, the Pacific and Africa before European spread after c.1500 AD. These religious traditions maintain some individual access to the supernatural, but also religious specialists with somewhat more formal roles than those of the shaman; the religious facilities and artifacts associated with communal religions, such as *shrines* and other ritual spaces, are significantly more elaborate than any of those found in shamanic religions.

The supernatural world is largely only accessible to highly-specialized *priests* and diviners in the *Olympian* religions. These religions are richer in detailed content and subdivisions and include *cults* (subdivisions) that specially venerate ancestors in the spirit realm, or specific supernatural beings. Such a religious tradition is well-known among many fully residentially sedentary cultures globally and through time, including Ancient Greece, Egypt, Babylon, the Shang civilization of ancient China and the civilizations of the Americas. Olympian religions are highly formalized and their works are carried out with many elaborate artifacts (including special clothing and substances such as incense) and architectural facilities as large, costly and elaborate as pyramids. A chief distinction of the Olympian religions is that they are *polytheistic*, *monotheistic* or *ecclesiastic* religions venerate a single god, though other supernatural beings, such as angels, may exist. These religions support a hierarchy of full-time religious specialists (the *clergy*) who have the most regular and richest communications with the supernatural world. These religions include the Judeo-Christian and Islamic traditions, organized as complex churches staffed by a clergy who have high social status. Non-clergy members of such religions have less direct access to the supernatural realm than in other types of religion; they look to the clergy to interpret many life and world events from the perspective of the supreme god.

Planners and predictors of the cultural dynamics of human attempts to settle beyond Earth may find interest in such a classification. For example, members of ecclesiastical religious origins on Earth might decide to modify their system to have more immediate access to their god when living in settler's conditions that for some time might now allow for the large religious facilities (e.g. cathedrals) of Earth. On the other hand, small, local churches and other facilities for religious worship have in many cases been among the first substantial buildings established in colonies, including those of New England.

I would urge caution on attempts at detailed predictions or prescriptions; it is hard to imagine predicting that a minor sect illegal at the time of Jesus would be the state religion just a few centuries later. Still, it may be interesting to consider

We may conclude here with the point that the human experience is cumulative, not static, and that adaptation is itself change to accommodate new circumstances. However a tradition or cultural institution resists change due to new circumstances, it is likely to fail in the long run because the universe is largely unknown to humanity and millennia-old Ultimate Sacred Postulates, devised on one planet, are unlikely to adequately explain the entirety of existence. As we have seen in other domains of adaptation, it seems here that we cannot usefully suggest exactly how circumstances will force change on cultures venturing to settle beyond Earth, but that those most adaptable, culturally and biologically, are most likely to succeed. It is entirely reasonable to surmise that religious traditions of any variety currently known on Earth could be flexible enough to mold itself to the conditions of environments beyond Earth. Religious traditions have been very resilient in the most technical sense, able to recover following disturbance. For example, in Central and South America, Pre-Columbian religious traditions have resurfaced in the past few decades and in fact were woven into the acceptance of Christianity, with the overwhelming arrival of Europeans in the 16th Century, rather than replaced entirely. We might otherwise put it that practitioners of many religious traditions should have nothing to fear of new circumstances and discoveries beyond Earth, however fundamentalism, of any variety, is least likely to be able to smoothly adapt to such new circumstances.

4.2.5 Styles of Bodily Decoration

Styles of Bodily Decoration	Human identity is often communicated by bodily decoration, either directly on the body or with clothing	Some cultures heavily tattoo the body while others communicate identity more with clothing styles

Elaborate bodily decoration is a behavior quite unique to humanity and has long been used to visually communicate identity and, sometimes, individual history. Methods of bodily decoration have included tattooing and a myriad of styles of the hair and clothing. Such 'trivial' differences, in sum, constitute the very differences between cultures that are critical to maintaining cultural identities. Bodily decoration is largely, then, more about social than physical survival, but social survival is not unimportant and in fact if one is entirely socially inviable, one will not be biologically viable. How styles of bodily decoration will be used to adapt to the new cultural conditions of environments beyond Earth is not readily predictable, though we may reasonably predict that:

- New forms of bodily decoration will utilize new materials used in off-Earth settlements
- New forms of bodily decoration may well be used to differentiate off-Earth populations from those of Earth as a form of expressing a new identity.

4.2.6 Family Structure

Family Structure	Concepts of kinship or relations between kin, and associated ideas such as inheritance	Some cultures are polygynous, where males have several wives, and some are polyandrous, where females have several husbands

All cultures have some variety of *family structure*, which normally includes mated pair of individuals and their most immediate offspring. Families generally have the following characteristics:

- *coresidence*: family members live, more or less, together
- *economic cooperation*: members work for the economic interests of their family in mind, assisting (and being assisted) in times of stress
- management of *reproduction* and *enculturation*: taking part in the process of having children, providing for them, and bringing them to adulthood
- *management of property*: orchestrating the movement—within the generation and from one generation to the next—of the family's property.

In some cultures the immediate *nuclear* family is of less significance than the *extended* family, which includes more distant relations such as cousins and uncles. The demographic structure of a population (its age and sex constitution) are important determinants of the family structure, as they factor significantly into issues including inheritance and marriage eligibility. In remote, small-population agricultural settlements there has historically been an emphasis on large and extended families, likely because the complex tasks of agriculture—though they do provide stored foods as a hedge against scarcity—require large numbers of people working often simultaneously; "more children meant more help on the farm" (Benedict 1972 [1934]: 83). Though such conditions may seem likely in some environments beyond Earth, I do not think this large-family structure pattern will be common for the same reason, if it is common at all. The large-family pattern was established in the noted communities largely for unmechanized and relatively simple farming methods, but such will not likely be employed beyond Earth for the forseeable future. Rather, early off-Earth agriculture will likely be highly technically efficient and while very carefully monitored and maintained will likely be far less labor-intensive than on early Earthly farmsteads. Still, family size and structure may well be adjusted from those common on Earth for other reasons, as noted, demographic structure being particularly important. Other reasons will include the spatial distribution of communities on, for example Mars, which might well differ significantly from the distribution in a closed, orbital colony.

Again we find that while we cannot with useful reliability estimate just how this variable of cultural adaptation (family structure) will be adjusted to the conditions of various environments beyond Earth, but we may expect that they will be adjusted. Again the message is that if we wish to make a success of space settlement we should expect change, and welcome it as a sign of healthy adaptation.

4.2.7 Course of Life Stages

Course of Life Stages	Lives of individuals are segmented across the lifespan	Some cultures have formal rituals commemorating leaving behind a stage of life and entering another, while others are less formal and may have different stages demarcated

All cultures recognize different rights and responsibilities among members of different chronological ages (or age categories) which serve to organize many aspects of life, from *enculturation* (learning the rules of a culture) during childhood to *active teaching* of the young (for an example among Hutterites, who live in somewhat isolated rural communities, see Hostetler 1995). These *course of life stages* largely organize (a) the period before adulthood, in which young are enculturated, (b) the period of adulthood, in which culture is protected and propagated and (c) the period of post-adulthood, when many responsibilities are relieved. It has been argued that a period of 'adolescence' as understood in modern Western civilization did not exist in some traditional, non-industrial cultures (Condon 1990): nonetheless a periodization of human life is universal, despite some important details.

What is responsible for cultural changes in life course periodization, and will such factors obtain when settling beyond Earth? As usual there are many answers to the first part of the question. Alternations in the lifespan may significantly change the chronological age of the life stages, but biological and cultural necessity it seems will always distinguish between infancy/childhood, adulthood, and post-adulthood. But complex cultural phenomena can drive and attend such alterations. In ancient Greece, for example, the life span of males and females typically did not exceed over 35 years of age (Angel 1947) with the effect that childhood ended earlier and adulthood was accelerated compared to that of today; regarding cultural values, infant and childhood mortality was somewhat higher and there was a cultural disvalue of infants as they could not hold particular value until 'proven' of worth by surviving into the age of walking and talking. More specific instances of the human variety in life stages will be explored in Chap. 7.

4.2.8 Reproductive Behavior

Reproductive Behavior	Regulation of reproductive behavior, including incest rules	Cultures differ in the age at which sexual activity is permitted

Each culture has rules managing the sexual behavior of its members. The most commonly-recurring such rules have to do with the prohibition of close-kin reproduction, arguably as a precaution against inbreeding. Inbreeding in small populations can indeed be responsible for a higher incidence of genetic maladies than we would otherwise expect, an issue explored in Chap. 3. In this way, a cultural tool (for example, *incest taboos*) is used to protect the biological health of the population. If conditions in habitats beyond Earth are significantly unhealthy for early-reproductive-age individuals (as has been argued as unavoidable on Mars, where only a life lived entirely underground could screen out all of the increased radiation dose there) it may be that culturally, sexual behavior may be accepted earlier than it currently is on Earth in many Western cultures. But only a century ago in the West, people married on average a several years earlier than today and the mother's age at giving birth has also increased as life span has increased. In the same way, allowed sexual activity and the time of marriage have been strongly correlated in England over the past few centuries, also somewhat driven by lifespan issues; males several centuries ago married at an age near the average lifespan of their fathers, reflecting rules of inheritance, the lesson being that many factors, including religious beliefs, and even inheritance rules, can subtly but significantly shape these behaviors as well (Stone 1977).

When considering the regulation of sexual behavior to include earlier mating in populations beyond Earth, it has been pointed out that the current problems of early motherhood, which may include reduced access to education and lower overall life income (for an example in modern New Zealand see Borden et al. 2008)—are consequences of a culture not currently established for early motherhood. That is, it is not an issue of a biological reality but of a cultural system (shaped significantly by economic and religious traditions) that does not support early motherhood, shaped not from a biological but a moral perspective. That moral perspective may change, however, when we consider the reproductive health of people living beyond Earth.

Cultural regulation of reproductive behavior often relates to population control. While it is true that foraging populations generally prefer smaller populations that can be supported by the relatively widely-distributed resources of their environments, and that farming people generally prefer larger populations because of the labor demands of farming, there are other considerations to population regulation. These may be economic and/or related to social ranking, as pointed out by cultural anthropologist Ruth Benedict (1972 [1934]: 82–83):

> [human population control includes a concern for] scare social resources, objects or behaviors which give status and prestige…In terms of economic production, more children mean an increased food supply and perhaps the production of surpluses for trade. Most of the world's religions place great emphasis on the production of children…they may have a high survival value where the mortality rate is high…but they are proving disastrous where the mortality rate is low…

The point is that population control will be exerted by cultures beyond Earth for many and complex reasons, as on Earth. Some of this will likely be carried out by regulating reproductive behavior. Specific means of such are outlined in Chap. 7. We may conclude here by reiterating that if such measures are required for biological viability, we should not find this alienating; our cultures have managed such issues before and should be expected to endure such changes for new existence beyond the home planet.

4.2.9 Food Preferences

Food Preferences	Concepts of what are appropriate food and drink in certain situations	Some cultures eat certain animals while others consider them unfit to eat

Each culture has food preferences or *foodways* determined in variable degrees by religious concerns, resource availability and other factors including, importantly, identity (Gabaccia 2000). What is considered 'fit to be eaten' is not always simply a matter of what is available; cattle are common in India but are not consumed due to religious taboo; and while pigs could be raised in the modern Near and Middle East, ancient taboos against pork continue to inform Muslims about what is appropriate to consume.

Again what appears superficially trivial is particularly important to cultural identity. In the migrant experience, one of the most important ways for relocated people to maintain their identity is to 'have the right food'. In the case of Chinese immigrants to the early United States, immigrants working on railways and other large projects worked particularly hard to obtain the 'correct' spices and other ingredients in order to maintain a 'proper' diet, and thereby their identity (for an example see Ma 2015). These observations indicate that attention to the particular foods that people will want in early settlements beyond Earth will be of tremendous importance. The very domesticated plants and animals taken as food or food supplements (e.g. condiments) will likely not be entirely representative of the whole cuisine of the founding cultures, so foodways will have to be adjusted to the available resources. And of course food innovation will occur beyond Earth, such that new palates, new tastes for and against certain foods and beverages, will arise.

For members of different cultures, adjusting to a new cuisine can take a great deal of work, and may be so difficult that it is practically impossible; such an inhibition is not trivial, as it may limit social interactions and even the prospect of finding marriage or business partners. The point here is as we have seen many times before: we cannot and perhaps have no business suggesting how foodways will change beyond Earth, only that they will and that this will be a significant change in the daily life and long-term cultural history of the new, off-Earth populations.

4.2.10 Aesthetics and Arts

Aesthetics and Arts	Concepts of ideals, beauty, and their opposites	Some cultures value visual arts more than song, and vice versa; artistic expression differs radically between cultures and among members of the same culture

Artistic expression is also rather distinctive to each culture and of course in many cases, to subdivisions of a given culture. *Art*, anthropologically defined, is *embellished expression*, a broad term encompassing music, dance, the visual arts and even literature. The essence for the purposes of this book is that art serves as a means of communication often of novel ideas, and has historically been a source of cultural variation. Such variations have not always been universally popular, and highly symbolic, artistic works have played significant parts in large cultural revolutions; the visual arts were intensively employed to normalize counter-culture messages in the West during the late 1960s AD, and to communicate communal labor ideas during the emergence of Stalinism in the U.S.S.R., for example. Considering the human proclivity for the novel association of ideas (significantly underwritten by the cognitive operations employed in the use of modern languages; see Sect. 4.2.1) artistic expression would undoubtedly flourish in the presence of new phenomena and conditions of off-Earth settlements. We can imagine the qualities of sky light on Mars, versus on Earth, for example, where salmon tones and sky-blue tones, respectively, differ significantly. We can imagine the possibilities of sculpture, or dance, in Mars' 0.38 gravity field, or in the 0-g centers of rotating orbital or Lagrangian habitats. We can imagine the possibilities of literary expression when new concepts and perceptions, including those of being separated, perhaps permanently, from Earth, arise. Certainly this is a field of tremendous potential for investigation, and certainly art will have the same potential power to shape culture as it has on Earth for millennia (which is exactly why authoritarian systems strenuously try to regulate expression).

4.2.11 Ultimate Sacred Postulates

Ultimate Sacred Postulates	Central, unquestionable concepts about the nature of reality	Some cultures consider time to be cyclic while others consider it linear

The Ultimate Sacred Postulates (USP's) that shape a culture's core perceptions and moral principles are have varied globally and through time, but have also been rather homogenous over large areas and populations and through long spans of time. The founding concepts of ancient Egypt were established over 5000 years ago and persisted for three millennia, a period occasionally beset with civil war and cultural breakdown; still by the time of Greek conquest of Egypt it still maintained very ancient, fundamentally-organizing concept. On the other hand, the European Renaissance was a time of radical change in even founding concepts, in which humanity was repeatedly shown to not be at the center of time, or the universe or even at a pinnacle of creation, but another of many life forms produced by a natural (rather than supernatural) process (Shrader 2015). The point for our interest is that USP's of any founding Earth culture are quite likely to be challenged by the radically new circumstances of life in habitats such as the surface or subsurface of Mars, or orbital colonies on which a life may be led entirely at ½ of one's ancestor's gravity field. We do not know in what way such conceptions will change, only that they are likely to be challenged. I have discussed this realm in more detail above, in Sect. 4.2.4 as they pertain to religious conceptions.

4.3 Mechanisms and Rates of Culture Change: The Process of Ethnogenesis

How does culture change? Considering that anthropology has recognized that cultures are not concrete, unchanging monoliths or types, it is surprising that the field of cultural anthropology has not worked to systematically build a theory of culture change. This is likely due to the field's broad rejection of evolutionary approaches to culture following the mistakes of the social-darwinist approach, but that is an historical matter that I do not wish to address here. For the moment, we may say that an evolutionary approach to culture change, as programmatically constructed by anthropologist Durham (1991), has been influential. This approach identifies significant similarities between genes (the units of information in biology) and memes (the units of information in cultural evolution) as tabulated in Table 4.6.

In Chap. 3 we learned that evolution occurs in populations of individuals and that four forces were responsible for this change. These also have significant analogues in cultural evolution, as tabulated in Table 4.7.

By innovation and selection, then, new concepts, that is, cultural guides to behavior, appear and are dispersed in populations of minds. The *mechanisms of culture change* then include all ways that innovations occur and are dispersed in a population, and then accepted, becoming commonplace guides to behavior. These include word-of mouth moving ideas (='memes') among contemporaries, and the transmission of legends or histories from one generation to the next, verbally, gesturally and/or with any external record medium. How and why ideas (='guides to behavior') are evaluated and accepted or rejected, mind by mind, on a moment-by-moment basis in the daily lives of all members of a given culture, are at present of interest, but beyond anthropological and even psychological investigation.

As these new concepts are used to condition new behaviors, cultural norms change through time. This can occur rapidly or slowly, and there is no generalized rate of such change. The *shift, through time, of the centers of cycles of activity*, is sometimes referred to as the *longue duree*, a useful concept. For example, over centuries, ancient Egypt became more warlike as they were constantly attacked from the North. The

Table 4.6 Genes and memes in biological and cultural information transmission

Unit of information	Nature of information	Mode of transmission	Principle	Example
Gene	Biochemical code	Sexual; vertical, from parent to offspring. Also horizontal information transfer among microbes, but uncommon in multicellular species	Genotype governs development and operations of the phenotype	Sperm and egg each contribute some genetic material forming the next generation
Meme	Neural linkages store information as memories, perceptions and integrated 'worldviews'; may include symbols and other information stored outside the body	Social; vertical, from parents to offspring, but also cultural transmission *among* offspring and indeed among any communicating people; thus transmission is also horizontal	Cultural guides to behavior condition the behavior of the individual. However, individuals are highly differentiated and display a broad spectrum of adherence to cultural norms	Parents, peers and others freely exchange information that includes symbols and guides to behavior

Table 4.7 Genes and memes in biological and cultural information dispersal and selection

Unit of information	Origin of variation	Spread of variation	Evaluation of variation (selection)	Example
Gene	Mutation	Migration	Natural selection	Sperm and egg each contribute some genetic material forming the next generation
Meme	Innovation	Diffusion	Natural and cultural selection	Choices among a pool of options guiding behavior. Promotion of ideas e.g. propaganda, restriction of ideas, e.g. censorship

shift towards militarism was a relatively slow shift of the centers of activity, from less-warlike to more-warlike. Examples of rapid and broad-scale culture change include the large changes in American culture between 1960 AD and 1970 AD.

By such methods, cultures change. Such change includes the active (and sometimes not active) reshaping of the Cultural Universals introduced above. It is by this reshaping that humanity adapts itself, culturally, to new conditions, both physical (e.g. encountering new resource distribution in a new territory) and cultural (e.g. encountering new cultural traditions when encountering new trade partners). This 'reshaping' is adaptation, which we reduce to *culture change* in the section below, leading to a close examination of *ethnogenesis*, the appearance of new cultures.

Rates of culture change are highly variable: some culture change occurs on the scale of just minutes, as when an influential individual reaches a global audience with a convincing message that conditions future human decisions. Other culture change occurs more slowly, for example the radical generational change in American culture noticeable in the late 1960s; compare the values of the US in 1960 and 1970. On the other hand, some aspects of culture may not change for a long time. In the American example, again, consider that while women's rights were advanced as part of culture change between 1960 and 1970, the basic American cultural concept of the household—a family unit cohabiting a structure—did not fundamentally change in this period.

It is by such processes that culture changes, which assists us in addressing the common question: *will space colonists become another culture or species*? As for species, the biological dimension of the question is addressed elsewhere in this book. Here we examine the phenomenon of cultural diversification.

The origin of culture change, just as with the origin of biological change, is some *novelty* or *innovation* in the replicating information system; in genes, these are DNA mutations on the chemical level, in cultures they are new memes or ideas. Specifically, they are *innovations*. Anthropologist Homer G. Barnett (1906–1985) carefully examined the roots of culture change, focusing on innovation, which he defined as "…any thought, behavior or thing that is new because it is qualitatively different from existing forms" (Barnett 1953: 7). The 'existing forms', in our case, may be thought of as the *pool of options* existing in a person's mind (and therefore more or less shared by members of a given culture) when they consider some course of action. An innovation is some new option introduced to that pool of options. For it to be adopted as a new guide to behavior, it must be seen valuable enough to overcome the inertia of current satisfaction with the established pool of options, or what Barnett describes more generically as a 'configuration', which consists of a "unified pattern of experience". Barnett clarifies the situation in this enlightening way, highlighting that innovation is disruptive (Barnett 1953: 202):

> The real adhesive between the member parts of any configuration is habituation. The innovator is the person who either consciously or unconsciously reconfigures the conventional associations that are known and accepted by his group. His rearrangements are dissociative from the standpoint of tradition. They are extrinsic or extrapolative, they diverge from the associative sequences of the customary thought responses to a given stimulus. They take off on their own divergent course from a point which until then has been embedded, hidden, or so firmly integrated in one context that the thought of it almost automatically leads to the thought of that context only.

The point for our purposes in this book, and its larger goal of establishing a science of human space settlement, is that understanding innovation—a root cause of culture change—should better our understanding of our adaptations to environments beyond Earth. Adapting is simply changing to accommodate new conditions, but those changes are often disruptive to an established pattern; this is why they seem strange and/or cause discomfort. But it is by such change that we will pave the way for comfort of later generations. It is interesting to note that in the past century of globalization, Western culture has largely influenced other cultures, such that cultural variation has reduced and we have seen a general homogenization of cultural styles and norms. Generally, Western culture has influenced others but has itself not much changed in the last century; but it will be exactly this culture that itself will have to change when it faces new conditions beyond Earth. Its members will be well-advised, then, to be accepting of novelty and innovation, or at least hear it out before rejecting concepts that run counter to established patterns or what Barnett referred to as 'configurations'.

At some point in the accumulation of changed guides to behavior, 'new cultures' will form among populations beyond Earth. This is the phenomenon of *ethnogenesis*, the evolution (*genesis*) of new ethnicities (units of cultural identity) and is broadly applicable to the concept of human space settlement over multiple generations. Cultural anthropologists identify a given culture as a group sharing many beliefs or cultural guides to behavior. The boundaries of cultures are fuzzy; consider the case of the Quebecois of Canada, many of whom identify themselves as distinctly

Quebecois (rather than Canadian) but who share more behavioral traits with Canadians than, say Floridians. Cultures in this sense are real, but fuzzy. Keeping this in mind we may consider how such cultural groups originate.

The process of *ethnogenesis* is known, to anthropologists, to occur, and in some circumstances it might be inevitable. For example, the case of early 20th-Century 'Italian Americans' adapting culturally to the new social conditions of North America (Glazer and Moynihan 1970: xxxiii):

> As the old culture fell away—and it did rapidly enough [for early Italians emigrating to the United States]—a new one, shaped by the distinctive experience of life in America, was formed and a new identity was created. Italian-Americans [shared] precious little with Italians in Italy, [and in] America they were a distinctive group that maintained itself, was identifiable, and gave something to those who identified with it.

Though evident enough, the ethnogenetic process is not structural to anthropology. That is, it is not a well-known or well-understood process, even though it is not rare. This appears to be because the events that lead to new culture formation are repeatedly found to be contingent or unpredictable. For example, in a study of ethnogenesis among native peoples of North America, we find a wide variety of factors resulting in a new cultural identity among the Blackfoot (Juneau 2007: 96–97):

> For the Inuck'siks band, the foremost epic transitions defined some of the conditions that fostered a widespread process of ethnic hybridization, and that contributed to the emergence of new cultural groups and ethnic identifications during the eighteenth and nineteenth centuries were:

- 1760s Inuck'siks acquires the horse from [other Native North Americans]
- 1781 smallpox epidemic decimates over one-half the population
- 1790s [significant reorganization of trade relations]
- 1800 began an ethnic bloc confederation with the Salish and Kootenai [tribes]
- 1830 actively seek a new powerful belief system in Christianity to add to their existing belief [adapting to an overwhelming inevitability]
- 1837 smallpox devastates the band in their wintering ground with a loss of over half the population
- 1840 [numerous demoralizing battle losses]
- 1845 Inuck'sik loose an estimated 50 lodges, and 150 women and children in a battle with the Crow…30 remaining bands split over new Christian belief system…
- 1855 Sign the Lame Bull Treaty at Judith Landing
- 1870 Heavy Runner band is decimated by U.S. Calvary."

In this case, the specific events taking place could not be predicted, even though perhaps some overarching themes might be predictable.

While the processes of ethnogenesis have eluded anthropological systematization, we may generalize productively on a few of its aspects. First, one study of the process indicated that it is not as simple as a culture 'accommodating to new conditions' in a general, unidimensional way (Fig. 4.3a) but can be observed to occur at least in four main domains of behavior (which will resonate when considering those mentioned in the previous sections), and in a recurring sequence of *domain-specific*

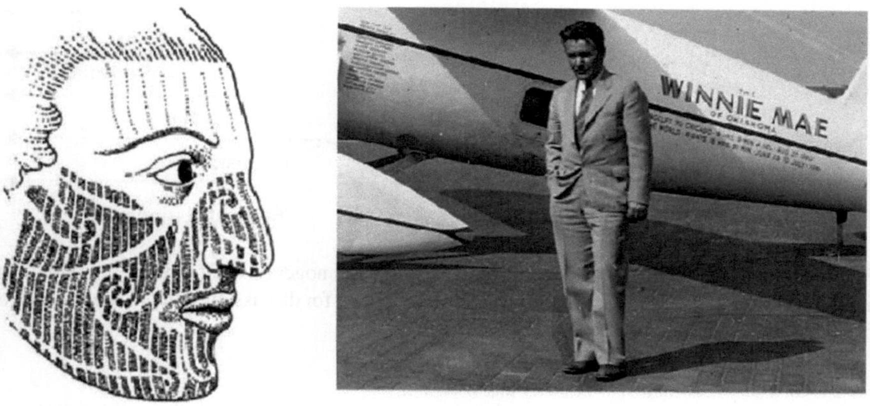

Fig. 4.3 Bodily Decoration Examples. Pictured are traditional Polynesian facial tattooing on a Maori chief and the 1930s aviator Wiley Post wearing a suit rather than technical flying clothing to emphasize the comfort of early passenger aviation

change; this is the transformation of the domains of language, social organization, economic organization, and civic organization (Fig. 4.3b). Note that these are (from language on) increasingly-integral domains of the culture, such that their change is increasingly difficult or, better-said, increasingly resisted by the conventional configuration, but do eventually occur. That is, language changes first, incorporating new vocabulary, for instance; this is easy to do as it only requires an individual to try out new words or phrases; of course, they must then gain acceptance by more individuals to become common. This is followed by social changes such as value adjustments that are *implied* by the new terminology (e.g. the difference between the Arctic seen in the 1880s as a 'wasteland' and in the 1980s as a 'cultural resource'). Economic changes are more difficult as economic systems such as manufacture tools & techniques represent a considerable investment and startup of new such tools and techniques can be costly. Finally, civic changes occur, being those associated with the essential cultural configuration, often shaped by such change-resistant foundations as the Ultimate Sacred Postulates described above. While the process specifically described in the study from which Fig. 4.4 was adapted was assimilation rather than ethnogenesis, I find the parallels significant and worth describing in this way.

By whatever methods and sequence, and on whatever timescale, we can be sure that new options for behavior will be innovate early in the adaptation to conditions beyond Earth, such as a Mars or Lagrangian colony. After some time, the divergence between the Earth and colony cultures will be palpable, noticeable, in communications between those diverging cultures. Anthropology also has no quantification for this phenomenon, but it is obvious that it occurs. At some point, anthropologist Patricia Albers has noted, there becomes clear an:

> emergent ethnic community [wherein] the process of ethnogenesis has reached its completion.... In the process, they not only form a political entity that is separate from their parent populations, but they also assume an ethnic identification that is distinctive as well. It is an

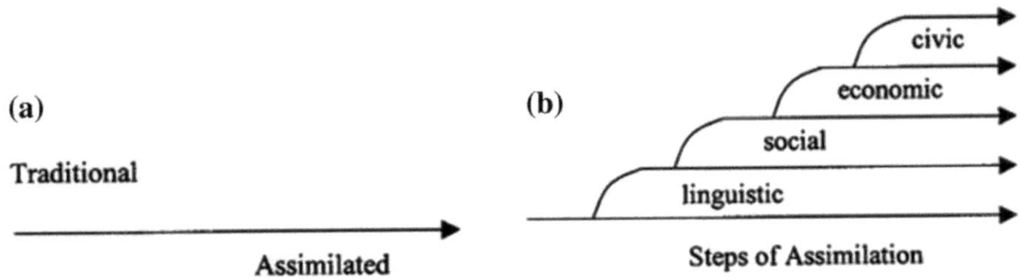

Fig. 4.4 Models of Culture Change that Can Result in Ethnogenesis. Time is on the horizontal axis, from recent (on the left) to future (on the right). See text for discussion. Adapted by the Author from Figs. 1–3 of Flannery et al. (2005)

> identity that emphasizes unity and solidarity over any differences from their ethnic pasts. (Albers is quoted in Corr and Powers 2012: 27)

This examination of how cultural innovations, if accepted and spread, become the very adaptations humanity will use to flourish beyond Earth has done the important work of at least beginning to demystify the process. We need not simply wave our arms and broadly say 'people will adapt' when asked just how people will live 'in space'. We can say that they will do so by the same methods we have used for millennia, processes we know and that we can study to give space settlers a better chance. This is the process of inventing new ways to live, particularly in the universal human domains of behavior outlined, including language, household organization and enculturation. And when it is asked, 'will humanity become a different species or culture?' we may say, of course, new cultures will emerge, and this is entirely natural and expected of healthy cultures.

We may also note, though this will be explored more in Chap. 7, that the bulk of adaptation will be the burden of the first and perhaps second generation of space colonists. Within a few decades, people will be born into a culture, just as they have been on Earth for millennia. They will not have a difficult adjustment process; they will just grow up, just as an Australian aborigine or a Thai villager's child is born into a culture and environment not alien to them, though it was alien to their ancestors.

Finally, we may note that there is nothing apparent in the behavioral range of humanity, as expressed in our Cultural Universals, that is so inflexible as to preclude space settlement. As far as we can see at present, informed though with a century's anthropological study, there are no cultural 'show stoppers' to human space settlement. It will simply be cultural adaptation, a phenomenon we have productively examined in this chapter.

4.4 Cultural Regulation of Biological Evolution

It is commonly argued that advances in genetics will enable humanity to settle beyond Earth, using techniques such as DNA repair promotion to overcome DNA damage due to radiation, or using genetically activating bone-mineralization pathways to overcome bone density loss that occurs in microgravity. Currently the Mason Lab (Institute for Computational Biomedicine, Weill Cornell Medical College) is actively seeking out the genes most likely to be useful in understanding how we can best biologically adapt to conditions beyond Earth as part of their "ten-phase, 500-year plan for the survival of the human species on Earth, in space, and on other planets"; the project includes the goals listed below (MASONLAB):

Improve the ability of humans and other species to survive in space-based environments or on other planets

- Genetic engineering
- [establish a] "do-not disturb" list for the human genome; regions that cannot

  or should not be altered

- Engineer mammalian insulators that can carry genetic payload safely
- Establish long-term safety, efficacy, and functional characterization of genomes which have been engineered
- Improved DNA damage detection and repair (vis-à-vis *Deinococcus radiodurans*)

The lab's timeline includes the following phases:

> **Phase 1** (2010–2020): Experiments that complete the functional annotation of the human genome, specifically experiments that delineate base-by-base examination of those elements which cannot be changed, or those that are most flexible (tolerant of mutations). We will use comparative, evolutionary functional genomics and an examination of those human diseases that constitute the most aggressive deviations from normal biological function (such as aggressive tumors) to understand the genome. Early stages of synthetic and systems biology.
>
> **Phase 2** (2021–2040)
>
> We assume that whole-genome sequencing and molecular characterization is common, cheap, and accurate. Thus, work is focused on methods for contextualizing common and rare variation and its affects on the human genome. Begin efforts on integrating new elements into mammalian genomes.
>
> **Phase 3** (Years 2041–2050)
>
> Begin long-term human trial on the genome engineering.
>
> **Phase 4** (Years 2051–2060)
>
> Test protected human genomes in space environments
>
> **Phase 6** (2060–2100)
>
> Begin settlement of other planets and genesis of synthetic genomes

Certainly the ethical issues of genetic engineering of our species are invoked here, and they will be decided by space-settling peoples themselves. From an anthropological perspective, my contribution to the discussion at this stage would be that the human genome is already quite plastic (see Chap. 2) and that we need not wait for genetic tailoring of our species to explore the possibility of settling beyond Earth. The nearly 50-year project of humans-in-space provides a cautionary tale, as referenced in Chap. 2; today, despite a quarter century of human habitation of LEO, we still feel unprepared to characterize low-gravity effects on the human body. Certainly we wish to be informed of the hazards, but we must also be ready to take some risks, as an expectation of zero risk will ensure that we never set out from Earth. I see, then, from the anthropological perspective, no pressing need for genetic engineering of spacefarers. If this becomes necessary later, it will be a discussion best held by the spacefaring people themselves, as with many of the ethical concerns about space settlement.

4.5 Artificial Intelligence and Cultural Evolution

The field of *artificial intelligence* (AI) has worked towards building 'thinking tools' analogous of the human mind for about two generations, or c.50 years. Currently AI research exists at least in part to emulate and then exceed the capacities of the human mind, or, more specifically, human cognition. I have often been asked whether AI will be a part of future human evolution, a question I address below.

Modern major subdivisions of AI are *Machine-*, *Reinforcement-* and *Deep-learning*. These are concerned with (respectively) pattern-recognition, goal-oriented behavior based on trial and error (an essentially evolutionary approach that selects for variations that approach the desired performance and against variations that do not) and neural network emulations that develop rich representations of the outside world with large memory matrices. Today's AI's are sometimes capable of rapid pattern recognition, and can quickly develop behavioral rules or algorithms that improve with time and experience to carry out certain activities. These 'thinking machines' are often highly capable.

All of these activities are very local and limited. That is, while AI's can call upon large memory arrays for precise information in response to a given stimulus, they do so in a small bubble of time- and space-awareness. In the study of the evolution of consciousness this is termed *episodic consciousness*, analogous to an infant's awareness of time and space (see Donald 1993 for other varieties of consciousness). Infantile consciousness is fully capable of processing information, making correlations, recalling specific memories and a few other functions, but all of this is done without a significant awareness of the scales of two universal variables: space and time. An infant does not know of the atom, or molecule or the size of Earth, or the galaxy, nor does it know that we estimate the Earth to be on the order of 4+ billion years old, the universe closer to 15 billion years old, or that we can observe phenomena that take place on a very short timescale of milliseconds and less. An infant does not know that its lifespan is estimated in the 70-year range and that this should be important for some decision-making. This space- and time-unawareness of context leads to a mind that is situated in the moment, with episodes occurring during life and only slowly being assembled into an integrated *worldview* that is capable of appropriately linking perceived phenomena in cause and effect relationships, for example, or generating useful metaphors (the worldview concept is introduced in works by Liane Gabora of the University of British Columbia e.g. Gabora 2001). Such higher cognitive functions eventually result in a more contextually-aware mind characterized by what anthropologists call *behavioral modernity*, analogous to the cognition evolution researcher's terms *cognitive fluidity* and *theoretic consciousness*, to name just two. Human behavioral modernity, as currently understood anthropologically, exhibits an array of differences in cognition from that of other intelligent life, as summarized in Fig. 4.5.

The point for our purposes is not an exploration of human intelligence but a reminder that that while AI will be an important tool for the forseeable future, at present AI's only perform human-mimicing functions (and many of these not very well) in stark isolation; isolation from the size and timespan of the cosmos, knowledge of which both tremendously enrich humanity's consciousness. It has been pointed out that language evolved as a mode of communication and that its evolution cannot therefore be understood as something that occurs in just one mind but is always influenced by other minds in the social environment. AI research will likely continue to generate episodically-aware, non-human cognitive devices until such social integration is devised.

For the forseeable future, then, I think of AI as a highly-capable tool with the capacity to influence human culture rather passively as have many of our other tools; consider the reshaping of the world after the invention of the gasoline-powered automobile. But I do not consider it as an active part of human cultural evolution, participating in humanity's endless reworking of our worldviews and understandings of the universe in the incessant, highly-malleable cultural information flux, textured with many layers of metaphor and informed by broad time- and space-awareness. This I feel, and have argued elsewhere in print, is in fact a distinctly human capacity resulting from our variety of consciousness.

4.6 Conclusions and Recommendations

The cultural adaptation of humanity to conditions beyond Earth is not a given, but it seems likely enough to succeed to attempt. That is, once the voyaging aspect of settling new habitats is overcome, as mentioned in Chap. 1, the cultural challenges to adapting to the new environments we may reasonably expect do not appear to be beyond the cultural plasticity of our species. It will require proaction, and cannot be expected or taken for granted, but humanity has a deep and rich reservoir of experience in cultural adaptation. And today we can identify the tools of such adaptation, and begin to think about how best to use them for space settlement.

Cultural adaptation will be most difficult for the first generation or two of space-settlers. Once the excitement of arrival at a new habitat dies down, the often-difficult work of adapting ad adjusting will begin. Culturally adapting to new conditions will require adjusting how people live, from Earth-modes to modes appropriate for new conditions. Such cultural change disrupts traditional ways of life, including established social and economic structures; a culture wishing to adapt, then, to new conditions, must be a culture willing to change. Not all must be changed, however, and some traditions will of course be preserved for some degree of cultural continuity,

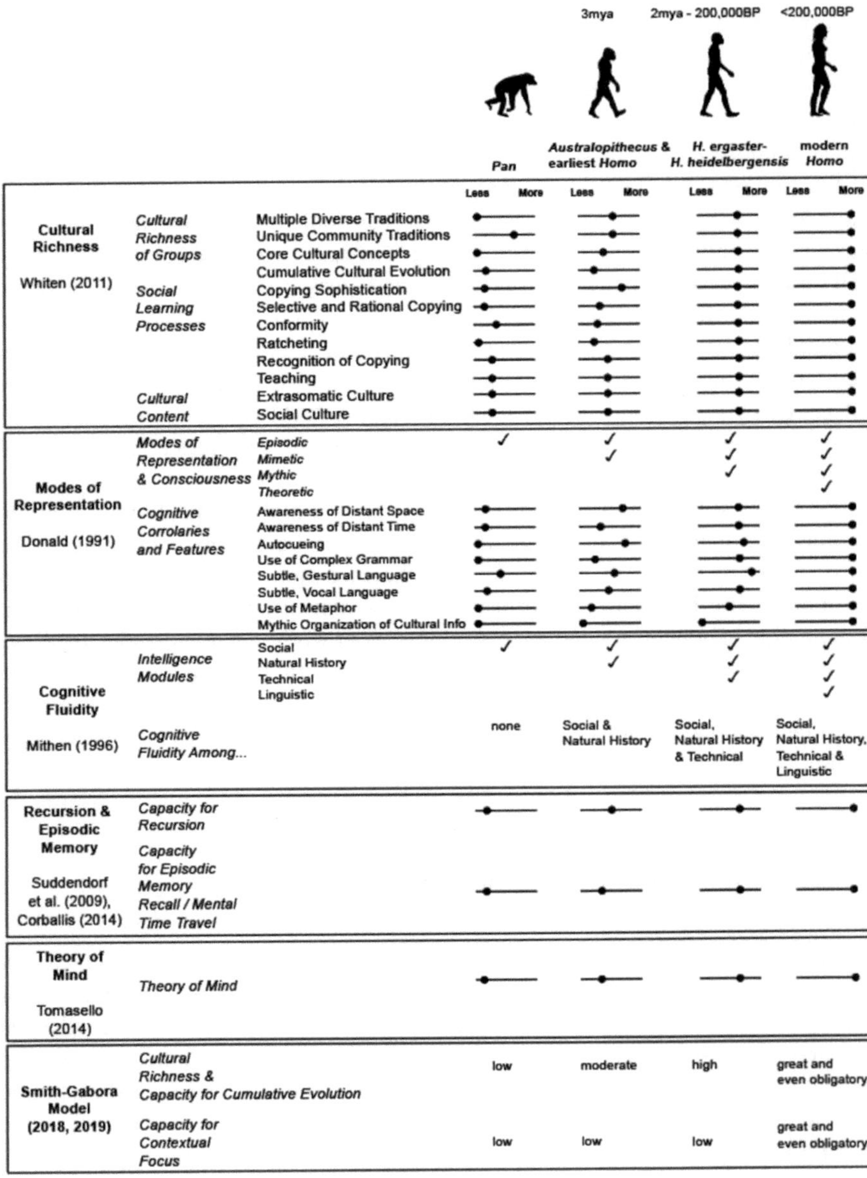

Fig. 4.5 Likkert Scale Indications of Cognitive Features in *Pan* and *Homo*. Each domain of cognitive action (e.g. "Modes of Representation") associated with a particular author is indicated. The Likkert Scale in each case per species indicates there is a greater or lesser degree of the cognitive feature (more on the left, less on the right). Figure by the Author, referencing work published in Gabora and Smith (2018)

important to identity. But over time, beyond-Earth cultures will take on their own identities. They will devise new patterns of life and new social rules.

In discussing space settlement I often encounter the view that establishing such new social rules would be 'meddling' or 'social engineering'. But, of course, every new law we pass or every tradition we begin to follow, is by this definition social engineering. We might as easily call it I take the view that rather than interfering with 'natural' culture change, such informed regulation of culture would make for a better future (especially for future human voyagers) than if we left things to 'fate', a vague notion at best. The cultural future is made by the moment-by-moment decisions of the members of a culture. Guiding those decisions are the values of the culture. When the decisions are made to better the cultural health of the culture, establishing the cultural rules or customs is taking care of the future, not meddling or social engineering.

Considering this general stance, I do not have significant, specific prescriptive suggestions for cultures of space-settlement pioneers, although I revisit these topics in Chap. 7 with more detail regarding possible future outcomes. Rather, I think the cultures of space settlers will shape themselves. I will point out, however, that they may best shape these cultures if they have a good understanding of the adaptive features of human culture, and that is what we can provide with space anthropology. The Cultural Universals are the clay, we might say, of the body of human culture and they are highly malleable. A familiarity with how they have been shaped by past and present human cultures will be useful to both the eventual settlers themselves, but also in the interim to space-settlement planners, as guides to action. What aspects of culture must we accommodate when we propose the settlement of space? At least, the Cultural Universals. This knowledge alone can help structure the discussion of space settlement by delineating the boundaries of the discussion, which I think is often necessary. In my experience, space settlement discussions rapidly become a discussion of the details of each discussant's personal belief of how human cultures *should* be organized (typically specific but based on a general and vague familiarity with cultures other than those of the modern, Western tradition). Personal opinions are fair enough, but this discussion form has two main downsides. First, it fragments a discussion. This can be remedied by 'containing' a given discussant's specific interest, e.g. economics or religion, into a broader Human Cultural Universal category. With the Cultural Universals outlined, doem basic terminology has been established, which should streamline a discussion. Second, the individual discussions of specific cultural interests is often unproductive because it is not well-informed, but rather a very personal view based on one's own cultural experience rather than an informed, profession understanding of the topic from a broad, cross-cultural appreciation. That specific knowledge is the domain of anthropology, and a significant reason for writing this book has been for me to bring this chapter's overview to the discussion. Understanding Cultural Universals should make serious discussions of space settlement more focused and productive.

4.7 Summary Points

- Human culture is highly malleable
- Human cultural adaptation is carried out by the adjustment of behavior in eleven broad but definable domains of behavior termed 'Cultural Universals'
- Human cultural adaptive behavior appears entirely capable of adaptation to the social and physical environments beyond Earth reasonably expectable of the earliest sites of settlement, including Mars
- Artificial intelligences will assist humanity in cultural adaptation to habitats beyond Earth, but at present are highly constrained in their capacity for genuinely-informed cultural innovation or even intelligence as it is expressed in humanity
- Humanity's adaptive 'Cultural Universals' should be further researched to increase our understanding of these as adaptive tools for space settlement
- Anthropological understanding of Cultural Universals can be used to better organize the often fractuous and disorganized discussions that occur about space settlement.

References

Angel, J. L. (1947). The length of life in ancient Greece. *Journal of Gerontology, 2*(1), 18–24. https://doi.org/10.1093/geronj/2.1.18.

Barnett, H. G. (1953). *Innovation: The basis of culture change*. New York: McGraw-Hill.

Benedict, R. (1972 [1934]). *Patterns of culture*. New York: Houghton-Mifflin.

Borden, J. M., Fergusson, J. M., & Horwood, J. L. (2008). Early motherhood and subsequent life outcomes. *The Journal of Child Psychology and Psychiatry, 49*(2), 151–160. https://doi.org/10.1111/j.1469-7610.2007.01830.x.

Brown, D. E. (1991). *Human universals*. Philadelphia: Temple University Press.

Condon, R. G. (1990). Adolescence and changing family relations in the Central Canadian Arctic. *Arctic Medical Research, 49*(2), 81–92.

Corr, R., & Powers, K. V. (2012). Ethnogenesis, ethnicity and 'cultural refusal: The case of the Salasacas in Highland Ecuador. *Latin American Research Review, 47,* 5–30.

Creanza, N., Ruhlen, M., Pemberton, T. J., Rosenberg, N. A., Feldman, M. W., & Ramachandran, S. (2015). A comparison of worldwide phonemic and genetic variation in human populations. *PNAS, 112*(5), 1265–1272. https://doi.org/10.1073/pnas.1424033112.

Donald, M. (1993). *Origins of the modern mind: Three stages in the evolution of culture and cognition*. Cambridge, Mass: Harvard University Press.

Durham, W. H. (1991). *Coevolution: Genes, culture and human diversity*. Stanford: Stanford University Press.

Edgerton, R. B. (1992). *Sick societies*. New York: Simon and Schuster.

Flannery, W. P., Reise, S. P., & Yu, J. (2005). An empirical comparison of acculturation models. *Personality and Social Psychology Bulletin, 27*(8), 1035–1045.

Gabaccia, D. R. (2000). *We are what we eat: Ethnic food and the making of Americans*. Cambridge: Harvard University Press.

Gabora, L. (2001). *Cognitive mechanisms underlying the origin and evolution of culture* (Unpublished PhD thesis). Center Leo Apostel, Brussels. Available online at https://www.vub.ac.be/CLEA/liane/thesis_gabora.pdf.

Gabora, L. (2008). The cultural evolution of socially situated cognition. *Cognitive Systems Research, 9*(1–2), 104–113.

Gabora, L. (2013). An evolutionary framework for culture: Selectionism versus communal exchange. *Physics of Life Reviews, 10*(2), 117–145.

Gabora, L., & Smith, C. M. (2018). Two cognitive transitions underlying the capacity for cultural evolution. *Journal of Anthropological Sciences, 96,* 1–26.

Gert, M. (1988). *Morality*. Oxford: Oxford University Press.

Glazer, N., & Moynihan, D. P. (1970). *Beyond the melting pot: The Negroes, Puerto Ricanns, Jews, and Irish of New York City*. Cambridge: MIT Press.

Hostetler, J. (1995). *The hutterites in North America*. New York: Wadsworth.

Juneau, L. M. (2007). *Small robe band of blackfeet: Ethnogenesis by social and religious transformation* (Unpublished MA thesis). Department of Anthropology, University of Montana Missoula, Montana. See https://scholarworks.umt.edu/etd/890/.

Lieberman, E., Michel, J.-B., Jackson, J., Tang, T., & Nowak, M. A. (2007). Quanitifying the evolutionary dynamics of language. *Nature, 449,* 713–716.

Ma, G. (2015). Food, eating behavior and culture in chinese society. *Journal of Ethnic Foods, 2*(4), 195–199.

Maffi, L. (2005). Linguistic, cultural and biological diversity. *Annual Review of Anthropology, 29,* 599–617.

MASONLAB. (2018). See masonlab.net.

Mencken, H. L. (1937). *The American language: An inquiry into the development of English in the United States*. New York: Alfred A. Knopf.

Mostafavi, H., et al. (2017). Identifying genetic variants that affect viability in large cohorts. *PLOS One*. https://doi.org/10.1371/journal.pbio.2002458.

Nettle, D. (1999). Is the rate of language change constant? *Lingua, 108,* 119–136.

Newman, G. (1976). *Comparative deviance: Perception and law in six cultures*. New York: Transaction Publishers.

O'Neill, P., & Petrinovich, L. (1998). A preliminary cross-cultural study of moral intuitions. *Evolution and Human Behavior, 19*(6), 349–367.

Palmer, S. (1997). Language of work: The critical link between economic change and language shift. *Teaching Indigenous Languages*, 263–286.

Schwartz, S. H. (1992). Universals in the content and structure of values: Theory and empirical tests in 20 countries. In M. Zanna (Ed.), *Advances in experimental social psychology* (vol. 25, pp. 1–65). New York, Academic Press.

Shrader. (2015). Shrader book or three scientific revolutions.

Stone, L. (1977). *The family, sex and marriage in England 1500–1800*. New York: Harper and Row.

Thornhill, N. W. (1991). An evolutionary analysis of rules regulating human inbreeding and marriage. *Behavioral and Brain Sciences, 14*(02), 247–261.

Vatican. (2018). See the Academy's website at http://www.vaticanstate.va/content/vaticanstate/en/altre-istituzioni/accademia-scienze.html.

Wallace, A. F. C. (1966). *Religion: An anthropological review*. New York: Random House.

Wenke, R. J. (2002).

Westermarck, E. A. (1906). *The origin and development of the moral ideas*. London: Macmillan.

Whiten, A., Hinde, R. A., Laland, K. N., & Stringer, C. B. (2011). Culture evolves. *Philosophical Transactions of the Royal Society B, 366,* 938–948.

Wichmann, S., & Holman, E. W. (2009). Population size and rates of language change. *Human Biology, 81*(2), 259–274.

UDHR. See online http://www.un.org/en/universal-declaration-human-rights/.

Chapter 5
Humanity and the Migration Experience Beyond Earth

Abstract A defining feature of animal life is that it moves. Early in life, with little knowledge of its environment, each life form is engaged in exploration. As information about its environment is accumulated by exploration, the animal may decide to remain where it is, or go elsewhere; various external 'push' and/or 'pull' factors also tend to move animal life across geographies. Repeated, cyclic movement is often termed migration, as when birds fly to distant feeding grounds for months at a time. Other life forms, in particular some aquatic species, move one-way, leaving the place where they were born to settle permanently elsewhere. Humanity is engaged in many such movements on Earth. The one most pertinent, however, to the case of space settlement is emigration, a permanent, one-way phenomenon, in the human case an intentional one. This chapter examines the emigration experience in general and proposes a four-stage model for permanent human space settlement. As in the other chapters of the book, I highlight the evolutionary aspects of our topic, identifying the main trends in animal migration and emigration so far identified, and applying, when appropriate, lessons from evolutionary biology to the prospect of human space settlement.

> The great arc along which all the possible human behaviors are distributed is far too immense and far too full of contradictions for any one culture to utilize even any considerable portion of it. Selection is the first requirement. Without selection no culture could even achieve intelligibility, and the intentions it selects and makes its own are a much more important matter than the particular detail of technology…that it also selects in similar fashion.—Ruth Benedict

5.1 Introduction

For the past two million years, populations of our genus, *Homo*, have been rather mobile; Fig. 5.1 displays our ancestors' global dispersals. In the past 10,000 years, when agriculture became more prevalent than foraging in a number of centers worldwide, such dispersal slowed, although both the Pacific and North American Arctic were thoroughly explored in the last five millennia. For the last 3000 years humanity

C. M. Smith, *Principles of Space Anthropology*, Space and Society,
https://doi.org/10.1007/978-3-030-25021-8_5

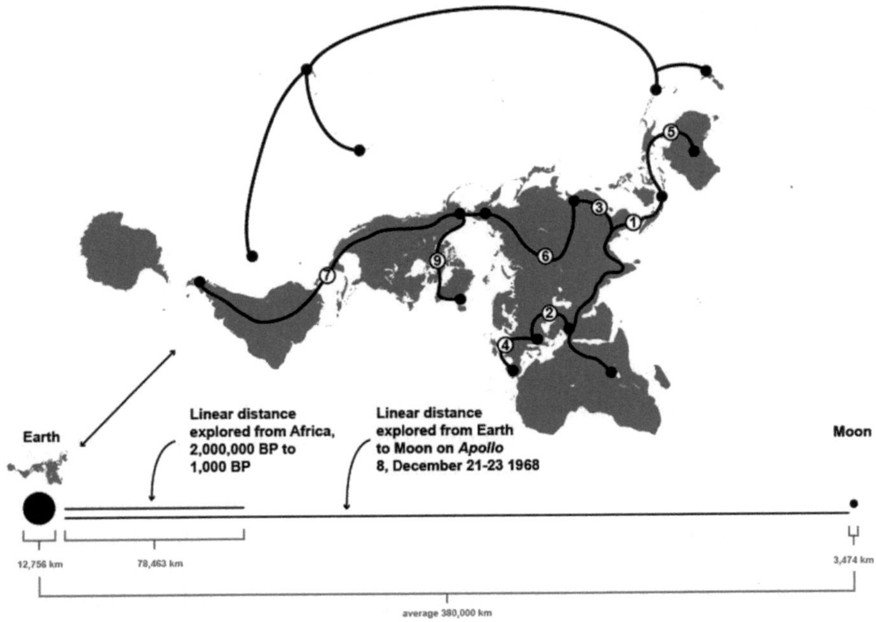

LEG	YEARS BEFORE PRESENT	LEG DISTANCE (KM)	CUMULATIVE DISTANCE from AFRICAN ORIGIN (KM)
1: Africa - SE Asia	2,000,000	14,247	14,247
2: Africa - W Asia	2,000,000	2,099	16,346
3: SE Asia - E Asia	1,500,000	2,555	18,901
4: W Asia - Europe	1,000,000	4,418	23,319
5: SE Asia - Australasia	50,000	6,018	29,337
6: E Asia - NE Asia	40,000	7,073	36,410
7: NE Asia - Americas	20,000	17,719	54,129
8: SE Asia - Oceania	3,000	18,482	72,611
9: NW Americas - NE Americas	1,500	5,852	78,463
10: Earth - Moon	50	384,400	462,863

Fig. 5.1 Human global dispersal and comparison with moon voyage. Over fifty millennia of modern human dispersals have included all manner of 'drivers' including simple range expansion with ecological opportunity, happenstance and proactive exploration and colonization. Figure and data by the author

at large has been sedentary on Earth in the respect of encountering and adapting to new environments. For the last 500 years environments beyond Earth have been observed by telescope, and in the last 50 years robotic explorers have gone to some of these environments. In just one case, the *Apollo* program expeditions of 1969–1972, have humans themselves gone to an environment beyond Earth.

In Chap. 1 we saw that on average, land mammals have a natal range of something on the order of 40 km, while many humans today, assisted by technology, routinely exceed this figure by some orders of magnitude, and could of course range even farther if the potential to settle beyond Earth is taken up. The evolutionary significance of these movements is that they place humanity into new *selective environments*—sometimes biological, sometimes cultural and sometimes both—that require adjustment and adaptation of the individual and population (respectively) for success (survival and propagation). In this chapter we will examine the prospect of *intentional*, proactive human population movement to new selective environments beyond Earth. In Chap. 6 we will examine in more detail the specific ways that humanity will adapt to these new regions, whereas in this chapter we will focus on the decision to move, and the movement and settlement more specifically. As usual, throughout this chapter we will draw on the lessons of evolutionary biology for guidance in the project of human space settlement.

5.2 The Biology of Species Dispersal

Figure 5.2 introduces useful summaries of animal postnatal movement. Note that *migration* refers normally to a seasonal or cyclic movement, which includes significant return to a given territory or species range; while some space settlers will plan to return to Earth (return to the home country after some years abroad is a common migrant experience), in this book my focus is on ensuring multigenerational success for permanent space settlers; this is more commonly termed *emigration*, taking up permanent residence in an environment different from the home, and in Fig. 5.2 would follow *conscious colonization*.

Figure 5.3 schematically illustrates the distribution of animal populations across a geography. We see that animal life occupies zones of various distributions, ranging from relatively densely-occupied zones of 'continuous distribution' (bottom left) to zones of 'sterile invasion' where the species occurs in very small, vulnerable exploratory populations. We see this in Earth life (see distribution of modern chimpanzee upper right) and may consider it in the case of the settlement of Mars, which for quite some time may be seen in evolutionary terms as a case of very small and vulnerable populations.

When undirected by a conscious choice, as in most Earth life, substantial, permanent movement of species is referred to variably in the literature as *range expansion*, *colonization*, *dispersal* and *ecological invasion*. It has been found in the study of each of these subtly-differing modes of species movement into new habitats to divide the process into several stages. One excellent review suggest three main terms for the process: "the decision to leave the current [habitat] (*emigration*), movement between [habitats] (inter-patch *movement*), and the decision to enter, and remain in, a new [habitat] (*immigration*)" (Bowler and Benson 2005: 207). In another, more recent review the authors describe *predeparture, departure, transfer, settlement* and *post-settlement* phases (Bonte et al. 2012: 292). In the study of invasive species, again several stages have been identified as real and worthy of study in themselves: *introduction* ("the transport of invaders from their native to exotic range, as well as transport within the exotic range"), *establishment* ("populations of introduced species increase from rarity") and, if the invasive species is successful "spread and impact [upon endemic species]" (Levine 2000).

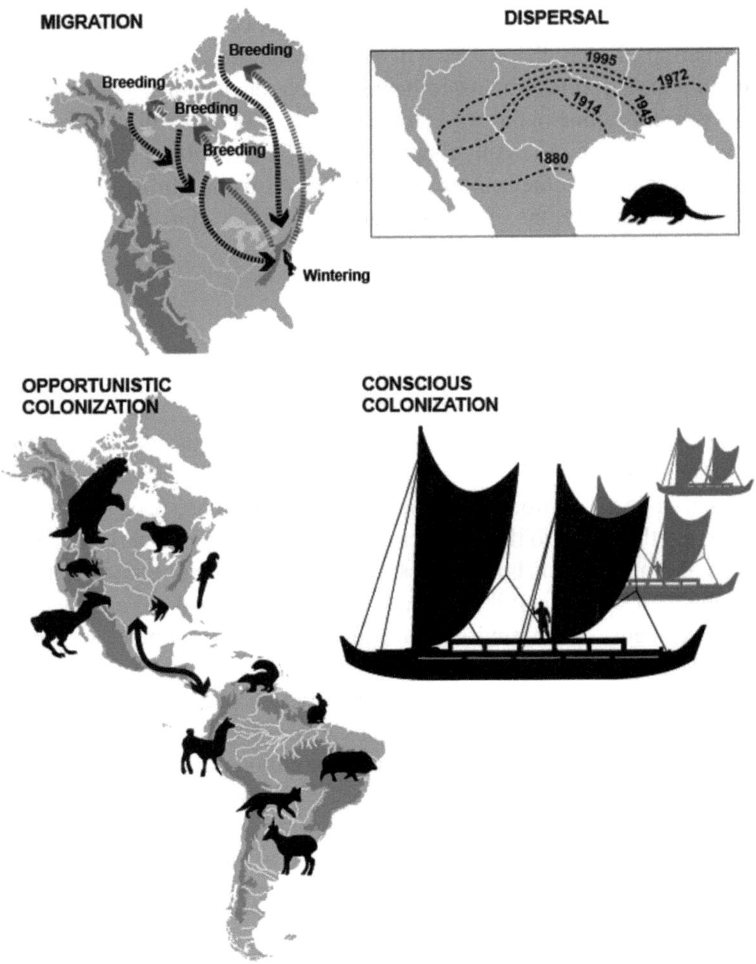

Fig. 5.2 Modes of animal postnatal movement. In *migration*, animals move rather seasonally from a home range to another range for such actions as breeding or feeding, and then return to the home range. In *dispersal*, animals extend their ecological range due to a variety of extrinsic factors; in this case the range expansion of the armadillo from North Mexico into some Southern US states resulted from the expansion of automobile roads at the turn of the century. In *opportunistic colonization* animals move into areas of new ecological opportunity, as when the North and South American continents connected via the Panama land bridge, c.3 million years ago. In *conscious colonization*, which appears to be unique to humanity, potential new habitats are actively explored and settled, as in the case of Polynesian peoples settling the islands of the Pacific in the last three millennia. Figure and data by the Author

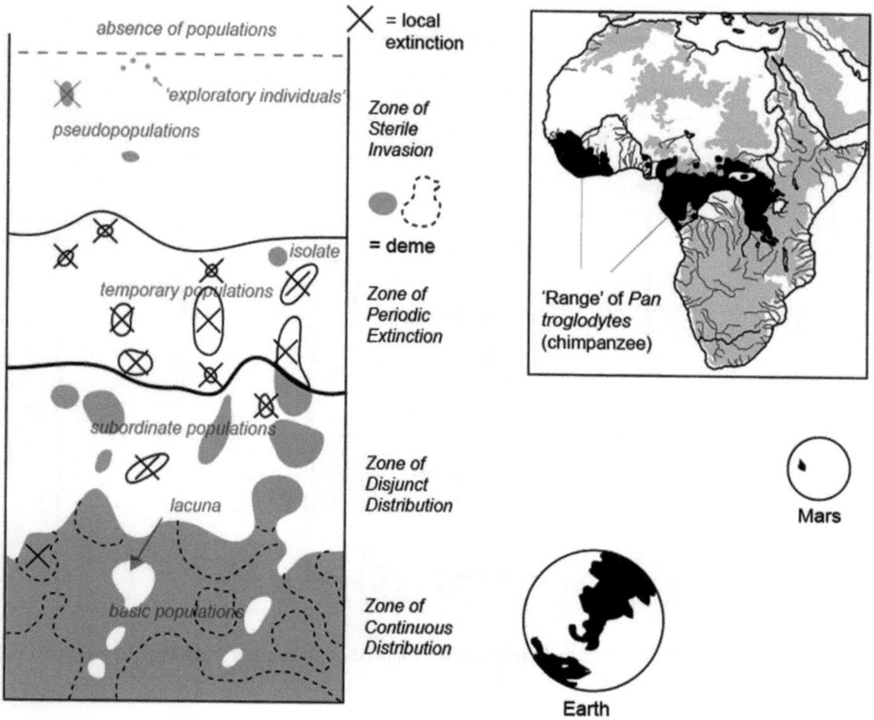

Fig. 5.3 Distributions of animal life across geographies. See text for important discussion and details. Figure and data by the Author, adapted from Smith (2018)

5.3 Stages of Human Emigration Beyond Earth

For our purposes of learning from these natural processes to make space settlement more likely to succeed, but tailoring our thought and concepts to account for distinctly-human purposiveness and adaptive capacities, I direct attention to Fig. 5.4, drawn directly from the field of animal dispersal ecology. Here, Travis et al. (2012) communicate the broad stages or phases of animal dispersal (*pre-emigration*, *emigration*, *transfer* and *settlement*), and indicate that each has specific time and energy costs (e.g. development of *means* of emigration, *information acquisition* at all stages, and *integration* [with resources at the emigration site] and *establishment*). The same periodization is evident in human emigration, which differs from that of other animal life in that it is consciously, proactively planned, as discussed below.

The stages and other variables outlined in Fig. 5.4 are useful as thinking device because they have evolved over time, in many varieties of animal life, as recurring modes of successful behavior in the emigration-based settlement of new habitats. A human conception of emigration to Mars may differ significantly, and to its detriment, if uninformed of such stages. For example it may be generally thought that an original, founding research institution will naturally evolve into a colony over time, but that

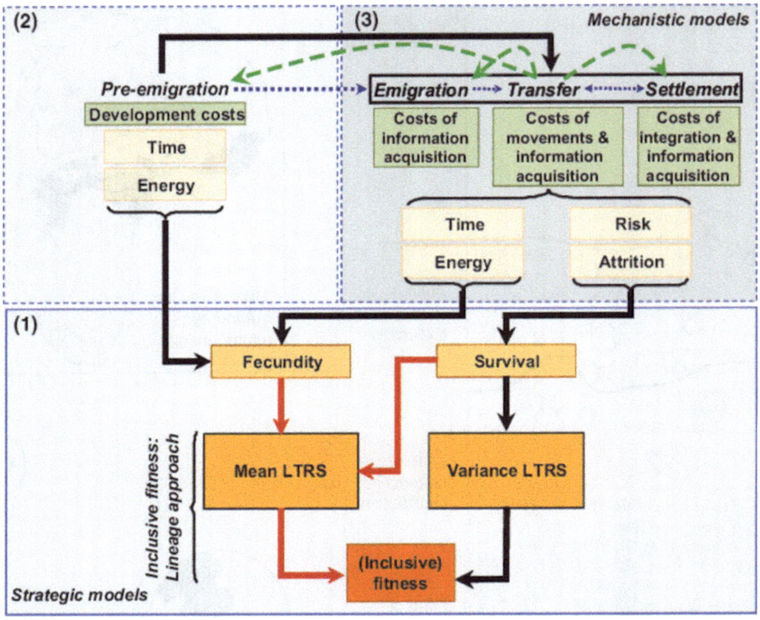

Fig. 5.4 Schematic model of animal emigration. Adapted by the author from Fig. 1 of Travis et al. (2012). See text for discussion

does not as specifically address the costs of integration and establishment of families and communities, which would be different from those of a research population; nor would the requirements of families and communities be as well attended to. Or we might over-emphasize the cost of transport, as it is the most exotic phase of the process, forgetting to invest in earlier or later costs. It may work, but my emphasis is to model our actions based on what has evolved already in natural populations and, where appropriate, at some scale of organization, emulate such evolved systems.

As a productive emulation of naturally-evolved systems of animal emigration as discussed, I propose four periods pertinent to space settlement: *preparation*, *travel*, *arrival* and *establishment*. As in the periodization of non-human animal dispersals mentioned above, each stage or period has several substages and implications for biology and or culture; these are summarized in Table 5.1 and discussed in the following sections of this chapter. Such stages and their subdivisions serve as chronological and activity guidelines for space settlement studies.

5.3.1 Stage 1: Preparation

The period of preparation in our case consists of the decision to settle space and then planning for such a project. We may currently be said to be early in the preparation

Table 5.1 Four stages of human emigration beyond Earth

Stage and provisional timescale	Substage	Biological implications	Cultural implications
Preparation (decades)	Study	Learn what is needed for multigenerational success	Learn what is needed for multigenerational success
	Planning	Establish a biological plan for multigenerational space settlement	Establish a cultural plan for multigenerational space settlement
	Preadaptation	Preparation of populations for space settlement	Preparation of cultures for space settlement
Travel (<12 months)	Movement	Sustenance and protection for duration of transit	Cultural accommodations for periods of *departure* and *arrival*
Arrival			
(first 24 months)	Settlement	First exposure of biology to new conditions	Introduction of emigrant culture to new conditions. Initial assimilation
	Adjustment	Short-term adjustments = acclimatization to new conditions	'Unpacking' and invention of new routines of life
Establishment			
(decades)	Proliferation	Establishment of families, reproduction and the life cycle; microevolutionary processes	Cultural maintenance of families, offspring and the life cycle
	Fixation of new variety	Heritable biological adaptations become widespread Biological divergence e.g. subspecies or species emergence; macroevolutionary processes and *cladogenesis*	Formation of new cultural identities accommodating new conditions Cultural divergence e.g. dialects and sects become new languages and cultures; *ethnogenesis*

Such periodization treats the prospect of human dispersal seriously in evolutionary terms, and each stage and substage offers a rich field of study to advance the project of human space settlement

stage, in the substage of *study*, with many worldwide expressing an interest in this concept. A review of current humans-in-space goals as publically stated is found below:

- NASA, ESA, JAXA, ISRO, CSA: *study of individual human health in non-Earth gravity field*
- SpaceX: *establish viable human settlements on Mars within decades*
- Blue Origin: *establish viable human settlements 'beyond Earth' within decades*
- National Space Society (USA): *settle space generally, with emphasis on orbital and Lagrangian-point settlements*

The focus on individual health (as opposed to the health of populations, a requirement of multigenerational space settlement) has been reasonable to present, as discussed in Chap. 2, and there seems no reason we would ever stop researching it; but for space settlement we must supplement this with the investigation of reproductive biology beyond Earth, including population genetics issues as discussed in Chap. 3.

An illustration of the current *study* period of space settlement is found in a NASA summary of the 2017 meeting of the 14-space-agency International Space Exploration Coordination Group in the January 2018 *Global Exploration Roadmap*. This document identifies the ICEG's collective goal of sending humans to 'the surface of Mars', though at this time only exploration, and not settlement, is considered. To plan for reaching Mars, the Global Exploration Roadmap contains the following list of *Critical Enabling Technologies* requiring development, an example of the delineation of prerequisites of the goal (CET):

- In-Space Cryogenic Acquisition and Propellant Storage
- Liquid Oxygen/Methane Cryogenic Propulsion
- Beyond-LEO Crew Autonomy
- Enhanced Reliability Life Support*
- Closed-Loop Life Support*
- In-Flight Environmental Monitoring*
- Life Support*
- Long-Duration Spaceflight Medical Care*
- Long-Duration Behavioral Health and Performance*
- Microgravity Counter-Measures*
- Deep Space Mission Human Factors and Habitability*
- Crew Health and Performance*
- Space Radiation Protection (GCR and SPE) Infrastructure and Support Systems High Data Rate (Forward and Return Links)*
- Adaptive, Internetworked Proximity Communications
- In-Space Timing and Navigation
- Low Temperature and Long-Life Batteries
- Comprehensive Dust Mitigation
- Low-Temperature Mechatronics
- ISRU: Mars In Situ Resources
- Fission Power (Surface Missions)

- Deep-Space Suit*
- Surface Suit (Moon and Mars)*
- Next Generation Surface Mobility
- Tele-robotic Control of Robotic Systems with Time Delay
- Robots working side-by-side w/crew

Note that in this list 12 of the items (marked with an *) involve human biology and/or culture and are thus potential subjects for anthropological consideration and preparation. Similarly, the ESA recently established *Thesesus*, a project to identify the 'research priorities for human exploration of space' with the following priorities: "*(i) to define reference scenarios for a European participation in human exploration and to estimate their influence on the life sciences and life support requirements; (ii) for selected mission scenarios, to critically assess the limiting factors for human health, wellbeing, and performance and to recommend relevant countermeasures; (iii) for selected mission scenarios, to critically assess the potential of advanced life support developments and to propose a European strategy including terrestrial applications; (iv) to critically assess the feasibility of existing facilities and technologies on ground and in space as testbeds in preparation for human exploratory missions and to develop a test plan for ground and space campaigns; (v) to develop a roadmap for a future European strategy towards human exploratory missions, including preparatory activities and terrestrial applications and benefits*" (Vernikos et al. 2016). The *Theseus* project early on identified the following fields of research to be particularly important within its domain: neurophysiology, immunology, cardiovascular, respiratory and renal health, and nutrition and metabolism.

While in April 2018 human sperm were sent to the ISS to investigate motility in microgravity, as mentioned, neither NASA nor the ESA (nor any federal space agency) has space settlement as a current serious focus, and are rather focusing on space exploration. This is probably as it should be, and the planning for space settlement can probably be carried out independently for some time, though some coordination in this field should be attempted to seek the benefits of collaboration and to reduce duplication in research efforts.

In Table 5.1 I specify that in this phase of study we should focus on what is required for multigenerational success in both biology and culture. These are largely the focus of this book, which is meant to sketch out and encourage a program of study informed by evolutionary and anthropological principles. May we sharpen focus here, rather than make these vague suggestions? I believe so. The list below indicates pertinent questions for study and may act as research guidelines, much as research guidelines in the larger field of evolutionary biology were outlined in Harvard University's 2009–2013 *Foundational Questions in Evolutionary Biology* programme, which included such questions as '*what are the differences between genetic and cultural evolution, and how can these differences be formalized?*' (FQEB 2014). Note that in the present book I have sketched out the preliminary answers to some of these questions, yet there is plenty of room for further work.

Questions for Researching Multigenerational Success in Human Biology and Culture

- What are the recurring features, if any, of successful human populations and cultures?
- What are the recurring features, if any, of unsuccessful human populations and cultures?
- What is the adaptive range of the human genome and phenotype, in particular regarding early development?
- What is the adaptive range of human behavior and culture?
- Can the adaptive range of human biology and culture be accommodated by existing or conceivable space settlement methods and technologies?

While I propose that some time should be spent in answering these questions and applying the answers to the project of space settlement, I advocate an action-oriented approach that does not become crippled by an ever-expanding research programme. Rather, I advocate the establishment of learning goals *embedded within a larger plan with a discrete timeline*; the goals should be achieved and these fields of study 'checked out' such that the rest of the programme may proceed. I note that subsequent to the *Apollo* programme, no such concrete direction has been adhered to on the generational timescale likely necessary for such a project.

Who should do this research and work? Absent any serious institutions dedicated to human space settlement, at present it will be carried out by researchers such as myself. The downside of this is that such research often is highly focused or biased by the specific interests and competencies of specific researchers, such that we have many detailed papers on the economies, legal structures, material cultures, population structures, and so on, of space settlements, all of which are interesting but none of which consider the larger picture, a task I am trying to address in the outlining of research questions and methods sketched out in this book. Until the large (read *wealthy*) institutions support such research, however, the space settlement concept could receive quite a boost from being at least minimally represented as a serious objective among, for example, federal agencies. Should and can the federal agencies at present shape the current solar system exploration program in a way to support the eventual goal of space settlement? If so, what steps could be taken to bests support this period of study in the stages of human space settlement? These are good questions that others will have to pursue.

The Stage 1 substage of *planning* should be the application of what has been learned in the study phase. This requires a rather detailed consideration of some space settlement destination and, if plans are to remain valid, a realistic consideration of the available technologies. At present, such detailed plans are not public, though I suspect some are being developed privately e.g. by entities such as SpaceX and Blue Origin. In any case, specific plans must be developed to apply the biocultural lessons directly to such plans. Currently it is impossible to say how specific such plans will have to be to productively make use of the lessons learned in the *study* phase, but it is likely that preliminary plans, e.g. a *reference study*, should be carried out even before the substage of *study*. For example, NASA's 1998 *Mars Reference Mission 3.0* (Drake 1998) outlines quanta for various system components, as shown in Fig. 5.5.

With modern computing it should be relatively easy to update such reference studies, and with rapidly-updating, multiagent system modelling, planning should in the forthcoming decades be radically improved. Currently my colleague Kai Staats, as part of Arizona State University's Interplanetary Initiative, is project lead for SIMOC, a "Scalable Model of an Isolated, Off-World Community" (Staats 2018). This computer-based model simulates an off-Earth community composed of many individual agents (each a system component with its own inputs, outputs and

Earth Return Vehicle	Reference Version 1.0	Final Version 3.0	
Habitat Element	51974 kg	26581 kg	
Life Support System	6000 kg	4661 kg	TransHab Study 7/97
Crew Accomm. + Consumables	22500 kg	12058 kg	TransHab Sizer from K. Kennedy
Health Care	2500 kg	0 kg	Included in Crew Acc. & Cons.
EVA equipment	4000 kg	243 kg	TransHab Sizer from K. Kennedy
Comm/info management	1500 kg	320 kg	TransHab Study 7/97
30 kw PVA power system	3474 kg	3249 kg	B. Cataldo 12/97 & TransHab Study 7/97
Thermal Control system	2000 kg	550 kg	TransHab Aerobrake Study 11/97
Structure	10000 kg	5500 kg	Scaled from Boeing 7.6 x 16.2 Mars Hab
Science equipment	900 kg	600 kg	12\97 Scrub
Spares	3500 kg	1924 kg	Rodriggs/Munk 12/97
SUBTOTAL	**56374 kg**	**29105 kg**	
Excess consumables	kg	-7392 kg	Jettison 560 days of food before TEI
Crew & Samples from Surface	500 kg	kg	Crew and Payload from surface
ERV MASS AT TEI	**63274 kg**	**27042 kg**	
TEI stage drymass	5200 kg	4806 kg	MSFC update 1/98
Propellant mass	52000 kg	28866 kg	MSFC update 1/98
Earth return RCS propellant	0 kg	1115 kg	MSFC update 1/98
Aerobrake	17300 kg	10180 kg	ARC update 1\98
TOTAL TEI MASS	**120474 kg**	**61829 kg**	
TOTAL MOI MASS	**131374 kg**	**74072 kg**	
NTR Propulsion System	28900 kg	23400 kg	
Shadow Shield	0 kg	0 kg	
TMI Propellant	86000 kg	50000 kg	
TOTAL INITIAL MASS	**246274 kg**	**147472 kg**	

Fig. 5.5 Selected quanta from Mars exploration reference studies 1.0 and 3.0. Planning details of Mars exploration missions depends on the production of such reference studies, as will planning space settlement. As discussed in the text, reference study revisions as seen here, from version 1.0 to version 3.0, will be best used when integrated into complex, multiagent simulations. Adapted by the author from Drake (1998)

interactions) who inhabit a settlement that has its own specific properties such as breathing gas pressure and composition, temperature, solar and GCR radiation flux and so on. At each computing cycle or desired time increment the status of thousands of variable states are updated, based on the properties of the entire system. The simulation is highly scalable and easily modified with regards to system variable rules (e.g. does a kilogram of rice yield *X* or *Y* calories? Such figures may be updated in the computer code to continually refine the model). The simulation may be run in parallel on many machines, in the manner of the distributed-computing project *SETI at Home*, radically increasing the number of simulations and quantifiable data. The goal of implementing such a model is (Staats 2018):

> to design a habitat that sustains human life through a combination of physio-chemical (machine) and bioregenerative (plant) systems, and then scales over time, with SIMOC Phase IV-V including options to grow the community with the addition of inhabitants and infrastructure…[based on]… an agent-based model (ABM), a class of computational models for simulating the actions and interactions of autonomous agents (both individual or collective entities such as organizations or groups) with a view to assessing their effects on the system as a whole.

Figure 5.6 illustrates a small section of the SIMOC model indicating just a few of the relationships between the model's many agents.

Recently the HERITAGE program has been devised by Frederic Marin of the University of Strasbourg, using Monte Carlo methods to derive general system effects from the specific instances of many system simulations (this program was also discussed in Chap. 3). The HERITAGE program was devised to simulate "a large number of free parameters such as life expectancy, age range allowed for procreation, percentage of infertility, unpredictable accidents, etc. to be investigated proactively in order to ensure a viable [interstellar voyage to an exoplanet]:" (Marin 2017). Though Marin has some suggestions that differ from my own, I am sure his approach will also be of value in space settlement planning. Figure 3.15 (Chap. 3) illustrated Marin's HERITAGE computer model and some of its variables. HERITAGE is currently being used to also model a closed ecosystem and its state over time, for example calculating the agricultural productivity required to sustain 500- or 1000-person 'villages' beyond Earth (Marin & Beluffi 2018, Marin et al. 2018). Such work will be fascinating to compare with Staats' SIMOC model, which models these systems as well and is being developed in tandem with empirical studies of closed ecosystems, to 'ground truth' the model (Fig. 5.7).

Earlier-generation space-mission and -settlement planning has used rather static reference studies that were laborious to update and were not integral, so that simulations or the effects of one variable on another would have to be (again, laboriously) computed individually, resulting in a fragmentation of efforts and interests in such simulation. As mentioned earlier in this section, the 'study' substage of the Preparation stage of human space emigration will have to be studied for multigenerational health in some certain set of circumstances. It seems likely that SIMOC and models that will derive from it will furnish us with far more realistic, rich, subtle and informative models and quanta than we have had to present, and will advance the goal of outlining what we need to study at present to best ensure the success of

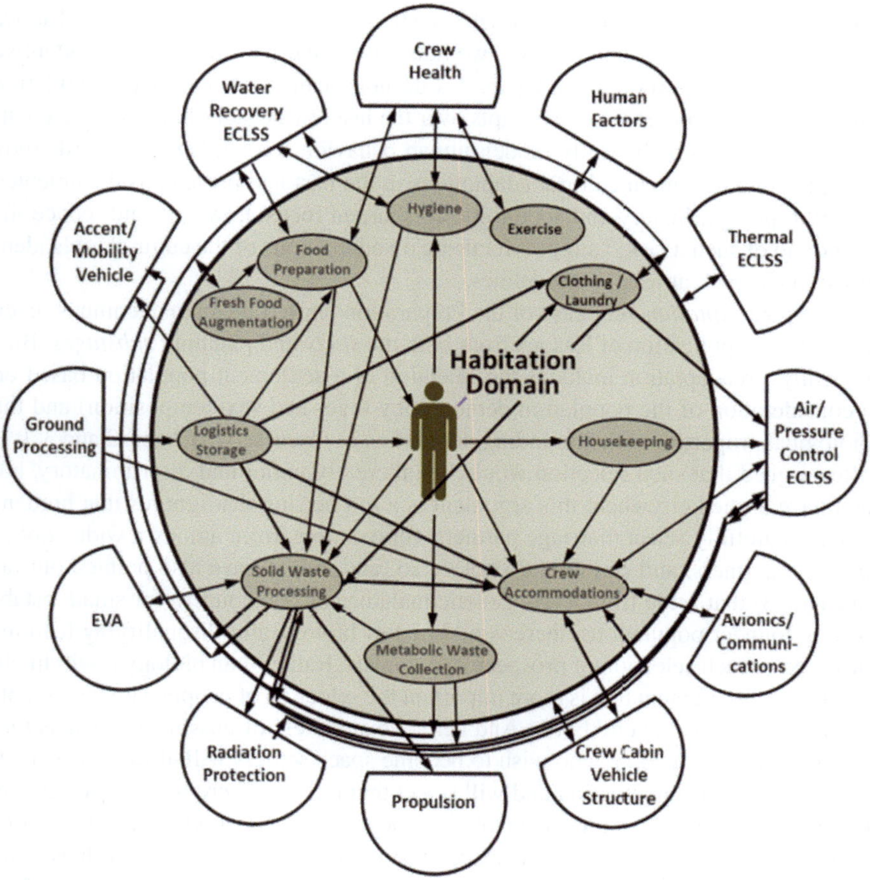

Fig. 5.6 Selected model agents and interactions in SIMOC. Planning for permanent space settlements will be radically improved by using dynamic models of variables, interactions and both expected and emergent properties. Image courtesy of Kai Staats, M.Sc.

Fig. 5.7 Barley being prepared for closed-ecosystem agriculture as part of Arizona State University's 'Interplanetary Initiative'. The SIMOC program mentioned in the paper will be ground-truthed and improved by comparison with empirical studies. Image courtesy of Kai Staats, M.Sc.

the goal of permanent human space settlement. Note that the models described above focus on consumables and biology, we must study and plan with cultural adaptive tools as well, as introduced in Chap. 4. I do not mean to use computer simulation of cultural dynamics, as such attempts over the last few decades have been fascinating but unable to realistically model human behavior. Rather, here I suggest study and planning of human cultural adaptations in the past and present, as documented in anthropology, history and sociology. To sharpen focus, however, and reduce the workload of such a task, I suggest focusing on adaptations of the cultural tools identified in Chap. 4, at least as guidelines.

The *preadaptation* subphase of the Preparation phase I mean to encompass even more direct application of lessons from both the *study* and planning *substages*. Biologically, preadaptation includes the selection of a settlement population based on a consideration of the population demography (age- and sex-composition) and the biological properties of that founding population as discussed in Chaps. 2 and 3. It is often argued that such selection would be inherently unfair and discriminatory, but as I have argued elsewhere this argument is a red herring that ignores that humans already carefully select marriage partners, for example, from among a wide pool of possible partners, and certainly do not do so randomly. I have also pointed out (as in Chap. 3) that aside from a few genetic maladies of real concern for small, establishing founder populations, there would be few biologically-disqualifying features involved in such selection of prospective colonists. Rather than biology it will likely be group temperament that is more important the selection of people who may set off as the first settlers beyond Earth. And here, we are likely discussing a *self-selecting group*, that is, the people who wish to become space settlers will always be a small group of the Earth population and will select themselves. There is no reason in any rationally-foreseeable future to invoke a draconian, Hollywood-style selection of individuals as fit or unfit to settle space. The reality I think will be that those who wish to go will volunteer to do so, and will always be rather small in number when the real risks of such a venture are realistically appreciated.

Cultural preadaptation would include establishing reasonable expectations for prospective beyond-Earth settlers. The pioneer of innovation research, H. G. Barnett, has noted that innovations occur in individual minds and then are either accepted or rejected by populations of minds following communication; naturally many variables make the specific outcomes of such generalizations very difficult to predict. However, generally speaking, Barnett suggests that innovations or new conceptions (such as conceptions of settlement beyond Earth) are most well-received when they address a discontent and are easily-understood. Furthermore, innovations are more likely to be accepted under the more dynamic social conditions outlined in Table 5.2, than the fixed. Communicating the conception of space settlement should be done more skillfully than in the past, if it is expected that wide social backing for the project is a requirement (currently some indications suggest that private entities will be first to attempt space settlement, rather than it being a federal project). Table 5.2 may assist in better communication of the space settlement concept.

Table 5.2 Social atmospheres of fixity versus dynamism

Fixity	Dynamism
Emphasis on structure	Emphasis on dynamics
Fixed conceptions	Investigation of concepts
Content with establishment	Discontent with establishment
Order	Disorder

Innovative concepts, such as those relating to settling beyond Earth, are more likely to be accepted in conditions of social dynamism than fixity

Adapted by the author from the comprehensive discussion in Barnett (1953)

5.3.2 Stage 2: Movement

This stage is of obvious importance as it is the means by which the beyond-Earth habitats themselves will be reached by human populations. However, aside from a few concerns discussed below, I find it largely the domain of the engineering needed to reach those destinations in good health and in this regard we must assume that such technologies be developed.

Having said all this we should remember that the project of permanent human settlement of space is a *human* project and that its function is to support humanity; the tools and mechanisms for space settlement will be just that, tools, and they should be designed to support humanity from the outset. Spacecraft designs have increasingly taken on such considerations, as we see in Fig. 5.8, contrasting vehicles of space *exploration*, *transport* and *habitation*. The Spartan, utilitarian features of exploration vehicles are well-known to explorers, but environments for living out entire lives are likely to be differently designed. While not a feature of this book, such design will be a lively and important field for creative expression. Some anthropological input to this field may be found in the anthropological subdiscipline of *proxemics*, the cross-cultural study of the use of space e.g. 'body space' and its associated fields, but these are not a large concern of this book.

Important to the cultural preparation for the movement phase of space settlement will be a *demystification* of the period of travel. Currently the voyage to Mars, for example, is highly exoticised; it will involve some months (perhaps three to six), conditions of increased risk of radiation exposure, relatively tight quarters and, sometimes, it is said, a lack of things to do. Some have even proposed that this voyage would be so unusual in the human experience that it should be carried out in hibernation (Bradford et al. 2014). I find this proposition unnecessarily complicating, largely because such voyages may be divided into realistic periods of *departure*, *cruise*, and *arrival*. During departure, which could consume a month or so, travelers will be able to anticipate and mark special occasions including (a) the departure from the Earth's gravitational well (on the first day), (b) the passing beyond the moon into truly interplanetary space (in the first few days), (c) passage beyond the Van Allen belt, and others including (d) the last day that Earth's continents are distinguishable by the

Fig. 5.8 Design considerations in human space exploration and settlement. Space exploration (upper left, astronaut in an orbital capsule) and short-term habitation (upper right, International Space Station) architectures emphasize life-supporting and scientific hardware. As a shift is made towards permanent space settlement, the design of vessels for transportation over long periods (e.g. bottom left, several months to Mars; courtesy of SpaceX) and habitats for living out entire lives (bottom right, biosphere analogue of the University of Arizona) will increasingly be built out from and to accommodate human bodies, families, communities and cultures on a multigenerational timescale

naked eye. In the cruise period of another month or more, many special activities of this once-in-a lifetime voyage may be carried out, including exotic-g field sporting events, marriages, ceremonies marking the solidarity of settlement subcultures and the settlement culture as a whole; time may also be spent in recreational astronomy, which will be superior in the space between Earth and Mars. And one can hardly imagine the arrival period (perhaps the last 30 days) not being marked at such events as (a) Mars' surface features becoming visible to the naked eye and (b) the glide into Mars' gravity well, as well as a host of preparations for the arrival and landing. Whatever the specifics of the occasions to be marked and activities carried out in the voyage to Mars (or other relatively nearby, reasonably-expectable habitats of the next 150 years), it seems excessive to consider the 3-6 month travel period intolerable. One may just as easily propose that it will be a period of great significance in the lives of the settlers, something to be enjoyed rather than slept through.

5.3.3 *Stage 3: Arrival*

This stage regards the actual arrival of early settlers at some environment beyond Earth. The substage of *settlement* is biologically distinctive as it refers to the first exposure of the body to non-Earth conditions of gravitation, breathing gas composition and pressure, and so on. Whatever plans have established whatever plans and preadaptations, the first such exposure will be unique and should be carefully attended. After some time, as the new habitat's unexpected properties are learned and adjusted to, the first exposure of arriving immigrants will likely be made decreasingly perturbing to the body. Still, it is unlikely that conditions anywhere beyond Earth will ever be precisely as they are on Earth, and in fact some months will likely have been spent in the journey from Earth to another habitat, in yet other conditions. In any case here we may recognize the transition from one set of bodily circumstances to another.

The immediate biological reactions to such changes include alterations in heart rate, thermoregulation, breathing patterns, metabolism, and so on, all under the short-term phenomenon of *acclimatization*, defined as "biological changes occurring within the lifetime of an organism that reduce the strain caused by stressful changes in the [organism's environment]" (Frisancho 1993: 485). This a phenomenon of bodily *homeostasis* involving changes that are generally *reversible* and are not *heritable*, that is they are expressions of a range of physiological responses to perturbance innate to the individual. It is a common misconception that such acclimatisations are subsequently passed to the offspring, but despite revelations in the field of *epigenetics* (rare, heritable changes in gene expression), what happens to the adult is rarely transferred directly to the offspring (if it were, blacksmith's offspring should be born with giant muscles, which is not the case). Rather what is inherited are the genetic programs for a range of such physiological responses, such that if there is selection for a certain range of physiological response, that range will become more common in the gene pool. But here we consider only acclimatization of the individual. It is safe to assume that space settlers will know enough about the various habitats they will experience in transit to a new settlement and at that settlement that planning for such acclimatization will not be an undue challenge. A general rule of thumb will be to minimize acclimatization requirements as is practical.

Culturally, immersion in a new culture can require somewhat longer for some variety of *assimilation*, a behavioral and psychological process we may consider a cultural analogue of acclimatization as defined above (Lazear 1998, 1999). Note that cultural assimilation is entirely behavioral, not genetic, and is therefore amenable to immediate transfer to peers (members of the same generation) and offspring (teaching to the next generation). Cultural adjustment, then, can be very rapid and sharply tailored to specific environmental features (consider both physical and cultural environments) and thus is a powerful adaptive tool. Still, some assimilation, such as the formation and internalization of a new identity to the extent that the intaking culture accepts the immigrant culture, can take some time, perhaps several years. The phenomenon of assimilation is well-known in cultural anthropology and the research literature in this field will be rich in useful lessons for space settlement planners concerned with cultural adjustment of immigrants to space settlements.

After settlement I propose a substage of *adjustment* which, biologically, is a continuation of acclimatization, and not very distinct here, but culturally is a period of deeper engagement with the new environment. Specifically here I refer to a period, on the order of up to a year or so in duration, of *unpacking* and the invention of *new routines of life*. By unpacking I mean both the transfer of material culture and concepts brought from Earth to the new habitat, and some cultural disengagement from Earth and to the new habitat. Cultural disengagement from the originating region includes such phenomena as ceasing to use certain words or phrases, or an entire language, to ease communication with people of the intaking culture; this is particularly common among children of second-generation colonists, who must rapidly adapt to the severe social environments of new schoolmates (Fillmore 1991). Establishing new routines of daily life is also important to successful assimilation, making behavior more predictable and thus integration with an intaking culture more palatable. The importance of routines and rituals in daily life is well-communicated in a 2002 review (Feise et al. 2002: 382):

> Routines involve a momentary time commitment and once the act is completed, there is little, if any, afterthought. Routines are repeated over time and recognized by continuity in behavior. Rituals, on the other hand, involve symbolic communication and convey "this is who we are" as a group. There is an affective commitment that leaves the individual feeling that the activity has a felt rightness and provides a sense of belonging. Furthermore, there is often an emotional residue where once the act is completed, the individual may replay it in memory to recapture some of the affective experience. Rituals also provide continuity in meaning across generations with the anticipation for repeat performance and an investment that 'this is how our family will continue to be.' When routines are disrupted, it is a hassle. When rituals are disrupted, there is a threat to group cohesion.

Again, anthropology has a rich research literature on ritual and routine, and this may well yield significant information for space settlement planners interested in this specific aspect of space settlement; how can the establishment of satisfying and effective routines be fostered, rather than hindered, in the design of beyond-Earth space settlements and habitats? Without prescribing or predicting, we may assist the overall effort by point out just what routines are. Important work on the concept of cultural routines was laid out by Pierre Bordieu, who defined the *habitus* as "A system of lasting, transposable dispositions which, integrating past experiences, functions at every moment as a matrix of perceptions, appreciations, and actions and makes possible the achievement of infinitely diversified tasks, thanks to analogical transfers of schemes permitting the solution of similarly shaped problems" (Bordieu 1977: 95).

The establishment of routines will involve both the construction of and be influenced by an overarching *worldview*, defined (as introduced in Chap. 4) by Liane Gabora as "a mind as experienced subjectively, from the inside; it is a way of seeing the world and being in the world that emerges as a result of the structure of ones' web of understandings, beliefs, and attitudes. A worldview reveals itself through behavioural regularities in how it expresses itself and responds to different situations" (Gabora 2018: 1–2). The combination of a crystallizing worldview and the establishment of novel routines, specific to the new habitats beyond Earth, will be central to success of the space settlers. Routines and a realistic, shared worldview are not trivial: regularities in behavior and one's environment provide a stable, referable structure to experience required for productive activity, communication and social life.

The period of arrival will likely require some rather rapid cultural accommodations. Here I mean accommodation to *unexpected* features of the extraterrestrial habitat, not features for which the habitats and their associated technologies have been designed and built by a *system* of emigration discussed in Sect. 5.4. Adaptation or accommodation to rapid changes has been common in human evolution, and of all species our proactivity and capacity for problem-driven invention give us significant survival skills. Table 5.3, for example, outlines indigenous North American behavioral adjustments to rapid (one-generation or less) climatic changes.

We may expect that such change will occasionally be necessary as aspects of the habitats we encounter and/or transform beyond Earth change from what we initially expect.

Table 5.3 Rapid changes to native North American Arctic environments and accommodating cultural adaptations

Environmental change	Cultural accommodation	Speed of cultural accommodation	Effectiveness of cultural accommodation
Thinning sea ice	Shift from ice-surface hunting to open-water hunting	Rapid behavioral change, but will take **a generation or more** to devise, test, evaluate and teach new methods	To be determined
Shortening Winter season, though more tundra snow cover in Winter	Increase use of male caribou to break ice crust so herd can access grazing below	Rapid behavioral change, but will take **a generation or more** to devise, test, evaluate and teach new methods	To be determined
Thawing permafrost including coastal erosion	Shoreline stabilization engineering; settlement relocation	Rapid; engineering and relocation methods simple and well-known	Effective, but not culturally preferred
Shorter winter river surface freezing period	Intensify Winter fishing period with improved infrastructure	Moderately slow: new infrastructure slow and expensive to build	Likely to be effective if observed freezing period remains through period of infrastructure development

For raw data (columns one and two) see Hovelsrud et al. (2011)

5.3.4 Stage 4: Establishment

This is a period of human space settlement I mean to delineate as the period of long-term success. To this point, rather shorter-tem biological and cultural adjustments to new conditions have been successful. However for multigenerational success these adjustments must be sustainable and tolerable.

The earliest substage, *proliferation*, indicates a rapidly-increasing population. All evidence of population genetics and demography instruct us that–particularly when far from the zone of continuous distribution (see Fig. 5.3) and the resources provided there—small populations are particularly vulnerable to both biological and physical calamities that could upset a breeding population structure in such as way as to be unrecoverable (see Chap. 3). Thus establishing a rapidly-expanding population, at least at first, is likely to be a good safety strategy for a space settlement. This occurs on the timescale of decades. To achieve proliferation will require the maintenance, both biological and cultural, of such entities as families and the long developmental life cycle leading from conception to child-rearing from one generation to the next. It is on this multigenerational timescale that natural selection will undoubtedly shape some generations of beyond-Earth settlers. That all individual offspring will have an identical likelihood of survival (and having their own offspring) even in such carefully-maintained environments as beyond-Earth space settlements is hardly credible. Naturally certain developmental phenomena will be unsuitable in these new conditions, others less suitable, and the resulting preadult mortality due to these fitness differences, which I think is inevitable, will be (if not technologically mitigated) textbook *natural selection*. But this should not persist for long, and within a few generations—I suspect a century, or about three, 30-year generations—the most unsuitable developmental genomes will have been selected against to extinction. This is no more a program of *eugenics*, as some might argue, than was allowed by countless of our ancestors in their adjustments to new global environments worldwide in the last 100,000 years. And it does not mean that people born with such nonlethal developmental deficiencies (read 'deficient in the new environment') would be unable to live out full lives. However space settlers wish to deal with the likely unavoidable phenomenon of higher-than-current (in the Western world) infant and pre-adult mortality, for about three generations, will be up to them. I do not think a culture wishing to settle space will be deterred by such challenges.

Such multigenerational change to the gene pools of viable off-Earth populations will lead to discrete, habitat-specific and heritable biological, genetic adaptations to humanity's new worlds. What these adaptations will be is naturally of great interest to humans speculation on the future, and perhaps the future of their own children, so that in fiction and science fiction we have seen all manner of projections; tall, lithe Martians, orange-skinned Venutians and so on. However the outward physical characteristics of these people will not be of the greatest biological concern for continued success any more than superficial variation is a concern of biology textbooks (these are not a concern in such books except as examples of racial thinking or slight, *clinal variation*). What will matter, from the perspective of multigenerational success, will be that the adaptations make for good health across generations. If so, they will become widespread and thus will proceed macroevolutionary processes of large-scale differentiation, or the *fixation of new variety*. This 'final' stage of space settlement is no more final than any process in nature, and we see that nature does not rest but is forever in flux. On the timescale of the human project of establishing successful, multigenerational, Earth-independent populations, however, it will be in this period of macroevolutionary differentiation from Earth populations by well-adapted Earth descendants that success may be considered achieved, at least for a given moment. Whether such change results in *subspecies* differentiation e.g. *Homo sapiens* ____, somewhat different than but still able to breed with Earth-based *H. sapiens sapiens*, or *speciation*, e.g. *H. cosmicus*, similar to but unable to breed productively with Earth-based *H. sapiens sapiens* is not really relevant here. The point is that successful long-term adaptation will not be considered complete (for a given set of circumstances) until such adaptations are genetically-transmitted and essentially *fixed* or universal in a population.

Finally, it seems that a cultural divergence from Earth-based cultures will also occur on the decadal timescale, as language, values, symbols and philosophies and religions are all altered, and some invented and some cast aside, in humanity's new worlds. How long such divergence may take may be studied in the anthropological field of *ethnogenesis*, the emergence of new cultural identities. While cultures are clearly real—Australians find it difficult to adjust to some aspects of American culture—they can be hard to delineate; in this case, Australians and Americans each speak English. Cultural identity formation is similarly fascinating but difficult (perhaps impossible) to quantify or generalize, as suggested in the enumeration of factors that shape a culture, "...ecology and economy...language...politics, and history," (Hornborg 2005: 593).

Still, it is possible to identify broadly the time required for unique ethnic identities to form. They do typically do not occur overnight, but on the scale of a generation or more as the core identity of a given generation is formed, largely during adolescence and early adulthood (Arnett 2000). Part of this identity-formation is facilitated by a variety of socialization processes whose content is difficult if not impossible to predict, however the timescale of formalized socialization and enculturation, as it plays out in educational institutions, has been examined; here we also find that cultural divergence can be expected on the timescale of some decades, the time required for "changes of curricula and new teaching strategies" (Riley 2007) (Fig. 5.9).

Just what kind of cultural change will occur is perhaps naturally an object of curiosity—what will our descendants be like? But as a topic for an anthropology of space settlement it is unsuitable for at least two reasons. First, we find that culture change in the past and present, which is well-documented and for which we can observe many contributing factors, remains difficult if not impossible to predict accurately. Until we can do this, anthropologist should not be predicting all but the broadest aspects of future culture change; namely that it will change, and in such domains as the Cultural Universals introduced elsewhere in this book. Second, as a technical point, it is not of particular interest to know what specific cultural features will arise in the future, just as evolutionary biologists spend little time contemplating what specific anatomical or instinctual-behavioral features will emerge at some

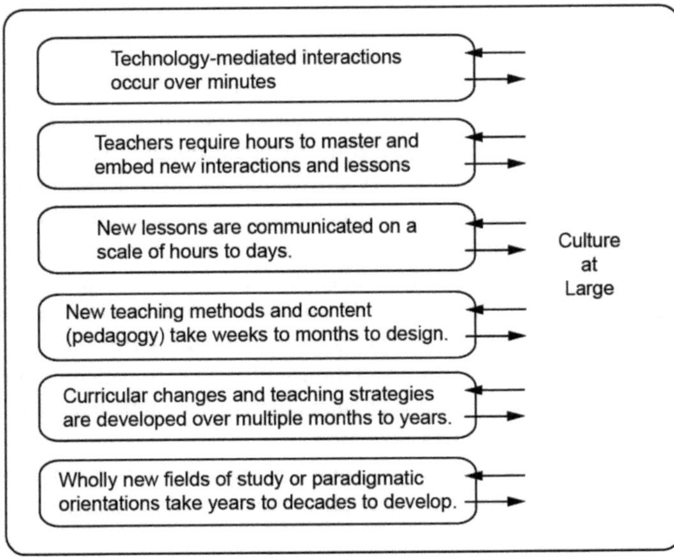

Fig. 5.9 A timescale for formalized educational change. The diagram, adapted by the author from Fig. 1 in Riley (2007), generalizes on the time required for educational institutional change. Arrows indicate interaction with members of the culture at large by members of those instituting and implementing new concepts and knowledge for transfer to the next generation. While culture can change quickly and unpredictably, we may expect the timescale of more gradual ethnogenesis in communities beyond Earth—also known as cultural divergence—to be similar

time in the future of a given species (if such detailed speculation were common and useful in biology it would appear in undergraduate texts, but it does not). Aside from assessing the adaptive capacities of endangered species, and identification of some essential long-term trends in the evolutionary process (e.g. a tendency towards increased diversity and complexity (McShea and Brandon 2010), such detailed projection does not feature in biology at large and is currently better left to science fiction (which I do deem to have legitimate value to space settlement) than science.

5.4 Developing an Evolutionary Human Space Emigration System

Migration biologists have developed a conceptual model of a generic animal 'migration system' based on a wide survey of animal movement worldwide. Because humanity has no existing model, based on any of the life or evolutionary sciences of animal movement over large time- and space scales, it may well be of use at least as a heuristic tool or thinking device for space settlement planners, to help structure their thinking on the matter. After introducing it in the section below I will describe, in a preliminary way, how it can be used to shape though about human space settlement.

This model is based on the concept of a migration syndrome, a special kind of a behavioral syndrome, defined as "a suite of correlated behaviors reflecting between-individual consistency in behavior across multiple (two or more) situations. A population or species can exhibit a behavioral syndrome" (Sih et al. 2004). Specifically, a 'migration syndrome' comprises "morphological, biochemical, physiological and behavioural traits that implement migration, and demographic traits that determine the migrants' schedules of reproduction and mortality" (Drake et al. 1995: 437).

For a great many migratory animal species on Earth, Fig. 5.10 capably expresses the pertinent variables and their interactions, but it can be challenging to interpret and requires some explanation and contemplation. The circular nodes in the diagram represent the *components* of the migration system, and the arrows the principal *processes* by which the components affect one another, shaping the system. The 'migration arena' refers to the sum of environmental properties of the migratory region, let us say between Earth and Mars, including interplanetary space. Such properties include both the abiotic (e.g. atmospheric gas composition and pressure) and biotic (e.g. gut bacteria, plants, other animals) environment. The 'population trajectory' describes the space- and time-components of a population's movement, for example the location in the arena of a given individual at a given stage of biological and or cultural development; it may be thought of in terms of a probabilistic map of the location in space (the migration arena) of a given member of the system at any time. The 'genetic complex' comprises the genetic underpinning of the system, including genes specific to biology and inherited or instinctual behaviors that facilitate animal movement. The 'migration syndrome' itself is the set of behaviors and biological features that facilitate migratory movement. Time, here, is indicated clockwise.

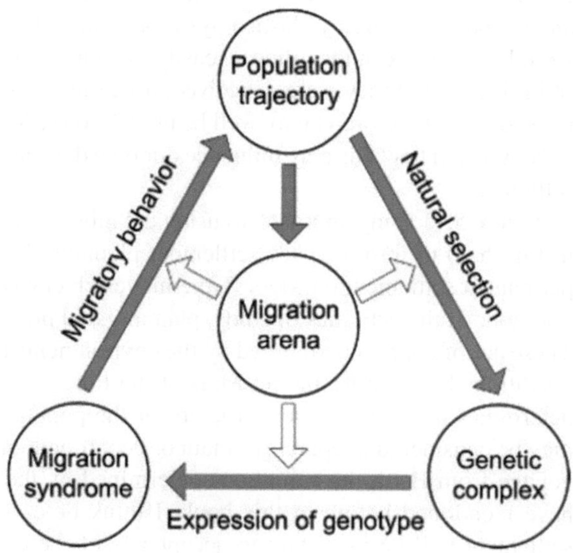

Fig. 5.10 An holistic model of animal migration. See text for important discussion and details. Adapted by the Author from Drake and Gatehouse (1995), Fig. 21.2

How can we use the conceptual model of a migration system to productively guide our thinking about human space settlement? I think two useful comments may be made at this time.

First, the model, which was developed to model insect migration specifically, has been influential since its introduction in the 1990s and has been extended to model migration in birds and other animals (Dingle and Drake 2007). It must therefore be a reasonably accurate representation of the organization of migration and its pertinent variables in many forms of Earth life, and therefore a reasonable candidate for consideration by the appropriative methods of biomimicry.

Currently a six-step process is advocated by the Biomimicry Institute to guide its students "which describes the six most important steps a design team should take when seeking biomimetic solutions to a design challenge" (BIOMIM 2018). Stage two of this process, *Biologize Function and Context* includes a literature review to identify how a certain process is carried out in Nature. In our case, the question would be 'how do Earth life forms migrate and emigrate, expand their range and disperse?' and a good part of the answer would be found in the descriptive model introduced above. We cannot yet be sure how the model will usefully or specifically guide thinking about human space settlement, but it is worth evaluating in these terms. As with any application of biomimicry, a spectrum of applicability, from very specific and useful to over-generalized and not of use, are likely to be found in such a research effort.

Second, the model has been useful in identifying the evolved features of long-term migratory success in multiple varieties of life; in birds, for example, the migratory syndrome includes the migration-specific features such as *precise timing of seasonal events* (e.g. premigratory fattening), *navigation* (e.g. stellar and solar navigation), *endurance performance* (e.g. reduction of digestive tract size), *contrasting environment coping* (e.g. highly plastic physiology), and others (Piersma et al. 2005). These would not have been identified so easily as a migratory complex evident in hundreds of bird species if they were conceived, for example, simply as independent traits of the species. Therefore, it may well be useful to consider the integration of such traits in Nature and how this may influence our own designs and processes in human space settlement.

Figure 5.11 is my approach to using the migration syndrome framework outlined in Fig. 5.10 to inform space settlement planning, here considering planning for a permanent settlement on Mars. Time moves clockwise, staring upper left. Here we find node 1, the activities of study, planning and preadaptation, as described above. These are of course conditioned by the environments of Mars and the interplanetary conditions between Earth and Mars at the time of travel. Planning of all varieties determines the age- and sex-structure of the populations to be transported, which may be considered a separate domain or department of planning. Travel is the actual voyage from Earth, its conditions determined by the interplanetary medium. As I have mentioned before in this book, I think de-exoticizing this portion of space settlement will be important to acceptance of the concept, and to success. Travel results in Arrival, which is immediately engaged in the next domain or department of protection of the biological lifecourse, through technological mitigation in the

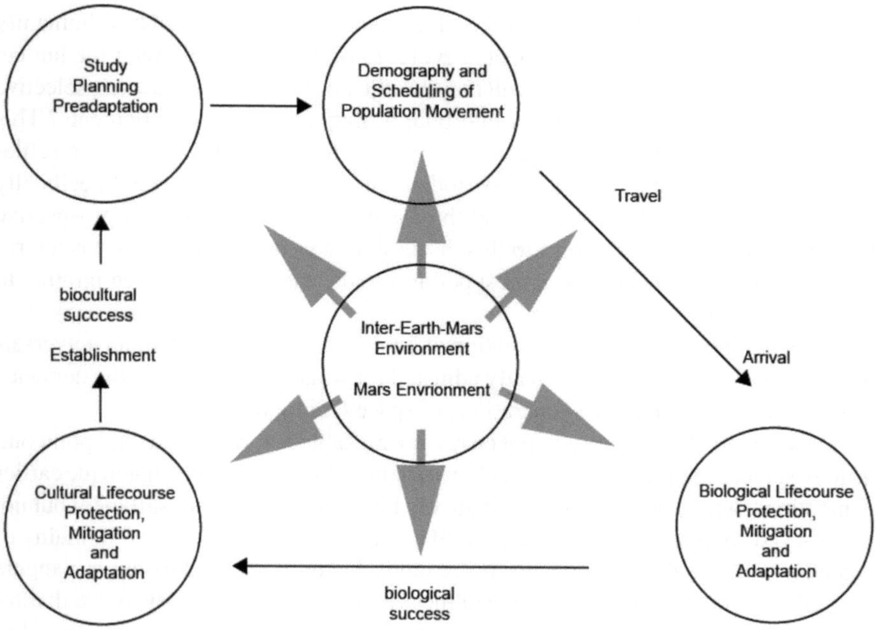

Fig. 5.11 A model of human space emigration. Emigration is one-way, as opposed to e.g. seasonal migration, which normally indicates a return to a given landscape. The diagram is informed by the framework of Fig. 5.8. See text for discussion

immediate term, physiological acclimatization in the shorter term and of course biological adaptation in the longer term. If this is biologically successful, cultural lifecourse success may follow, allowing behaviors, including socially-critical acts such as ceremony and the establishment of daily routines, to be developed. If this is successful, the population may be considered established, and its experiences are of course fed back into the preparation phases of planning for more arrivals from Earth. All aspects of the model are of course conditioned by the Inter-Earth-Mars and Mars environments, demanding high-fidelity understanding of those environments and technologies required to support human life among them.

5.5 The Emigration Experience and Human Universals

As discussed in Chap. 4, human universals address a variety of human requirements globally and through time; I argue that their variable manifestations are at least partially adaptations to local (in time and space) cultural and material selective pressures and that they thus constitute a minimal 'cultural adaptive toolkit' for humanity in the same way that the widely-conserved (in time and space) *homeobox genes* (AKA 'Hox genes') serve as the essential 'genetic adaptive tooklit' for multicellular animal life. Proceeding from this argument it is clear that to examine in detail how humanity will adjust culturally to environments beyond Earth we may begin with the human universals as adaptive tools. How will they be adjusted to accommodate new selective pressures on Mars, for example, or in orbital or Lagrangian-point settlements? This section tackles this critically-organizing approach to details of human space settlement, specifically with regard to the periodization discussed in Sect. 5.3. Specifically I wish to address the first, second and third generations of settlers in a prospective Mars settlement. We will return to this discussion later in this book, but it is appropriate to address it firs here, as it most pertains to the earliest stages of emigrating to environments beyond Earth.

We may proceed with an examination of Table 5.4, in which the human universals introduced in Chap. 4 are specifically adjusted for what I expect will be the demands of the first, second and third generations of space emigrants.

Three general lessons are evident on reviewing Table 5.4, which, I would point out, is more science than science (or speculative) fiction. **First**, it is likely that in the earlier generations, where populations are relatively low and resources are sufficient but not yet stocked to abundance, a variety of adaptations to human universal domains of behavior are likely to be required to ensure safety. **Second**, after some time, as dangers are buffered out, it is likely that some of these adaptive means could be relaxed; after this time it is evident that 'what will happen' is often times unpredictable, as it has been in Earth history. Though there are some longer-term expectable changes, the details of daily life as manifest in a myriad domains such as the clothing that one wears or the symbols cherished, will change unpredictably. And I would add, that such details are not our business, but will be the business of the descendants of the Founder Demes. **Third**, a consideration of the timescales involved suggests that by the second generation, many novel behavioral features will be underway or fixed (widespread and established), each subtly shaping the culture of the settlers. We see the general character of such change, which is not radial but in sum is substantial. We see that children born into the second and then third generations will not be specially aware of being settlers, but will simply be people living out their lives.

We may well ask, then, why try to predict specifics about how the second- and third-generation descendants of the Founder Demes will change culturally? I propose that, indeed, we do not try to predict or prescribe too specifically in these instances. As mentioned from the start, my objective is to give the project of human space settlement the best chance of success by supplying its planners with the lessons of evolutionary adaptation at large, and human evolutionary adaptation specifically. There is one reason, however, to speculate in an informed way on the succeeding 2nd and 3rd generations, and it is to give us a glimpse of what we may reasonably expect to occur, or the magnitude of what we may expect to occur. The expectation not of specifics, but at least of the scale of change that is likely to occur, throws a little light into the project of space settlement, importantly to make it more real in the mind, the author of action. For these reasons, Table 5.4 has its uses. In thinking about a second generation of immigrants to Mars or some other space settlement we may consider a recent enumeration of the chief difficulties faced by immigrants to Canada, listed in order of severity: *language barrier*, *employment*, *housing*, *access*

Table 5.4 Time and cultural adaptation by first- to third-generation space emigrants

Human universal domain	First generation Years 0–30 'Founder Deme' Corresponds to population *travel* and *arrival* as discussed in the text	Second generation Years 31–60 'Fixing Deme' Corresponds to population *establishment* as discussed in the text	Third generation Years 61–90 'Adapted Generation' Corresponds to biocultural *adaptation* as discussed in the text
Language	New vocabulary items Innovation of new communication patterns (vocal and gestural) based on effects of new gravity field, chemical composition of breathing gas and habitat pressure	New phrases and accents Fixation of gestural communication based on effects of new gravity field, chemical composition of breathing gas and habitat pressure	New dialects derived from Founder Deme languages; these may be strongly differentiated from the former if the culture wishes to communicate difference rather than continuity
Ethics	Founder Deme ethics will in some ways constrain Earth-typical freedoms, especially regarding resources and thus property	Unknown and unpredictable	Unknown and unpredictable
Social roles	Social roles likely to be more formalized than Earth-normal in higher-risk environment of early settlements, to offset danger. Formalization likely to manifest in more explicit assignment of social rights and responsibilities than in Earth-normal life today	Social role adherence may ease somewhat as dangers are buffered out with technology	Unknown and unpredictable

(continued)

Table 5.4 (continued)

Human universal domain	First generation Years 0–30 'Founder Deme' Corresponds to population *travel* and *arrival* as discussed in the text	Second generation Years 31–60 'Fixing Deme' Corresponds to population *establishment* as discussed in the text	Third generation Years 61–90 'Adapted Generation' Corresponds to biocultural *adaptation* as discussed in the text
The supernatural	Adaptations may be rapid or slow depending on Founder Deme conceptions and religious traditions. Largely unknown and unpredictable, but they will have to meet spiritual requirements or a crisis may occur	Unknown and unpredictable	Unknown and unpredictable
Styles of bodily decoration	Some traditional styles likely to be retained as links with Earth; rapid innovations will signal distinction from Earth	Likely fixity of styles that will differentiate subpopulations of (a) Founder Deme descendants and (b) newly-arriving migrants	Unknown and unpredictable
Family structure	Whatever family structure is originally, it will likely be more formalized to best protect the next generation. Formalization likely to manifest in more elaborate and memorable family structure reiterations (e.g. forms of interaction) than in Earth-normal life today	Family structure adherence may ease somewhat as dangers are buffered out with technology	Unknown and unpredictable

(continued)

Table 5.4 (continued)

Human universal domain	First generation Years 0–30 'Founder Deme' Corresponds to population *travel* and *arrival* as discussed in the text	Second generation Years 31–60 'Fixing Deme' Corresponds to population *establishment* as discussed in the text	Third generation Years 61–90 'Adapted Generation' Corresponds to biocultural *adaptation* as discussed in the text
Course of life stages	Whatever life stages are originally, they will likely be more formalized to best organize the activities and success rate of the first and second generations. Formalization likely to manifest in more explicit stage-of-life rituals than in Earth-normal life today	Course of life structure adherence may ease somewhat as dangers are buffered out with technology	Unknown and unpredictable
Sexual behavior and reproductive rights	Sexual behavior and reproductive rights likely to be more constrained than in Earth-normal life today in order to maintain population in line with resources while, early on, sufficient resource buffers are established to protect subsequent generations	Likely easing of some sexual and reproductive restrictions, with emphasis on rapid growth of population to buffer against catastrophe	Likely continued emphasis on larger population
Food preferences	Food preferences will be tailored to what is available locally, though exotics imported from Earth may be desirable or may be shunned if separation is more valued. In either case, innovation of a new cuisine is inevitable	Fixation of new cuisines by second generation	Unknown and unpredictable

(continued)

Table 5.4 (continued)

Human universal domain	First generation Years 0–30 'Founder Deme' Corresponds to population *travel* and *arrival* as discussed in the text	Second generation Years 31–60 'Fixing Deme' Corresponds to population *establishment* as discussed in the text	Third generation Years 61–90 'Adapted Generation' Corresponds to biocultural *adaptation* as discussed in the text
Aesthetics	As with styles of bodily direction, whether aesthetics are used to differentiate founders from Earth or maintain links to Earth, or to some variable degree each of these, innovation is inevitable	Fixation of new styles to communicate identity by second generation	Unknown and unpredictable
Ultimate sacred postulates	As with concepts of the supernatural, it is difficult to say how these will be altered to promote survival and propagation; but however it is done it will have to meet the requirements of Ultimate Sacred Postulates, which include provision of simple and broad explanations for how the universe is organized and how to behave considering this arrangement	Unknown and unpredictable	Unknown and unpredictable

Noted are entirely likely adaptive changes and/or commentary on the magnitude of expected change, in whatever direction, based on the timescale of human cultural adaptations so far known to history and anthropology. The terms *innovation* and *fixity* are used chronologically, to connote the appearance of a biological or cultural feature (a variation), and its widespread establishment as a norm in a population, respectively

to services, *transportation*, *cultural differences*, *integrating children to new schools*, *prejudice*, *isolation* and unfamiliar *weather* (Immigration Canada 2018).

Overall what we see in consideration of culture change on the three-generation time-schedule is that we may well expect culture to change on this timescale, and for new identities and cultures to form. While the pace of culture change is highly variable and has not been successfully characterized in anthropology or any associated field, cultural information certainly changes more rapidly than biological information, at least in part because it is easily modified [by various cognitive mechanisms and the 'fuzziness' of metaphor, which allows receiving one message and then transmitting something similar but subtly- and importantly-different (Gabora 1999)] and in part because cultural information may be transmitted or shared among members of a generation ('horizontally') rather than just from one generation ('vertically') to the next (Perrault 2012).

If we define 'success' of a space settlement to be persistence over some generations as an entity biologically, culturally, and economically independent of Earth, and we know that both culture formation and the average human life course run into the decades, rather than, for example, days or years, we may be confident in saying that by the time human space settlements are entirely successful and independent, they will be inhabited by cultures different from those of Earth. This should not be seen as a problem or something outlandish, but a natural result of the adaptation of humanity to conditions beyond Earth. Indeed, if a culture were to attempt to bring intact an Earth culture to just about anywhere conceivable beyond Earth, and change nothing, it is likely that that culture would fail. The lesson is to expect change, and welcome it as a sign of healthy human adaptation. Details of the timeline for human biological and cultural differentiation over time are presented below, in Chap. 7.

5.5.1 Screening Recommendations for Early-Generation Human Space Settlement

Recommendations for early-generation human space settlement may be screened for their appropriateness by evaluating their positions regarding the following criteria. Few recommendations will be perfect and satisfy all criteria, but the idea is to promote and allow adaptive evolution to proceed, rather than interfere with it.

Does the recommendation:

- Promote Replication or at least not impede it?
- Promote Variation, or at least not dampen it until it is known to be immediately deleterious?
- Promote Selection of adaptive features that themselves promote survival, wellbeing and propagation of human biology and culture?
- Interfere with the adaptive process of selective agent identification and accommodation?

- Interfere with the reorganization of system-functional information required for a true evolutionary transition?
- Interfere with the adaptive process of novel feature dissemination and fixation?
- Interfere with adaptive processes, at large, of the second generation?

5.6 Conclusion and Recommendations

A review of the adaptive systems ('syndromes') and periodization of long-term animal movements on Earth, including permanent, one-way movement, has revealed useful structuring concepts for human space settlement planning. Three main lessons are most evident.

First, space settlement should be planned to best accommodate the well-known stages of emigration. Serious, evolutionarily-informed consideration should be given to each stage, to give it the best chance to succeed. We may be said to be early in the stages of the emigration process, the process of *study*. Second, across multiple Earth lineages, *migration syndromes* have evolved, integrated genetic and behavioral complexes fitted to certain ecosystems; we may learn from their organization and integration to best tailor our own long-distance and permanent movement to environments beyond Earth. Finally, the timescales involved in making a success of human space settlement (broadly measured as long-term survival of human settlements independent of Earth) virtually guarantee cultural differentiation from Earth, and *ethnogenesis*, the emergence of identities distinct from Earth. Such change would be natural, healthy human adaptation.

5.7 Summary Points

- Humanity has a long history of dispersal from known into unknown geographies
- Human space settlement may be productively modelled on dispersal mechanisms of other Earth life
- Human space settlement may be studied, planned and carried out in discrete 'Emigration Phases' of *preparation, movement, arrival* and *establishment*, leading to multigenerational success
- Each Migration Phase has a knowable duration and reasonably-expectable cultural and biological implications
- The *preparation* phase of emigration is underway and should incorporate an adaptive, evolutionary paradigm
- The *movement* phase of emigration is relatively brief in the course of a human life and should not be exoticized but rather normalized (in the case of expectable solar system settlements)

- The *arrival* phase of emigration will be the most challenging of all, requiring sacrifice for one or two generations as the human biological and cultural adaptive systems are exposed to, and adjust to, beyond-Earth conditions
- The *establishment* phase of emigration, likely occurring late in the first and certainly by the second generation, will involve an easing of adaptive shocks and, importantly, the establishment of habitat-specific routines, habits and rituals; daily life will be normalized

References

Arnett, J. J. (2000). Emerging adulthood: A theory of development from the late teens through the twenties. *American Psychologist, 55*(5), 469–480.

Barnett, H. G. (1953). *Innovation: The basis of culture change*. New York: McGraw-Hill.

BIOMIM. (2018). See https://toolbox.biomimicry.org/methods/.

Bonte, H. V. D., et al. (2012). Costs of dispersal. *Biological Reviews, 87,* 290–312. https://doi.org/10.1111/j.1469-185X.2011.00201.

Bordieu, F. (1977). *Outline of a theory of practice*. New York: Cambridge University Press.

Bowler, D. E., & Benson, T. G. (2005). Causes and consequences of animal dispersal strategies: Relating individual behaviour to spatial dynamics. *Biological Reviews, 80,* 205–225. https://doi.org/10.1017/S1464793104006645.

Bradford, J., Schaffer, M., & Talk, D. (2014). *Torpor-inducing transfer habitat for human stasis to mars*. NASA Innovative Advanced Concepts (NIAC) Phase I Final Report. Online at https://ntrs.nasa.gov/archive/nasa/casi.ntrs.nasa.gov/20180008683.pdf.

CET. See https://www.nasa.gov/sites/default/files/atoms/files/ger_2018_small_mobile.pdf.

Dingle, H., & Drake, V. A. (2007). What is migration? *BioScience, 57*(2), 113–121.

Drake, B. G. (1998). *Reference Mission Version 3.0 Addendum to the Human Exploration of Mars: The Reference Mission of the NASA Mars Exploration Study Team*. NASA S/P-6107-ADD. Lyndon B. Johnson Space Center, Houston, Texas.

Drake, V. A., & Gatehouse, A. G. (1995). *Insect migration: Tracking resources though space and time*. Cambridge: Cambridge University Press.

Drake, V. A., Gatehouse, A. G., & Farrow, R. A. (1995). Insect migration: A holistic conceptual model. In V. A. Drake & A. G. Gatehouse (Eds.), *Insect migration: Tracking resources through space and time* (pp. 427–457). Cambridge: Cambridge University Press.

Fiese, B. H., et al. (2002). A review of 50 years of research on naturally occurring family routines and rituals: Cause for celebration? *Journal of Family Psycghology, 16*(4), 381–390.

Fillmore, L. W. (1991). When learning a second language means losing the first. *Early Childhood Research Quarterly, 6,* 323–346.

FQEB. (2014). The full list of questions of the FQEB follows: * *What is the transition from chemical kinetics to evolutionary dynamics, and can the transition from non-evolving to evolving systems be defined precisely and formally?* * *Are there different kinds of evolution, and is there evolution in the context of evolution itself?* * *What are the natural laws of evolutionary change, and can we derive a complete mathematical theory of evolutionary laws?* * *How does population structure affect evolutionary dynamics, and what role might coordination/cooperation play?* * *What are the differences between genetic and cultural evolution, and how can these differences be formalized?* * *What is the role of chance (stochasticity) in evolutionary dynamics?* * *What are the typical limits of evolvability, and how are these transcended?* * *Why does evolution (sometimes) lead to increasing complexity?* * *What are the evolutionary origins of emergence?* * *How does cooperation affect construction?* * *Can we build precise models for the evolution of cells, multi-cellular organisms, animal societies and human language?* * *What is the role of multi-level selection in the larger evolutionary process?* See https://www.trdf.co.il/eng/kolkoreinfo.php?id=1742.

Frisancho, A. R. (1993). *Human adaptation and accomodation*. Ann Arbor: University of Michigan Press.

Gabora, L. (1999). Weaving, bending, patching, mending the fabric of reality: A cognitive science perspective on worldview inconsistency. *Foundations of Science, 3*(2), 395–428.

Gabora, L. 2018. The making of a creative worldview. Book chapter in press, see https://arxiv.org/abs/1811.11236v1.

Hornborg, A. (2005). Ethnogenesis, regional integration, and ecology in prehistoric Amazonia. *Current Anthropology, 46*(4), 589–620.

Hovelsrud, G. K., Poppel, B., van Oort, B., & Reist, J. D. (2011). Arctic societies, cultures and peoples in a changing cryosphere. *Ambio, 40*, 1–110. https://dx.doi.org/10.1007%2Fs13280-011-0219-4.

Immigration Canada. (2018). *Top ten problems faced by immigrants* at https://www.immigroup.com/news/top-10-problems-immigrants.

Lazear, E. P. (1998). Diversity and immigration. NBER Working Paper no. 6535. *National Bureau of Economic Research*, Cambridge, MA

Lazear, E. P. (1999). Culture and language. *Journal of Political Economy, 107*(6), S95–S126.

Levine, J. M. (2000). Biological invasions. *Current Biology, 18*(2), 57–60. https://doi.org/10.1016/j.cub.2007.11.030.

Marin, F. (2017). Heritage: A Monte Carlo code to evaluate the viability of interstellar travels using a multigenerational crew. *Journal of the British Interplanetary Society, 70*, 184–195.

Marin, F., & Beluffi, C. (2018). Computing the minimal crew for a multi-generational space journey towards Proxima Centarui b. *Journal of the British Interplanetary Society, 71*, 431–438.

Marin, F., Beluffi, C., Taylor, R., & Grau, L. (2018). Numerical constraints on the size of generation ships: From total energy expenditure on board, annual food production and space. *Journal of the British Interplanetary Society, 71*, 382–393.

McShea, D. W., & Brandon, R. N. (2010). *Biology's first law: The tendency for diversity & complexity to increase in evolutionary systems*. Chicago: University of Chicago Press.

Perrault, C. (2012). The pace of cultural evolution. *PLoS ONE, 7*(9), e45150.

Piersma, T., Perez-Tris, J., Mouritsen, H., Bauchinger, U., & Bairlein, F. (2005). Is there a "migratory syndrome" common to all migrant birds? *Annals of the New York Academy of Sciences, 1046*, 282–293.

Riley, D. (2007). Educational technology and practice: Types and timescales of change. *Educational Technology and Society, 10*(1), 85–93.

Sih, A., Bell, A., & Johnson, J. C. (2004). Behavioral syndromes: An ecological and evolutionary overview. *Trends in Ecology & Evolution, 19*, 373–378.

Smith, C. M. (2018). *An atlas of human prehistory*. San Diego: Cognella Academic Publishing

Staats, K. (2018). *SIMOC: A foundational design document*. See https://www.kaistaats.com/?s=simoc.

Travis, J. M. J., et al. (2012). Modelling dispersal: An eco-evolutionary framework incorporating emigration, movement, settlement behaviour and the multiple costs involved. *Methods in Ecology, 3*(4), 628–641.

Vernikos, J., Walter, N., Worms, J. C., & Blanc, S. (2016). THESEUS: The European research priorities for human exploration of space. *Nature Microgravity, 2*, 16034. https://doi.org/10.1038/npjmgrav.2016.34.

Chapter 6
Adaptive Lessons from Ancient Technologies and Cultures

Abstract We have seen a wide range of humanity's adaptive tools, from biology to behavior. A significant aspect of that behavior is the making and using tools—technology—as our adaptations. The field of archaeology has over a hundred years' history of investigating these tools, from chipped stones to the finest metalwork. The archaeological record, then, shows us material adaptations of our species for some millions of years, and we may examine that record for lessons useful to the project of permanent space settlement. We can also learn quite a bit about how our ancestors arranged themselves, socially, demographically, and culturally, as adaptively-successful (most of the time) large, bipedal primates originating in Africa but in the last 100,000 years adapting to nearly every Earth biome. Many lessons here will be of great value to those planning space settlement, as it is our ancestors who carried out humanity's greatest project to date, becoming numerous, global and diverse.

[Material culture] is the extrasomatic means of adaptation.—L. R. Binford

6.1 Introduction

The study of *archaeology*, reconstructing ancient cultures from their material remains such as artifacts and skeletal evidence, has a roughly 100-year history of investigating our origins and evolution. This corpus of archaeological knowledge informs us about how humanity has adapted both biologically and culturally for the last 100,000 years. It also gives humanity a context for answering such gravid questions as "why are we here?" and "should we explore space?", but I have provided that context in my previous book, *Emigrating Beyond Earth*. In this chapter I would like to begin with an examination of some of humanity's cultural adaptations, which may serve to inspire space settlement planners with the adaptive genius of our collective ancestors. Following this, in the second half of the chapter, I would like to examine some specific aspects of the human past that may be very useful in planning the human future beyond Earth. These will be, first, the independent farming villages of our past, and

C. M. Smith, *Principles of Space Anthropology*, Space and Society,
https://doi.org/10.1007/978-3-030-25021-8_6

what they might teach us about early settlements beyond Earth, and second, the key aspects of all civilizations, and how they will play out beyond Earth.

6.2 Lessons of Material Culture: Material Adaptation

Archaeology's focus on the material objects made by humans and our relatives has resulted in a study of *material culture*, objects used as tools for human purposes. Material objects are best appreciated visually, so in the following pages I present a number of images of such objects, starting with small objects used close to the body and ranging outward in scale to shelters, vehicles and other large objects. In photo captions I describe the adaptive advantage of the artifact. This section of this chapter has three functions. First, it is a reminder of human adaptive ingenuity. This can be inspirational, and inspiration is important in challenging projects. Second, it is a reminder that human adaptation by material culture has a long history and that space settlement has a context in this history; it will simply be a continuation of an ancient way of adaptation. Third, while space settlement planning will require planning in the technological domains, those will only constitute a part of space settlement, a significant argument of this book. Therefore many other aspects of life can be productively studied, and studied by the many people who have an interest in space settlement, but no training in traditional space-settlement technologies. But, I argue, nearly everything we make and use beyond Earth will be some kind of adaptation, and so those with an interest in, say, clothing design, or architecture, do in fact have an expertise in a field important to space settlement. This third point is important because, as I have mentioned before in this book, when the voyage element of space settlement is completed, space settlers will begin living out normal, daily lives, and their surroundings, food, arts, and so on will all be just as important to their lives as they are to us today.

6.2.1 Clothing

See Figs. 6.1, 6.2, and 6.3.

6.2.2 Hand Tools

See Figs. 6.4, 6.5, and 6.6.

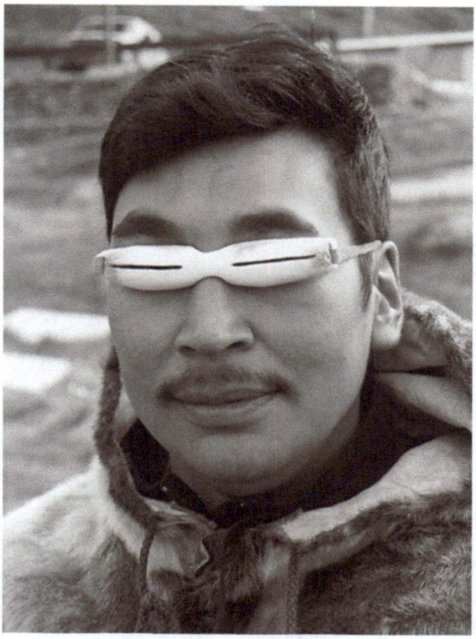

Fig. 6.1 Inuit (Arctic) snow goggles. Arctic winter conditions require eye protection from both sunlight, reflected by snow and water, and blowing grains of hard snow. Photo by Julian Idrobo from Winnipeg, Canada—Inuit Goggles, CC BY-SA 2.0, https://commons.wikimedia.org/w/index.php?curid=10787841

Fig. 6.2 Maasai woman's necklaces and cropped hair. All cultures have clothing and other bodily décor traditions that serve to mark identity. Successfully communicating identity, a social concern, can be important to socialization, mate-finding, alliance-making and other important cultural behaviors. By javic—Flickr, CC BY-SA 2.0, https://commons.wikimedia.org/w/index.php?curid=3732513

Fig. 6.3 Scandinavian fineskoe. Traditional boot of Arctic Scandinavia, with fur on the outside and stuffed inside with certain hay-like grasses as insulation and padding. This drawing shows the boot lashed to snowshoes, which were also made for horses. Naturally, making and using such tools will be critical beyond Earth, and will require the expertise of people with many different skills

Fig. 6.4 Northwest Coast native bent-wood chest. Here a natural resource (cedar wood) has been harvested and shaped into an artifact with both utilitarian function (container) and social functions, including a display of the owner's wealth, as indicated by the seal teeth embedded in the wood, and the fine carving and painting

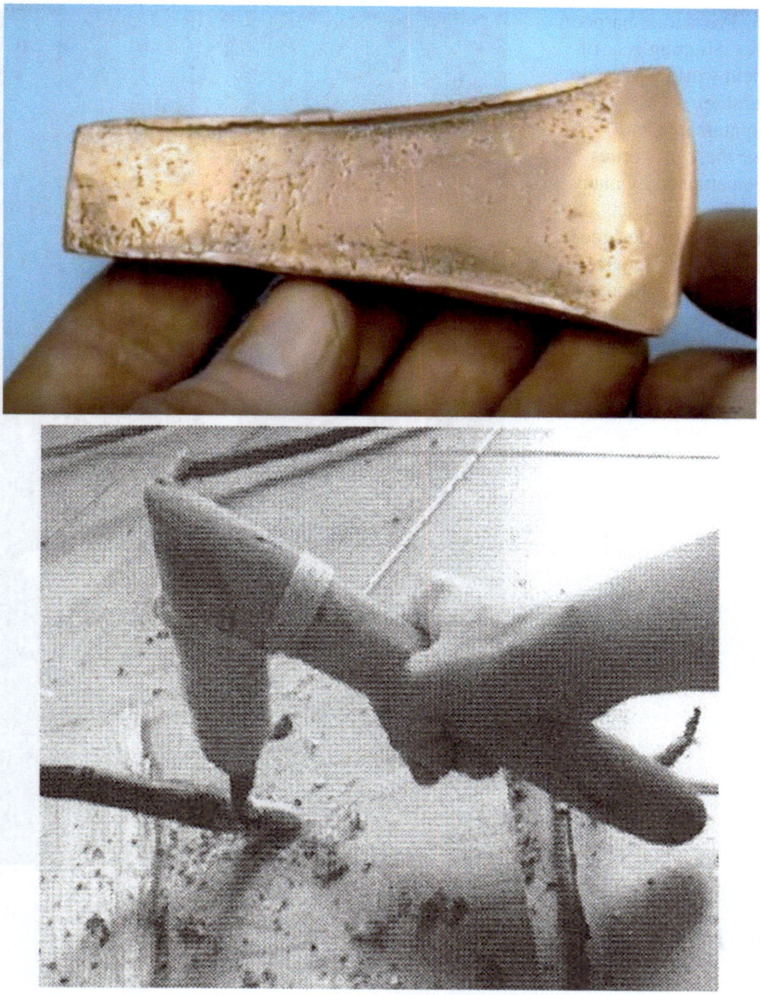

Fig. 6.5 Prehistoric axe replicas. Top, copper alloy replica of the axe carried by 'Otzi the Iceman', a prehistoric villager who perished in the Alps about 5200 years ago. Below, replica of a stone axe from early farming times, used experimentally by an archaeologist. As with the bent-wood box in Fig. 6.4, the object is both utilitarian and social, and in settlements beyond Earth, design to support both functions will also be employed

6.2.3 Housing

See Figs. 6.7, 6.8, and 6.9.

Fig. 6.6 Prehistoric harpoon armatures. Hunting gear of the ancient world was made to be durable, field-maintainable and effective; these properties have been attained by many design philosophies, any of which may inform designers of technologies and material culture for space settlement By Didier Descouens—CC BY-SA 4.0, https://commons.wikimedia.org/w/index.php?curid=10727062

6.2.4 Facilities

See Figs. 6.10, 6.11 and 6.12.

6.2.5 Transportation

See Figs. 6.13, 6.14 and 6.15.

6.2.6 Large Constructions

See Figs. 6.16, 6.17 and 6.18.

Fig. 6.7 Northwest Coast lodge. Native people of the Pacific Northwest lived in large, multi-family dwellings termed 'lodges' or 'plankhouses' such as this replica on the Columbia River. A village consisted of up to ten of these plankhouses, each house with a population of up to 100 people. Villages traditionally had no single leader, and were loose confederations of plankhouses. Plankhouses each had a royal family, who managed, but did not command, economy and social activity

6.3 Production Systems

A final note in this field of material culture studies is to note that all material culture is generated as part of a production system. As we have seen many times in this book, recognizing this usefully breaks down a larger topic for consideration in the prospect of space settlement. Figure 6.19 indicates a generic production system's components and flow. Such elements and their flows will have to be tailored specifically to the resources and products of any off-Earth economy.

6.4 Lessons of Ancient Culture: Cultural Adaptations

Ancient human cultures adapted to their environments in countless ways, and it is worth surveying what we know of these adaptations that may be of use in space settlement.

Fig. 6.8 Mandan lodge. Mandan loges, in the Dakotas of the United States, were particularly effective in remaining cool in summer and warm in winter, due to their thick walls and semi-subterranean design. Such structures are well-known worldwide where the temperature difference between winter and summer are particularly high

6.4.1 The Independent Neolithic Village and the Extraterrestrial Farming Village

Because for the foreseeable future, human populations beyond Earth will be domesticators (using farming as the mode of subsistence) but in relatively low populations compared to Earth's urban populations we may have something to learn from our ancestor's experience of relatively small farming populations, in particular the phenomenon of the *Independent Neolithic Village* (INV).

Independent Neolithic Villages flourished in each core domestication area of the Earth in the period of about 12,000–5000 years before present (Fig. 6.20). *Neolithic* refers to the New Stone Age recognized by archaeologists as a set of tool types (stone axes for cutting forested areas for farmland, and stone sickles for harvesting) and ways of living concomitant with early farming. More important here is the term *independent*, indicating an economy—production and consumption—on the scale of the community or settlement itself rather than for a global or other market economy. Finally, *village* refers to a demographic structure characterized by less 10,000 people.

In each core region indicated in Fig. 6.20, early in postglacial times, somewhere between 14,000 BP and about 10,000 BP, some foragers effectively abandoned for-

Fig. 6.9 Iglu snow house winter village. By Drawn by unknown artist based on sketches by C.F. Hall and photographed from the book by User: Finetooth—*Arctic Researches and Life Among the Esquimaux: Being the Narrative of an Expedition in Search of Sir John Franklin in the Years 1860, 1861, and 1862 by Charles Francis Hall* (1865), New York: Harper and Brothers. Public Domain, https://commons.wikimedia.org/w/index.php?curid=3648025

aging and took up some form of *domestication*, generally defined as the *capture of plant and animal species from the wild and holding and raising them for human purposes*. In some areas this transition was slow, in others rather faster, but at whatever rate, in the long run the transition was complete (note that foragers did not all drop their hunting tools, however, for the plow: but farming cultures emphasize larger populations and these grew, consuming landscapes for yet more farmland, and foraging peoples were 'pushed' to farther and farther out from the agricultural core areas). In inventing this mode of subsistence, deemed *horticulture* if low-intensity or *agriculture* if more intensive, these populations transitioned from a more mobile foraging culture, where resources were pursued, to a more sedentary culture, where resources are produced. A review of the features of these Independent Neolithic Villages, which evolved across the globe, though without coordination or communication, indicates some striking similarities which must be understood as *convergent evolution*, not just happenstance:

- Wattle-and-daub, adobe (mud-brick) and timber buildings
- Structures normally less than three stories high
- Houses generally of the same size (no housing hierarchy)
- Open areas for animal and plant processing
- Lack of monumental architecture or public buildings, including churches or city halls
- Simple defensive village architecture, but no evidence of military, e.g. barracks, stockpiled weaponry

Fig. 6.10 Archaeological replica of a Bronze Age Kiln. The local production of goods for daily life, such as making pottery for early villages, as reconstructed by archaeologists below, will be a feature common in early off-Earth settlements. There will for some generations be no global market or mass production, and focus will be on local-scale production and consumption. Thus, there will be many similarities with early farming village life on Earth, as discussed elsewhere in this book

Photograph by unattributed studio.Tyrrell Collection, Powerhouse Museum, Sydney. 85/1286-721, 00g0003

Fig. 6.11 Aboriginal fish traps, Australia. All human cultures have extracted calories from their environment, sometimes with passive technologies that transform landscapes, such as these fish traps made of rock alignments in a river. Like windmills or solar panels, they are passive but require maintenance, work that often develops into the specialty of specific community members. Public domain photo

- Villages generally of the same size (no settlement hierarchy)
- Economic independence
- Population in the 600–1000 individuals range

Figure 6.21 is a visual representation of three such villages, one from Demircihuyuk, Turkey (dating to over 8000 years ago) and one from Chaco Canyon in the United States (dating to over 1000 years ago).

What does the evolution of these convergent features tell us about the cultures of the INV people? We can infer some basic social information from the material culture, proceeding down the list.

The simple architecture indicates that non-specialized labor was used to construct the village structures. That is, the labor of construction and maintenance was that of non-activity-specialized people. Mud brick and minimally-worked timber, and stacked stone without mortar, require a bit of training to make and use, but not the specialized training needed for, say, complex masonry, or even kiln-fired clay bricks. This is the first of many suggestions that most village inhabitants were significantly

Fig. 6.12 Facilities include cultural landscape markers, such as the distinctive stone figures of the native Canadian Arctic, known as 'inusuk'. These may mark a landscape feature such as a cache of goods, or represent a symbolic claim to some resource, material or spiritual. Public domain image of an inusuk on Hudson Bay, photographed by Nicholas M. Perrrault

involved in food production, leaving time enough for such important work as house construction and maintenance, but not intensive training for specialized techniques in house construction and maintenance. In early communities beyond Earth, the pressing need to ensure a stable food supply, grown locally, might well for some generations require a renewed engagement with agriculture among the general populous. The domestic architecture will certainly be fabricated from more complex materials than wood and clay brick, but to free up time for agriculture it might well be kept as simple as possible to build and maintain. Here my focus is not so much on the architectural materials but what they imply about the time invested in various activities. In particular, with the capacity to fabricate many household goods with such technologies as additive (3d) printing from pre-prepared design files, time that might otherwise be engaged in making such objects—in the case of ancient people of the INV—would in beyond-Earth settlements be freed for other activities.

Fig. 6.13 Chinookan canoe. Native people of the Lower Columbia river had as many varieties of watercraft as modern Americans have different kinds of automobiles. Here a ceremony canoe waits to be invited to land ashore in another tribe's territory. While space settlement is often associated with giant rockets, settlers beyond Earth will be heavily engaged in making, using and repairing craft scaled to transport just a few people. Photo by the author

For these reasons, early beyond-Earth settlement architecture might have significant similarities with that of early, Earthly INV's.

In the same way, the fact that the architecture is nearly always undecorated and essentially utilitarian, just large enough to comfortably house nuclear families and their possessions, suggests that there was minimal investment in using the architecture to communicate economic rank. This and other archaeological phenomena suggest an essentially rank-free social organization, in which there was apparently equal access to resources by all community members. Note that while it is often said, offhandedly, that all humanity's problems began when agriculture started and people began to be economically segregated and ranked, this really is not archaeologically apparent; INV's worldwide normally continued for several millennia as apparently largely egalitarian, rather than ranked, before evidence of urbanism, economic ranking, and civilization. The lesson for space settlement is simply that somehow, such sociopolitical systems can indeed occur, and have, and they have remained stable or at least have persisted for multiple millennia. This can be encouragement for a pioneer population, and encouragement can be important to carrying out a goal.

The large open areas in INV's were largely used for economic purposes, such as communal processing of agricultural products (e.g. large areas for winnowing). Again I don't suggest that the same will be present in the same way in beyond-Earth communities, but rather that this tells us something important about the organization

Fig. 6.14 Alaskan dog-sled team. All cultures have domesticated some variety of animal, for many reasons. It is difficult to imagine settlement beyond Earth without some of our closest domesticates, dogs. Photo by Denali National Park and Preserve—Corrie Mile 9 Landscape—Jacob W. Frank. Uploaded by Albert Herring, Public Domain, https://commons.wikimedia.org/w/index.php?curid=29598849

of the INV cultures. What it tells us is that large open, communal areas were used for reasons much different from the same such large areas in later, socio-economically ranked cultures. In ranked cultures and state-level societies, large public areas were often used for formalized political and religious gatherings, which seem absent from the INV. This absence is also signaled by the lack, in INV's, of monumental architecture, again often used in state-sponsored religious and political activities, again marking the INV and the state-level city as quite different. Also common in the INV's were simple walls around the settlement, indicating some provision for defense; however the villages do not have evident storage facilities to provision standing military forces, but rather their storage facilities, for example for grain or corn, were scaled to the individual household. These both indicate that while there may have been occasional conflict, warfare extended in time or space by supporting large, full-time military forces was not a part of their world. We also see that while settlements of course varied in size, in a given region they typically do not vary in size; that is, there is a lack of settlement hierarchy, suggesting an independence and insulation of these communities rather than other socio-economic organizations, such as those using centralized decision-making, which normally have large urban centers and small, hinterland villages.

Fig. 6.15 Hokulea, a traditional Polynesian Voyaging canoe replica. Nearly all cultures have devised tools for transportation. Arguably the greatest explorers and settlers of human history were the Lapita people who explored and settled the islands of the Pacific ('Oceania') in the last three thousand years. Their vessels have been reconstructed and sailed globally in the last few decades, introducing the world to this ancient seafaring culture. Photo courtesy of Prof. Ben Finney

Another key feature of the INV was economic independence rather than interdependence. Although villages in a given region did have contacts with other villages—trading obsidian and other kinds of stone in the case of Southern Turkey over 9000 years ago, for instance—in most cases villages were largely self-sufficient and independent of others. In short, if one wanted a new pair of shoes in the independent agricultural village, one would make them. This is in great contrast to the pervasive economic interdependence we see in civilizations (discussed later in this chapter), in which villages began to have specialized functions, such that cobblers were concentrated in some villages, potters in another, and so on. In this case, if one wanted a new pair of shoes, one would trade for a pair made by someone in another village.

Finally we note that INV populations were small compared to the populations arranged in states—characterized by cities—that evolved later. Table 6.1 indicates population estimates of INV's known in the archaeological record. They average around 1000 individuals, a figure likely similar to later-stage early settlers in colonies beyond Earth. Hamlet and villages are 50–1000 persons with towns and cities having populations over 5000. Some centers, such as Mesoamerica's Tika, would have housed up to 30,000–50,000 but this was an unusual, large outlier.

Fig. 6.16 Great Ziggurat of Ur. Ancient civilizations built large, visible structures for many reasons discussed in the text. Many were carefully designed to be architecturally impressive, like this ziggurat. There is no reason that settlers beyond Earth would not also invest heavily in such cultural displays. By Hardnfast, CC BY 3.0, https://commons.wikimedia.org/w/index.php?curid=3544015

Figure 6.22 graphically indicates the INV population estimates presented in Table 6.1. The horizontal axis indicates time in BP or years Before Present. The vertical axis indicates the settlement population, the settlement type, and the political organization normally associated with such populations. For example, agrarian *hamlets* are the smallest of the settlements, often composed of a single, or several clustered, houses; when more, and more substantial, domestic structures are clustered, they form a *village* into the population of many hundreds to about 1000 people, grading into *towns* when populations exceed about 1000. Since at least 8000 years ago, then, there have been settlements in the 1000-person population range. Anthropological study of the political organization attending different population sizes and subsistence modes makes it clear that *tribes* or small *chiefdoms* are most common the organization types for populations less than about 2500, whereas larger chiefdoms, *superchiefdoms* and *protostates* are more common in the 2500–10,000 populations. Not seen here is the *state* or *civilization* variety of political organization, discussed in Sect. 6.5.

The patterns evident in the Earth's INV's are global and long-lasting. Prehistorians Jacquetta Hawkes and Sir Leonard Wooley wrote, in an erudite treatment of the phenomenon (Hawkes and Wooley 1962: 349–350).

Fig. 6.17 Stonehenge, Southern England. Ancient monuments are often aligned with celestial phenomena, such as Stonehenge, dating to over 4000 years ago, which was used in many ways but at least to mark the summer solstice. There is no reason that cultures would stop building such structures beyond Earth. Space Artist Sarah Jane Pell has recently proposed a 'Moonhenge', discussed later in this book. Public domain image

> …it is remarkable with what uniformity…we find the Neolithic communities from western Europe to China [led] lives remarkably close to the subsequent peasantries that have supported all civilizations…[there was] little essential difference in the way of life and outlook between those pioneers and their inheritors…The underlying uniformity of the primary [read 'Independent'] Neolithic village cultures, whether they were represented by the wooden houses of Europe, the matted huts of Egypt, or the mud-walled buildings of south-west Asia, is reflected also in their religious forms…the furnishing of the grave with goods and …weapons…the intense…preoccupation with [domesticated] animals…[and their] domestic and economic preoccupations.

That is, the INV phenomenon occurred globally without communication or coordination, and persisted normally (in each region) for millennia. A global study of early farming villages has shown that in many regions they persisted for 1000 years to c.3000 years before significant transitions in the settlement pattern (often a rapid growth of the villages) occurred (Bandy 2007).

My own data indicates that the duration of the independent farming villages, worldwide, was about 5000 years between establishment and the fundamental transition to much larger state-level societies referred to as *civilization* (discussed further in Sect. 6.5.2).

Fig. 6.18 Tenochtitlan replica. Modern civilization's urbanism, bustling markets and large structures are not unique. Conquistador Bernal Diaz, writing in the 1st quarter of the 16th Century, stated that the Aztec capital was far larger and better-organized than the largest city he had ever seen prior, which was Seville, Spain. Public domain image

6.4.2 *Learning from the Independent Neolithic Village*

I must emphasize that I do not take the patterns evident in the Earth's INV's to be a template for people living beyond Earth; it will be for the people who settle beyond Earth to arrange themselves. I also do not propose that the INV, or an extraterrestrial analogue, is some kind of utopia. What I do intend to show, though, is that there are significant similarities between the INV and what we may expect for at least some of the early generations of migrants to habitats beyond Earth. Table 6.2 indicates some of these similarities, and what I suggest they mean in terms of space settlement.

That people on Earth flourished in such arrangements as the INV for millennia, in multiple locations around the world, suggests that it approximates a stable, viable socio-cultural-technological arrangement. Some hundreds or a thousand people or so, most of them spending some time each day on farming subsistence activities, can survive and thrive as healthy communities for long, multi-generational time. They can have social and economic connections with neighboring communities, though need not directly compete with those communities, and indeed for millennia in each early farming region there appears to be no centralized, authoritative hierarchy, but rather, settlement independence. This independence, though with the capacity for

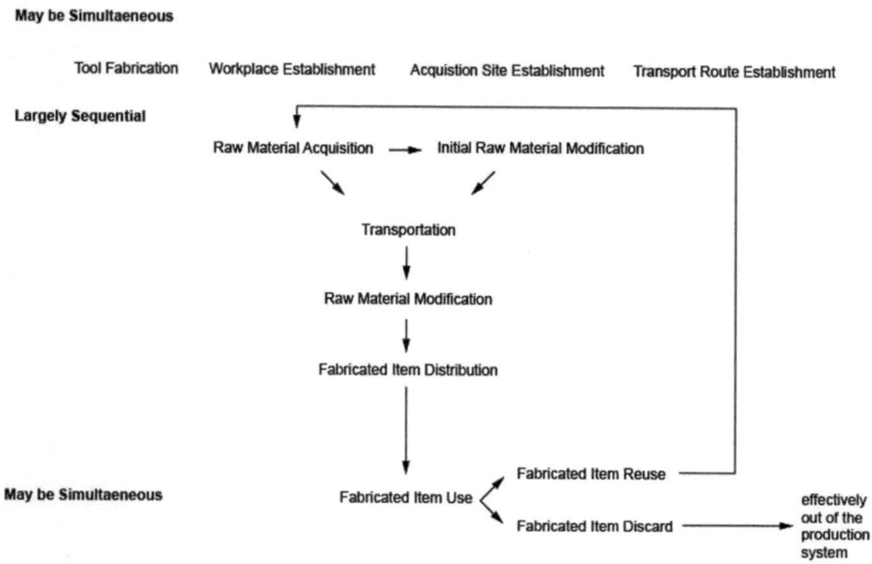

Fig. 6.19 A generic material culture production-distribution system. How this plays out differs for each item of material culture, but the framework is useful for organizing thought about how technology will be produced and distributed in habitats beyond Earth. Diagram by the author

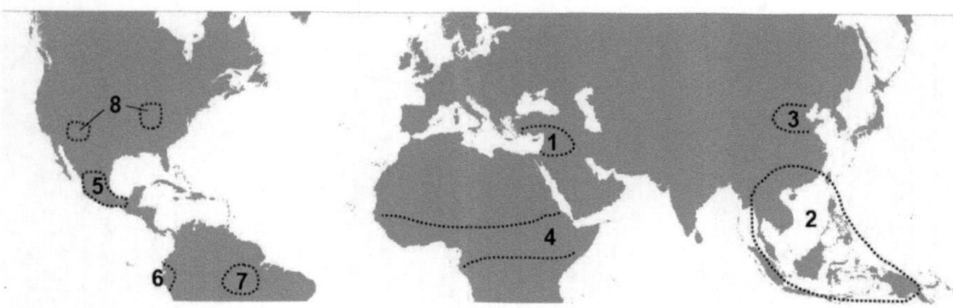

Fig. 6.20 Distribution of Early Neolithic Village complexes. See text for important discussion and Table 6.1 for village population estimates per numbered region. Diagram by the author

temporary confederation or integration with neighboring communities if there is some commonality in their activities and materials, would provide *both* the benefits of modularity (e.g. collapse of one does not necessarily mean the collapse of all) *and* the benefits of communal effort (e.g. resource pooling, task rotation-of-effort). So, this highlighting of the INV is not a prescription; it is a *description* of something that has worked before, and repeatedly, successfully, on Earth. Space settlement planners contemplating how to arrange their economies, architecture, agriculture and so on would be wise to at least know of the features of the INV.

Fig. 6.21 Reconstructions of two independent Neolithic Villages. **a** Demircihuyuk, Turkey and **b** Chaco Canyon. **a** Drawn by the author, **b** US National Park Service

Table 6.1 Independent Neolithic village population estimates

Region	Site	Farming origins	Village origins	Population	Reference
1. SW Asia	Jericho	10,000	10,000	225	Kujit (2000)
1. SW Asia	Netiv Hagdud	10,000	10,000	135	Kujit (2000)
1. SW Asia	Gilgal I	10,000	10,000	90	Kujit (2000)
1. SW Asia	Dhra'	10,000	10,000	41	Kujit (2000)
1. SW Asia	Nahal Oren	10,000	10,000	18	Kujit (2000)
1. SW Asia	Ain Ghazal	10,000	8900	405	Kujit (2000)
1. SW Asia	Tell Aswad	10,000	8900	360	Kujit (2000)
1. SW Asia	Jericho	10,000	8900	225	Kujit (2000)
1. SW Asia	Yiftahel	10,000	8900	135	Kujit (2000)
1. SW Asia	Kfar Hahoresh	10,000	8900	45	Kujit (2000)
1. SW Asia	Catalhoyuk	10,000	8600	6000	Cessford (2005)
1. SW Asia	Basta	10,000	8250	1260	Kujit (2000)
1. SW Asia	Ain Ghazal	10,000	8250	900	Kujit (2000)
1. SW Asia	Wadi Shu'eib	10,000	8250	900	Kujit (2000)
1. SW Asia	Beisamoun	10,000	8250	900	Kujit (2000)
1. SW Asia	Es-Sifiya	10,000	8250	900	Kujit (2000)
1. SW Asia	Ain Jamman	10,000	8250	630	Kujit (2000)
9. Europe	Cyprus	9000	6000	2000	Price (1977)
9. Europe	Serbian sites	8500	6000	1740	Porčić (2011)
3. East Asia	Xinglongwa	8300	7730	100	Shelach (2011)
3. East Asia	Cishan	8300	7700	100	Schleach (2011)
3. East Asia	Zhaobaogou	8300	7034	100	Shelach (2011)
9. Europe	Germany	7500	6000	135	Ebersbach and Schade (2004)
4. Africa	Merimda Beni Salama	7000	6000	1650	Hawass et al. (1988)
4. Africa	Hierakonpolis	7000	5500	1750	Hassan (1981)
6. South America	Real Alto	6400	5250	175	Damp (1984)

(continued)

Table 6.1 (continued)

Region	Site	Farming origins	Village origins	Population	Reference
6. South America	Loma Alta	6400	4680	175	Damp (1984)
2. South Asia	Ban Non Wat	4000	4000	700	Higham (2003)
5. Mesoamerica	Oaxaca sites	3500	3300	325	Bandy (2005)
5. Mesoamerica	Oaxaca sites	3500	2900	1973	Bandy (2005)
5. Mesoamerica	Oaxaca sites	3500	2770	1782	Bandy (2005)
5. Mesoamerica	Oaxaca sites	3500	2600	1828	Bandy (2005)
5. Mesoamerica	Oaxaca	3500	2600	1000	Bandy (2005)
5. Mesoamerica	Basin of Mexico sites	3400	3050	685	Bandy (2005)
6. South America	Titicaca basin sites	3150	3250	693	Bandy (2005)
6. South America	Titicaca basin sites	3150	2900	1752	Bandy (2005)
6. South America	Titicaca basin sites	3150	2500	3507	Bandy (2005)
8. North America	Snaketown	3000	1000	300	Lekson (2004)
8. North America	Galaz	3000	1000	300	Lekson (2004)
8. North America	Montezuma Valley	3000	800	2500	Lekson (2004)
8. North America	Yellowjacket	3000	800	2000	Lekson (2004)
8. North America	Zuni	3000	800	1600	Lekson (2004)
8. North America	Sand Canyon	3000	800	725	Lekson (2004)
8. North America	Marana	3000	800	700	Lekson (2004)
8. North America	Paquime	3000	600	4700	Lekson (2004)
8. North America	Sapawe	3000	600	2770	Lekson (2004)

(continued)

Table 6.1 (continued)

Region	Site	Farming origins	Village origins	Population	Reference
8. North America	Pueblo Grande	3000	600	1750	Lekson (2004)
8. North America	Los Muertos	3000	600	800	Lekson (2004)
7. Amazonia	Rio Negro Sites	2500	2300	1250	Heckenberger, Petersen and Neves (1999)
7. Amazonia	Upper Rio Xingu Sites	2500	1000	1250	Wüst and Barreto (1999)
7. Amazonia	Central Brazil	2500	1000	964	Wüst and Barreto (1999)
8. North America	Chaco Canyon	2200	1300	600	Wills and Windes (1989)
8. North America	SW USA	2200	1300	400	Kohler (1992)
8. North America	Mesa Verde	2200	1100	100	Wilshuen (2002)
8. North America	Chaco Canyon Main	2200	1000	3500	Bandy (2005)
8. North America	Chaco Canyon hamlets	2200	1000	200	Bandy (2005)
8. North America	Moundville	2000	1000	1200	Steponaitis (1998)
8. North America	Snodgrass	2000	1000	350	Holley et al. (1993)
8. North America	Lunsford-Pulcher	2000	950	1000	Pauketat (2004)
8. North America	Cahokia	2000	950	1000	Pauketat (2004)
Average		5676	4444	1088	
St dev		3292.81	3446	1163	

The date of 'commitment' in the region to a residentially-sedentary settlement mode is noted (Farming Origins) in years Before Present (BP); the date of the earliest recognizable villages in the region is similarly noted (Village Origins). Population is an estimate, a representative mean derived by the author from studies cited in the Reference column

LBK Linearbandkeramik culture of Danubian and Central Europe

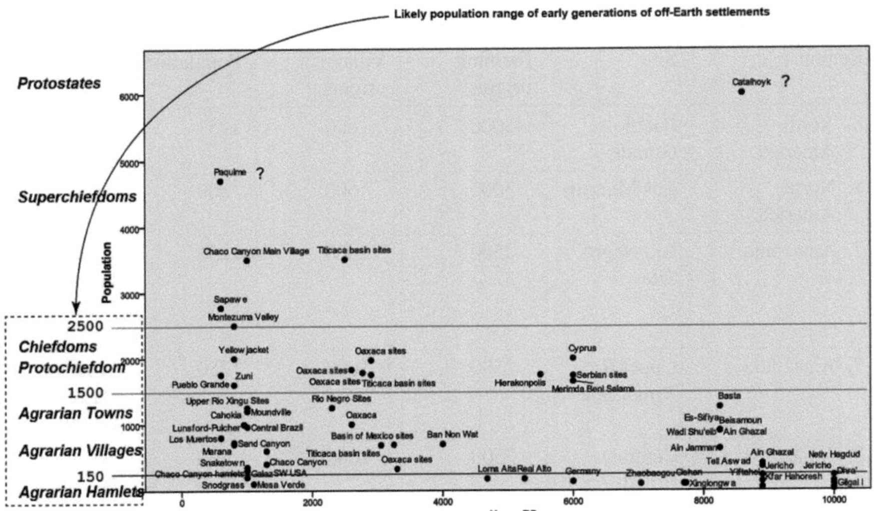

Fig. 6.22 Independent Neolithic Village population estimates. Note both implied political organization and observed archaeological population are indicated. Figure by the author

6.4.3 The Polis

If we wish to learn from humanity's past experience in arrangements—demographic and economic—similar to what we can reasonably expect of early settlements beyond Earth, the villages mentioned in the prior section are worth some study. Perhaps a slightly more contemporary example is seen, however, in the ancient Greek *polis*, which was fully developed by the 5th Century BC in the Eastern Mediterranean. Polis refers to a 'city' or 'city state' but this translation even specialists in the field find somewhat misleading (Howarth 1978). Today at least, the term 'city' of any kind invokes populations tens of times larger than those of most Greek *poleis* (plural). Poleis had a number of features reminiscent of the Independent Neolithic Village just described, but Table 6.3 indicates significant differences that make the polis somewhat more useful as an analog for, perhaps, a maturing early settlement beyond Earth.

The polis, then, had some features that make it slightly more urban in character and slightly more externally-involved than expressed in the highly-rural nature of the Independent Neolithic Village (for a fuller treatment, including discussion of confederations of polei see Ehrenburg (1969) and Morgan (2003)). The population of the polis was somewhat higher than the INV as well, into the several thousands, though commonly less than 5000 and so not approaching the character of cities in the modern experience. The polis was also somewhat more formally organized, as evidenced by its architectural uniformity and some areas of communal activity and even worship, though there was also fierce religious independence of the constituent population. The polis would seem to be a useful analog for early human settlements

Table 6.2 Features of the archaeological and beyond-earth independent farming village (see text for discussion)

Independent Neolithic Village feature	Essential inference of feature	Suggested parallels with early-generation beyond-earth settlements
Wattle-and-daub, adobe (mud-brick) and timber buildings	Unspecialized fabrication methods	Architecture likely to be technologically complex but easy to assemble
Structures normally domestic, and less than three stories high	Most architecture domestic	Most habitable space will be for small populations and agriculture
Houses generally of the same size (no housing hierarchy)	Surpluses small and domestic	Agricultural surpluses will be stockpile on both household and community scales
Open areas for animal and plant processing	Some simultaneous labor	Unknown; organization of labor likely to include simultaneous tasks
Lack of monumental architecture or public buildings, including churches or city halls	Fundamental independence of population sub-units	Unknown; socio-cultural and religious organization unpredictable; perhaps emphasis on individual worship due to cost of monumental architecture
Simple defensive village architecture, but no evidence of military, e.g. barracks, stockpiled weaponry	Lack of standing military forces, conflict limited in space and time	Unknown; cost (material and time) of standing military forces likely too high for some time
Villages generally of the same size (no settlement hierarchy)	Fundamental independence of populations; de-emphasis on polity interaction	Unknown; emergence of political interactions difficult to predict
Economic independence	Fundamental production and consumption scaled to the household	Likely independent, settlement-level production and consumption for some time
Population in the 600–1000 individuals range	Populations larger than foraging groups, smaller than urbanized states	Likely relatively small populations for some time, with later emphasis on population growth (once surplusses are established)

beyond earth for these reasons. Figure 6.23 shows the basic plan of the polis of Miletus, which housed from 1000 to 5000 individuals at a time over its centuries of activity (Morris 2005:10). Figure 6.24 shows a reconstruction of a polis, with the architectural cluster at the center and the supporting farmlands beyond. My current work includes more investigation of the polis as a useful analogue for consideration and design in early settlements beyond Earth, though it is too specific a topic to discuss more in the present book.

Table 6.3 Features of the independent Neolithic Village and the ancient Greek *Polis* (see text for discussion)

	Independent Neolithic Village	Polis
Population	c.1000	Low 000s to c.10,0000
Economic character	Internal	Both internal and trade
Specialist manufacture	Unemphasized	Emphasized
Formality of long-distance trade	Low	High
Architectural character	Defensive	Defensive
Martial character	Defensive	Defensive but may include vigorous displays of force
Settlement hierarchy	Little = most villages of roughly equal influence	Some = some nucleation of power though general settlement equality
Domestic architectural hierarchy	Little = most families of roughly equal influence	More = some significant nucleation of influence by specific households or lineages
Ritual and religious formality	Low	Low to moderate

6.5 The Features of Civilization

The main difference between the INV and civilization as we experience it today is the population, geographical distribution and global interconnectedness of modern civilization. However, the size of early off-Earth settlements will be comparatively small, and they will have far fewer significant and structural connections with distant communities than we do today; there will not be a global but a local economy, for instance. Hence the focus I have just paid to the INV phenomenon and the polis. There are aspects, however, of *civilization*, not the INV, that will be in play among early off-Earth settlements, so we may consider that those will be hybrids of the INV and civilization. To best understand this we require a brief review of civilizations.

Archaeologists use the term *civilization* to refer to a very specific kind of human social and economic organization and not just any organized group of people. Civilization is not considered the apex of a hierarchy or ladder of perfection in human affairs, as we know that some types of social organization, such as the tribal, has persisted for millennia in the Arctic and Australia (for instance) and these organizations continue to function (note that in Table 6.4 in the final row, living groups are indicated in bold. This is meant to emphasize that different kinds of socio-economic organization still exist today and, though touched by civilization, are not 'relics' of an ancient time, but still viable in their areas and with their variety of subsistence). This is because (Smith 2013: 212):

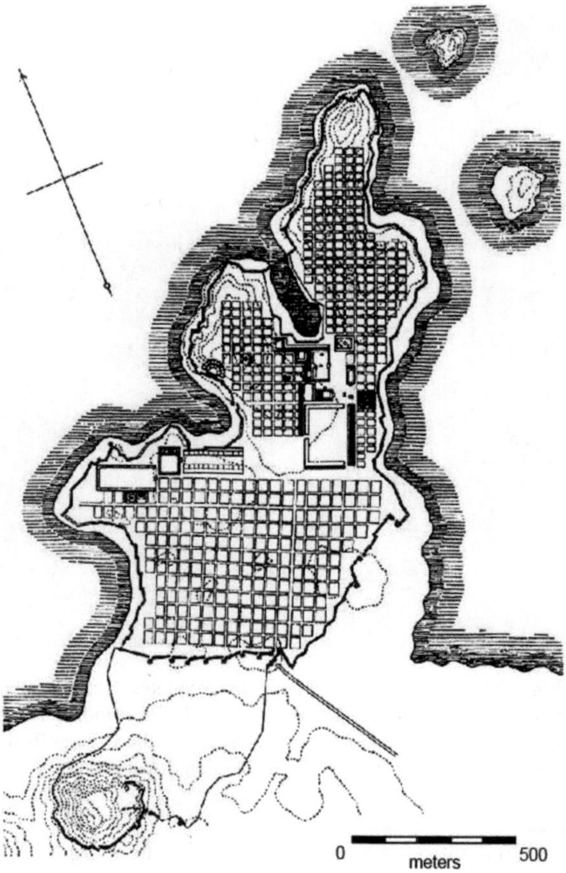

Fig. 6.23 Architectural plan of the Polis of Miletus. Public domain image from https://commons.wikimedia.org/wiki/File:Miletos_stadsplan_400.jpg

anthropology has found no internal drive for humanity to single-mindedly move 'through' [certain 'stages'], from one to the other. For example, agriculture simply will not work in the permafrost of the Arctic (or many other areas of the world), so in the Arctic people never developed agriculture or the highly populous, urban-dominated civilizations found where agriculture is practiced. Today, entirely cognitively modern human beings live in a foraging subsistence mode in many places worldwide, though their numbers are relatively low compared to the high populations of civilizations supported by a completely different mode of subsistence. The idea that all humanity would rise from one kind of existence to another—to use the archaic terms, from Savagery (foraging) to Barbarism (horticulture and pastoralism) to Civilization (agriculture)—has been shown by a century of anthropology to be false. While in some areas precisely this trajectory was in the past followed (foraging did give way to horticulture, and then agriculture and civilization), the point is that it is not an inevitable, internally-driven 'ascent'. It is simply how subsistence modes—and their attendant cultural and political correlates–played out in various environments worldwide in the last 10,000 years.

Fig. 6.24 Reconstruction of the Polis of Miletus. Note the suburban character, though there is some public architecture in the harbor-facing wall. These indicate both the household focus of the economy but also the slightly more urban character of the Grek polis than that of many earlier independent Neolithic Villages. See text for discussion. Image courtesy of Foundation of the Hellenic World

Civilizations are characterized by much larger populations and the use of agriculture as their subsistence base, so they are quite different from the tribes and chiefdoms of other and typically smaller populations, as summarized in Table 6.4.

Civilization is characterized by many traits, including cities, agricultural subsistence and military forces (all of which we will review below). The most influential set of these characteristics was published by archaeologist V. Gordon Childe (1892–1957) in 1950, and it remains, with some adjustment here and there according to research interest, a widely-used standard (e.g. see Trigger (1992) and Wenke and Olszewski 2006). Childe's essential characteristics of civilization are presented in Table 6.5. In the first column are the original descriptions based on Childe (1950); these are rephrased more concisely in the middle column, and finally (in the third column) I indicate the essential equivalents used in this book.

Table 6.4 Socio-economic organizations in the human experience

	Band	Tribe	Chiefdom	State or civilization
Subsistence	Foraging	Foraging/pastoralism	Horticulture	Agriculture
Residential mobility	High; seasonal moves across resource landscape	Medium/cyclic; may stay multiple seasons at a given resource area	Low; tied to farming plots	Lowest; tied to farming plots
Food storage	Little: days to months	Little: weeks to months, or 'meat on the hoof' (among pastoralists)	Medium, seasons to a handful of years (some stored food crops)	High, with reliance on staple foods stored in large quantities for years at a time
Property ethos	Low but present	Medium: among pastoralists, herded animals are property of individuals	Strong: elite classes own special objects unavailable to most of the population	Very strong; highly ranked members own objects forbidden to those of lower ranks
Social ranking ethos	Little: stratification; generally equal access to resources for all members	Medium: among pastoralists, families with more animals have higher rank	Strong: hereditary elite class exists, but has more power to coerce than command	Very strong: high rank can be achieved or ascribed, and access to resources depends on social rank
Population	10–150	Less than 200	Low hundreds to 1500	Tens of thousands to millions or billions
Example	Ba'Aka of Central Africa, Paiute of North American Great Basin, **Inuit** of Arctic Canada	**Maasai** of East Africa (cattle herders), Saami of Arctic Scandinavia (reindeer herders), Cheyenne of North American plains	**Maori** of New Zealand, Vikings of Medieval Scandinavia	Egypt, Greece, Shang (China), Maya (Mexico and Guatemala) United States, **Europe**

Subsistence indicates the main method of obtaining food and nutrients. *Residential Mobility* indicates the average amount of subsistence-related movement made during the course of a lifetime. *Food Storage* indicates the degree of reliance on stored food for subsistence. *Property Ethos* indicates the degree of emphasis on ownership of goods. *Social Ranking Ethos* indicates the degree of emphasis on differences in the availability of resources to population members. *Population* indicates broad and commonly-occurring group sizes

Table 6.5 Essential characteristics of early civilizations

Childe's comments (and page numbers from Childe 1950)	Essential point	Equivalent in this book
'In point of size the first cities must have been more extensive and more densely populated than any previous settlements.' (p. 9)	Population densities greater than can be supported by immediate environment	Urbanism
'In composition and function the urban population already differed from that of any village … full-time specialist craftsmen, transport workers, merchants, officials and priests.' (p. 11)	Some members of culture completely disengage from food production and are fully employed as artisans	Non-Food Production Specialists
'Each primary producer paid over the tiny surplus he could wring from the soil with his still very limited technical equipment as tithe or tax to an imaginary deity or a divine king who thus concentrated the surplus.' (p. 11)	Reliance on agricultural stores for subsistence; taxation of domesticated products from citizenry for	Taxation and Tribute Agricultural Subsistence Base
'Truly monumental public buildings not only distinguish each known city from any village but also symbolize the concentration of the social surplus.' (p. 12)	Use of permanent, highly-visible, costly structures as symbols of the supernatural domain and state power	Monumental Architecture
'But naturally priests, civil and military leaders and officials absorbed a major share of the concentrated surplus and thus formed a "ruling class". (pp. 12–13)	Centralized decision-making with delegation of administration to an hierarchy of permanent officers, involving political, military, and religious elites all led—most often—by a divine monarch	Hierarchical, Centralized Authority Social Ranking and Political Organization State Religion
'Writing.' (p. 14)	Use of high-fidelity non-biological memory devices to record business, administrative events and religious concepts in great detail	Durable Record-Keeping
'The elaboration of exact and predictive sciences – arithmetic, geometry and astronomy.' (p. 14)	Increasing detail noted in workings of celestial bodies leading to investigation of natural cause and effect, and quantification.	Astronomy and Mathematics

(continued)

Table 6.5 (continued)

Childe's comments (and page numbers from Childe 1950)	Essential point	Equivalent in this book
'Conceptualised and sophisticated [artistic] styles.' (p. 15)	Elaboration of material culture used to signal social rank	Concept embedded in Non-Food Production Specialists (artisans); see above
'Regular "foreign" trade over quite long distances.' (p. 15)	Formalized trade routes with full-time traders and standardized trade-related units	Long-Distance Trade and Standardized Measures
'A State organisation based now on residence rather than kinship.' (p. 16)	Relationships less organized by kin than by social position (class, status, rank)	Social Ranking and Political Organization
Not mentioned by Childe	Increased intensity, distance and duration of conflict. Maintenance of formalized borders	Standing Armies and Territorial Sovereignty

Each varies somewhat in expression in each civilization (including the modern) but these are real phenomena. See text for discussion
Tabulation by the Author based on Childe (1950)

6.5.1 Civilization Feature Manifestation Beyond Earth

Table 6.6 indicates the features of ancient (and most modern) civilizations, and suggests how they will play out in the early generations of space settlement, when populations will be relatively low, production and consumption will take place mostly on the household and community level, and political organization (referring to extension of power over large distances) will be minimal. Conditions of each possible beyond-Earth colony will slightly alter the manifestations of these structural variables.

What is most evident from this discussion and Table 6.6 is that early space settlements will be unique, not direct copies of civilization as it is experienced on Earth. They will be unique manifestations of human adaptation and arrangement. They will not have all the features of the Independent Neolithic Village, but will share its local scale and independence, reflected in its design and the influence of that design on culture at large. They will have many characteristics of civilizations, such as a high degree of specialized non-food-production activity, but there will also, at least early, be a large focus on food production; and there will not be a global, interconnected economy, but a rather smaller-scale economy that will include much household-scale production with somewhat automated technologies such as 3d printing. The lesson is that there is no good analog today, or in the past, for the precise features of early beyond-Earth settlements, but there are some analogues from the past that may be investigated to better inform early space settlement.

Table 6.6 Features of ancient and modern civilizations and likely manifestations in early space settlement (see text for discussion)

Feature of ancient and modern civilization	Required for space settlement?	Alternatives in space settlement
Urbanism	Large population, no; specialized activity, yes	New variety of specialized-activity center, perhaps similar to pre-Roman *oppidae* of central-Northern Europe
Non-Food Production Specialists	Yes	Yes; fabrication from hands-on to 'printing' from ISRU
Taxation and Tribute	Taxation likely, tribute (conquered vassal taxation) not required.	Some form of agricultural product stocking for emergency redistribution likely
Agricultural Subsistence Base	Yes	For some generations, growing area = small = local
Monumental Architecture	No	Infinite
Hierarchical, Centralized Authority	No	Wide variety of socio-political organizations applicable
Social Ranking and Political Organization	No	Wide variety of socio-cultural organizations applicable
State Religion	No	Wide variety of cultural organizations applicable
Durable Record-Keeping	Yes	Certain archiving and use of information as adaptive tool
Astronomy and Mathematics	Astronomy, no; Mathematics, yes	Certain use of mathematics as adaptive tool; astronomy likely but not mandatory
Long-Distance Trade and Standardized Measures	LD-trade, no; Standardized Measures, yes	Standard measures mandatory for efficiency; most trade quite local for generations
Social Ranking and Political Organization	Political Organization, yes; Social Ranking, no	Wide variety of socio-cultural organizations applicable
Standing Armies and Territorial Sovereignty	Standing Armies, no; Territorial Sovereignty, probably	Early settling cultures will not be able to afford prolonged conflict; protection of investment in settlements likely

6.5.2 *Lessons from the Durations of Ancient Civilizations*

It is often thought that space settlement is a proposition to save humanity from civilization collapse. While I would not use this argument in an urgent manner, suggesting that space settlement can be done quickly in an 'emergency mode', there are good reasons to think of space settlement as 'civilization insurance'. A brief look at the raw data for civilization and protocivilization durations in Chap. 1 indicated their repeated collapses or disintegrations. Here we look at a bit more of the raw data to identify patterns in the durations of ancient civilizations (and a few other varieties of social organization).

Table 6.7 indicates the durations of Independent Neolithic Villages, Civilizations and Superchiefdoms, Protostates worldwide; I also include the phenomenon of the representative democracy for reasons discussed below. We see that the large and better-know ancient civilizations have endured for some millennia, averaging in my estimation c.1700 years (though most experienced at least one temporary disintegration in its lifetime). The 'superchiefdoms' or 'protostates', which exhibit a mosaic of characteristics and are not considered full civilizations, average somewhat less, about 1500 years, and they are known to prehistorians to be rather unstable, with many collapses and reintegrations. I do not imply that global civilization is 'due for a collapse' based on these figures. Global, interconnected civilization is likely to be too distributed for any one catastrophe to disintegrate it completely. However, a vast amount would be lost in such a catastrophe. 'Civilization' may continue if Europe, America or East Asia were substantially disorganized by collapse, and much would be lost. And, of course, a few truly catastrophic scenarios for all of civilization are reasonable to consider, such as pandemic, earth impact by space debris, supervolcanism and a few others (for a discussion of these, see Smith and Davies 2012). For these reasons I think use of space settlement as insurance for the benefits and progress of human civilization is a good motivation, though not in a 'panic' mode. Nor should it ever be suggested that humanity means to abandon a resource-exhausted and burned-out Earth. That is a dystopian trope I have only ever encountered in science fiction and popular culture. In the world of serious space settlement planning, the discussion is always about small; populations of resource-responsible settlers beyond the home planet. I have never heard any serious suggestion that Earth would ever move more than a tiny percentage of its population from its surface, so 'relieving population stress', another science-fiction theme, is a nonstarter as well.

What is the use of pointing out these figures, then? As mentioned, to at least have an informed view when 'civilization collapse' becomes part of the discussion of space settlement. Also, it is interesting to note that the INV, as described, is a variety of human organization that has a long history. For example, Table 6.7 indicates that humanity's cumulative, global experience of the independent farming village has been about 66,000 years, or 66 millennia. By contrast, our entire experience of civilization has been 19,600 years or 19.6 millennia and our entire experience of somewhat-democratic state organization is on the order of 1300 years or 1.3

Table 6.7 Durations of independent farming villages, civilizations and superchiefdoms, protostates and representative democracies worldwide

	Civilization's political organization	Total years of independent farming villages before civilization in this region	Total years of civilization in this region	Total years of warring chiefdoms and feudalism (commonly after or between civilizations) in this region	Total years of representative democracy in this region
Civilization					
Mesopotamia/Sumeria	Dynastic Autocracy	5000	3000	2300	0
Egypt and North Africa	Dynastic Autocracy and Representative Republic	5000	2800	2300	300
Indian Subcontinent/Harappa	Unknown	5000	1200	3000	0
East Asia	Dynastic Autocracy	5000	2000	2000	0
Europe	Dynastic Autocracies and Representative Republics	5000	1500	1800	1000
Mesoamerican Complex	Dynastic Autocracy	5000	2000	1000	0
Andean Complex (e.g. Inka, Moche)	Dynastic Autocracies and Oligarchies	5000	1800	800	0
Mexico	Dynastic Autocracies and Oligarchies	5000	700	0	0
Subtotal		40,000	15,000	13,200	1300
Superchiefdom/Protostates					
SE Asia (e.g. Angkor)	Superchiefdom/Protostate	8000	800	300	0

(continued)

Table 6.7 (continued)

	Civilization's political organization	Total years of independent farming villages before civilization in this region	Total years of civilization in this region	Total years of warring chiefdoms and feudalism (commonly after or between civilizations) in this region	Total years of representative democracy in this region
Civilization					
Sub-Saharan Africa and Arabia (e.g. Great Zimbabwe)	Superchiefdom/Protostate	8000	800	300	0
NE Asia (e.g. Mongols)	Superchiefdom/Protostate	7000	1500	500	0
Pacific Complex (e.g. Maori)	Superchiefdoms	2000	1000	300	0
Mississippian Complex (e.g. Cahokia)	Superchiefdoms	1000	500	0	0
Subtotal		26,000	4600	1400	0
Grand total (years)		66,000	19,600	14,600	1300
Grand total (human generations)		2200	653	487	43

Chronologies are based on Wenke and Olszewski (2006). While archaeologists continually reshape their conceptions of the workings of civilizations, significant chronological revisions are rare

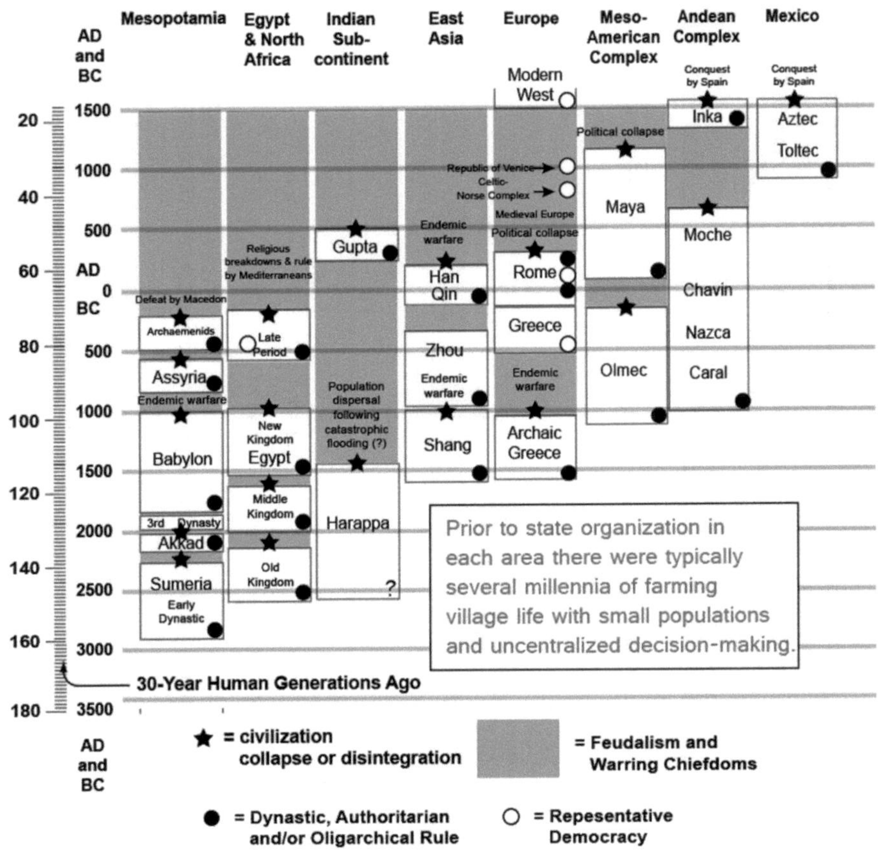

Fig. 6.25 Chronologies of ancient civilizations (see text for discussion). Figure by the author

millennia. For these reasons alone it seems worthwhile to explore the Independent Neolithic Village for its possible relevance to human space settlement planning.

Figure 6.25 reiterates Fig. 1.7d, but with more detail, indicating chronologies for the eight major ancient civilizations complexes. Stars indicate significant collapses, better termed *disintegrations* to reflect nonfunction of a previously-integrated socio-economic system. While the reasons for collapse are many, and sometimes debated (see Tainter 1990), it is plain that all prior civilization systems, across the globe and spanning several millennia, have collapsed multiple times (see also Weiss and Bradley 2001).

6.6 Conclusion and Recommendations

The lessons of the ancient world for improving the prospect of human space settlement are many. They may be found on nearly any scale at which one chooses to intelligently investigate the ancient cultures who so often adapted well to their own unique, new, challenging physical and cultural selective pressures. Material culture studies exhibit the human genius for proactive invention to overcome sever selective pressures. And studies of such phenomena as the Independent Neolithic Village and the polis may help us constructively imagine and plan future settlements beyond Earth, with lessons learned from such long-lived human organizations. Structured by the adaptive approaches described elsewhere in this book, I believe investigation into the ancient world will prove useful to the goal of permanent human space settlement. The only caveat is that, of course, one must not fall into the trap of considering people of the ancient world to have been perfectly adapted, and that we must return to ancient ways if we are to survive. Rather, I would use the ancient world's lessons to structure (not necessarily shape directly) our thoughts of the future beyond Earth, and to inspire rather than to blueprint. This is because, of course, life beyond Earth will approximate some ancient conditions, but not duplicate them. But utilized responsibly, the past provides much for the future.

6.7 Summary Points

- Human material adaptations to Earth habitats have been many and varied, particularly in habitation and settlement architecture and distribution, respectively.
- Ancient technologies may be studied for design principles expectably useful for early settlers beyond Earth; these include making conscious distinctions between expedient and curated artifacts, and always considering.
- Early space settlers will benefit from the study of the preindustrial exploration of Earth, which, though using simpler technologies, required skill in navigation and ingenuity in addressing a variety of terrains.
- The Independent Neolithic Village is a prehistoric variety of human settlement that has many productive analogues with early settlements we may expect beyond Earth, for example on Mars, the Moon, or other solar system moons.
- The distinctive features of human civilization will be carried to habitats beyond Earth, but early human civilization will differ in overall size and relative density from that of Earth.
- Specifics of how settlements grow and develop politically and economically are unlikely to be productive areas of detailed planning at this stage, even though some regularities of human economic and political change though time are evident in the archaeological and historical records.

References

Bandy, M. (2005). New world settlement evidence for a two-stage Neolithic demographic transition. *Current Anthropology, 45*(55), S109–115.

Bandy, M. (2007). Global patterns of early village development. In J-P. Bocquet_Appel & O. Bar-Yosef (Eds.), *The Neolithic demographic transition and its consequences* (pp. 333–357). New York: Springer.

Cessford, C. (2005). Estimating the Neolithic population of Çatalhöyük. In A. I. Hodder (Ed.), *Inhabiting Çatalhöyük: Reports from the 1995–99 seasons. Çatalhöyük research project 4* (pp. 323–326). Cambridge/London: McDonald Institute for Archaeological Research/British Institute of Archaeology at Ankara.

Childe, V. G. (1950). The urban revolution. *Town Planning Review, 21,* 3–17.

Damp, J. E. (1984). Architecture of the early Valdivia village. *American Antiquity, 49*(3), 573–585.

Ebersbach, R., & Schade, C. (2004). Modeling the intensity of linear pottery land use: An example from the Morlener Bucht in the Wetterau Basin, Hesse, Germany. In: K. F. Ausserer, W. Borner, M. Goriany, & L. Karlhuber-Vöckl (Eds.), *Enter the past: The E-way into the four dimensions of cultural heritage. CAA 2003. Conference Proceedings. BAR Int Series* (Vol. 1227, pp. 337–348). Oxford: Archaeopress.

Ehrenburg, V. (1969). *The Greek State*. New York: W.W. Norton & Company.

Hassan, F. A. (1981). Demographic archaeology. In M. B. Schiffer (Ed.), *Advances in archaeological method and theory: Readings from Volumes 1–4* (pp. 225–279). New York: Academic Press.

Hawass, Z., Hassan, F. A., & Gautier, A. (1988). Chronology, sediments and subsistence at Merimda Beni Salama. *Journal of Egyptian Archaeology, 74*, 31–38. https://www.jstor.org/stable/3821745.

Hawkes, J., & Wooley, L. S. (1962). *History of mankind; Cultural and scientific development: Vol. I. Prehistory and the beginnings of civilization.* New York: Harper and Row.

Heckenberger, M. J., Petersen, J. B., & Neves, E. G. (1999). Village size and permanence in Amazonia: Two archaeological examples from Brazil. *Latin American Antiquity, 10*(4), 353–376.

Higham, C. F. W. (2003). The origins of the civilization of Angkor. *Proceedings of the British Academy, 121,* 41–90.

Holley, G. R., Dalan, R. A., & Smith, P. A. (1993). Investigations in the Cahokia Site Grand Plaza. *American Antiquity, 58,* 306–318.

Howarth, I. (1978). *History to 404BC.* Milton Keynes: The Open University Press.

Kohler, T. A. (1992). Field houses and the tragedy of the commons in the early Northern Anasazi Southwest. *American Antiquity, 57*(4), 617–635.

Kujit, I. (2000). People and space in early agricultural villages: Exploring daily lives, community size and architecture in the late Pre-Pottery Neolithic. *Journal of Anthropological Archaeology, 19*, 75–102.

Lekson, S. H. (2004). Pueblo IV in the Chihuahuan Desert. In E. C. Adams & A. Duff (Eds.), *The protohistoric Pueblo World: A.D. 1275–1600.* Tucson, University of Arizona Press.

Morgan, C. (2003). *Early Greek stages beyond the Polis*. New York: Routledge.

Morris, I. (2005). *The growth of Greek cities in the first millennium BC.* Princeton/Stanford Working Papers in Classics. https://www.princeton.edu/~pswpc/pdfs/morris/120509.pdf.

Pauketat, T. R. (2004). The economy of the moment: Cultural practices and Mississippian Chiefdoms. In G. Feinman & L. M. Nichols (Eds.), *Archaeological perspectives on political economies* (pp. 25–39). Salt Lake City: University of Utah Press.

Porčić, M. (2011). An exercise in archaeological demography: Estimating the population size of Late Neolithic settlements in the Central Balkans. *Documenta Praehistorica, 38,* 323–332.

Price, N. P. S. (1977). Khirokitia and the initial settlement of Cyprus. *Levant, 9*(1), 66–89. https://doi.org/10.1179/lev.1977.9.1.66.

Shelach, G. (2011). The earliest Neolithic cultures of Northeast China: Recent discoveries and new perspectives on the beginning of agriculture. *Journal of World Prehistory, 14*(4), 363–413.

Smith, C. M. (2013). *Ancestors: An introduction to human evolution and prehistory.* San Diego: National Social Sciences Press.

Smith, C. M., & Davies, E. T. (2012). *Emigrating Beyond Earth: Human Adaptation and Space Colonization.* New York: Springer

Steponaitis, V. P. (1998). Population trends at Moundville. In V. J. Knight & V. P. Steponaitis (Eds.), *Archaeology of the Moundville Chiefdom* (pp. 26–43). Washington, D.C.: Smithsonian Institution Press.

Tainter, J. A. (1990). *The collapse of complex societies*. Cambridge: Cambridge University Press.

Trigger, B. (1992). *Understanding early civilizations*. Cambridge: Cambridge University Press.

Weiss, H., & Bradley, R. S. (2001). What drives societal collapse? *Science, 26*(291), 609–610.

Wenke, R., & Olszewski, D. I. (2006). *Patterns in prehistory*. Oxford: Oxford University Press.

Wills, W. H., & Windes, T. C. (1989). Evidence for population aggregation and dispersal during the Basketmaker III period in Chaco Canyon, New Mexico. *American Antiquity, 54,* 347–369.

Wilshusen, R. H. (2002). Estimating population in the Central Mesa Verde Region. In M. D. Varien (Ed.), *Seeking the center place: Archaeology and ancient communities in the Mesa Verde Region* (pp. 101–122). Provo: University of Utah Press.

Wüst, I., & Barreto, C. (1999). The ring villages of central Brazil: A challenge for Amazonian archaeology. *Latin American Antiquity, 10*(1), 3–23.

Chapter 7
Human Adaptation and Permanent Human Space Settlement

Abstract Chapters 2 and 3 have shown us aspects of biological adaptation necessary to consider in the project of human space settlement, while in Chaps. 4 and 5 we have examined some cultural aspects of such a project. In those chapters we were introduced to humanity's adaptive range and capacities, and some of the tools of human adaptation, from genes to complex behavior including culture and technology. Chapter 6 gave us some examples of the use of tools in humanity's deep past, showing that they are not unusual but should be considered entirely common and 'natural' to use. In this chapter we sharpen focus on these tools, refining our concepts as needed, to arrive at a practical, preliminary set of cultural and biological adaptive variables we deem to be necessary for human biocultural survival and sufficient to sustain it in environments beyond Earth. These tools may be thought of as the minimum adaptive toolkit of our species, significantly and informatively analogous to the HOX gene family common to all multicellular animal life. In these ways this chapter establishes a minimum set of adaptive tools for space settlement that may be studied as part of space anthropology and applied to the adaptive programme of space settlement established throughout this book.

> All growth originates with creative individuals or small minorities of individuals, and their task is twofold: first the achievement of their inspiration or discovery, whatever it may be, and secondly the conversion of the society to which they belong to this new way of life—A.J. Toynbee

> Designing spacecraft and space and planetary habitats for humans requires knowledge spanning a range of disciplines: engineering, medical sciences, psychology, human factors, life support systems, radiation protection/space weather, and other extreme space environments, at a minimum. These disciplines must result in an integrated human centered system, which should also be reliable, safe, and sustainable. This is space architecture—Bonnie Dunbar

7.1 Introduction

In Chap. 5 we discussed how space settlement may be initiated in a way structured by what we know of animal movement biology and evolution, looking to our knowledge of migration and dispersal for cues to a successful dispersal of our species beyond

C. M. Smith, *Principles of Space Anthropology*, Space and Society,
https://doi.org/10.1007/978-3-030-25021-8_7

Earth. We noted that once people are established in a new habitat (the bulk of the Chap. 5 material), they will need to *adapt* to it both biologically and culturally, and for multiple generations, before we may consider the project a success. Such adaptation is the topic of this chapter, which details the *adaptive domains of humanity* and, by sketching out their main features and functions in a culture, and how they change (adapt), suggests how they may be adapted to conditions beyond Earth. In this way I take an adaptationist and functionalist approach to culture, as I have through this book. These approaches—seeing culture's components as having specific functions—has a long history in anthropology, where it was most widely applied in the 1940–1970 timeframe (Eriksen and Nielsen 2013). One of functionalism's pioneers, Bronislaw Malinowski, tabulated the functions of cultural domains of behavior as summarized in Table 7.1. The concerns indicated here structured much functionalist anthropology, including the search for human universals. While it is clear that not every aspect of a culture, or every behavior of an individual, is functional in such a sense, many such behaviors are so functional, and many aspects of cultural organization are genuinely functional.

Through the 1970s and 2000s anthropology largely disfavored this approach, moving away from large-scale studies of culture to more consider specific cases, the individual, and a more insider's (*emic*) than outsider's (*etic*) perspective. This approach was taken to the extreme with postmodernism, which suggested that all knowledge was socially-constructed (a rather self-defeating claim) and that the search for large-scale (including evolutionary) patterns among humanity was sterile. This postmodern solipsism continues today, but I discount it but for a few of its contributions about the experiences of people and individuals in culture. Thus my approach does see cultures as having adaptive functions in maintaining the health of individuals and groups of individuals. But not every aspect of culture, of course, is easily mapped to some adaptive function, and of course cultures can be maladaptive or have maladaptive features—certainly we are not short of pointing this out about Western culture, with its many self-destructive features, and we should not be surprised that such may be found in other cultures as well.

The upshot of all this is that for our purposes in this book we must assume that yes, cultures have some recurring, important structural features that at least in part function to promote the health of the population, and are therefore adaptive. Cultural adaptations are sometimes obvious, and sometimes obscure. One evolutionary approach to culture suggests that *behavioral adaptations* they may be indentified when found to be *economic*, *efficient*, *complex*, *precise*, *specialized* and *reliable* (Cosmides and Tooby 1992). Identifying and evaluating each of these variables for a given behavior in a given culture is likely a work of many lifetimes. At the least they are suggestions for research approaches.

Throughout this book I have used evolution as our context for human space settlement. Table 7.2 reviews the cultural and biological varieties of evolution, and places adaptation in its context as one of several products of the evolutionary process.

While other chapters suggested what aspects of biology and culture might change in human adaptation beyond Earth, and occasionally sketched out the possible 'shape' or outcomes of such adaptation, the aim of this book has not been to make such

Table 7.1 Malinowski's functional aspects of culture

Basic needs (Individual)	Direct responses (Organized, i.e., Collective)	Instrumental needs	Responses to instrumental needs	Symbolic and integrative needs	Systems of thought and faith
Nutrition (metabolism)	Commissariat	Renewal of cultural apparatus	Economics	Transmission of experience by means of precise, consistent principles	Knowledge
Reproduction	Marriage and family				
Bodily comforts	Domicile and dress	Characters of behavior and their sanctions	Social control		
Safety	Protection and defense			Means of intellectual, emotional, and pragmatic control of destiny and chance	Magic Religion
Relaxation	Systems of play and repose	Renewal of personnel	Education		
Movement	Set activities and systems of communication				
Growth	Training and Apprenticeship	Organization of force and compulsion	Political organization	Communal rhythm of recreation, exercise and rest	Art Sports Games Ceremonies

For the functionalist, such broad concerns as feeding the individual and ensuring their safety, and that of offspring, are carried out by activities carried out by members of a culture, guided by some basic domains of behavioral guidance, such as religion. At this basic level, culture is indeed adaptive. See text for important discussions

Table 7.2 Aspects of adaptation pertinent to human space settlement

	Biological manifestation	Cultural manifestation	Biological rate and specificity of innovation	Cultural rate and specificity of innovation
Adaptation as noun	Developmental, anatomical or physiological trait	Learned behavioral trait including the fabrication and use of artifacts	Slow (multi-generational in most cases); specificity can increase only through generations of selection. Can be rapid with human genetic engineering	Rapid (intragenerational–cognitive innovation is nearly continuous and artifact prototyping for testing and refinement are rapid); artifacts may be immediately specific to a given environmental feature to be manipulated
Adaptation as verb	A process of natural selection for and against largely randomly-arising developmental, anatomical and physiological traits, resulting in better fit of organism to environment over generations	A process of cultural selection for and against courses of action (behavioral adjustments) consciously and proactively evaluated against performance criteria	Necessarily multigenerational in non-humans, and necessarily undirected regarding the origin of variations, which may not be beneficial. In humanity however these may be rapid and specific, as in the immediate effects of genetic engineering	Potentially very rapid (intragenerational) and precise; cognitive innovation is nearly continuous as noted above, however note that actors wishing to preserve cultural norms may significantly dampen or slow the spread of innovations and must be considered

predictions. Rather, I wish to outline a field of study, to indicate what needs to be studied, and provide guidance for ordering our study such that better plans for space settlement might be made.

The main adaptive tools of humanity so far reviewed have been:

- individual physiology and genetics/genetic therapies (Chap. 2)
- population genetics/genetic therapies (Chap. 3)
- Human Cultural Universals as Adaptive Tools (Chaps. 4 and 5)
- Human Material Universals as Adaptive Tools (Chap. 6).

The sections that follow below sharpen our understanding of adaptation and suggest a variety of tools for studying human adaptation to beyond-Earth conditions.

7.2 Focus on Adaptation

The term *adaptation* has been variously used and rather poorly-defined—or at least variably defined—for its history in the life- and social sciences. The American naturalist Ernest Lankester (1847–1929) outlined the principles of adaptation early in the 20th century AD, indicating that adaptation was the *adjustment of internal living system properties in relation to external conditions* (see for example Lankester 1915). More specifically, adaptation may be considered as a noun, *a thing or process that allows life where it previously was not able to survive*, or as a verb, a *process of adjusting internal states to external conditions*.

For our purposes the point is that humanity will not simply 'adjust' to conditions on Mars, for example, or an orbital or Lagrangian colony, or an asteroid belt habitat. The processes of adjusting to these new places can be better understood and planned when specified as adaptation, an evolutionary phenomenon well-known to biology. The thrust of this entire book is that we would be wise to learn from over three billion years of adaptation Earth life to better the chances of success in the project of extending live beyond Earth. We have done this to an extent with biological evolution, but here we must sharpen focus on cultural evolution.

7.2.1 Cultural Evolution and Adaptation

We have used the term culture throughout this book in a rather general sense, but it is appropriate at this time to indicate subtleties of the use of this term. Table 4.1 indicated multiple uses of the term in this book and in much of the research literature that readers of this book might encounter. For our purposes, the definitions in the first three rows are appropriate.

While biological adaptation is well-studied, a modern body of theory of cultural adaptation is only just being sketched out and investigated. Early, 19th-century approaches to culture change and adaptation crudely misinterpreted Darwin's theory of biological evolution, resulting in a highly-structured conception of cultural evolution. In this model, all human cultures were internally-driven by an adaptive process that led up through predetermined stages, from Savagery to Barbarism and then Civilization. When it was, decades later, found that this conception was false, and that human cultures adapted to a myriad of environments, each taking its own course with no internal 'striving' to become 'civilized', anthropology at large dropped evolutionary attempts to explain culture change. Still, some anthropologists persisted with an evolutionary framework for culture change that lacked the 'social Darwinism' misunderstandings just mentioned. This was because they noted that while a 'cultural striving' to 'climb the ladder' to Civilization was not apparent, there were significant patterns in cultural histories that suggested parallels with biological evolution. For example, while human cultures (numbering perhaps several thousand in the early 20th Century) varied widely in details of their political organization, that variation was clearly variation on a pattern, a recurring set of modes of political organization. That is, whatever the details, all human cultural political systems could be legitimately classified into four main varieties, as was presented in Table 6.4 (which also indicated concomitant patterning in *subsistence mode*, the manner of obtaining water, calories and nutrients).

Because such broad patterning was evident in human culture, some anthropologists persisted with evolutionary attempts to explain them, though not using the earlier and inaccurate conception of cultural evolution through predetermined stages. George P. Murdock, for instance, characterized and studied human cultures as adaptive phenomena, as summarized by medical anthropologist A. Alland, Jr.: "Murdock's statement of evolution is totally nonlinear. The process is seen as a series of adjustments to internal conditions which occur within particular types of social systems…his major focus…has always been on [cultural] adjustments which develop [in cultures]…" (Alland 1970: 38). Another anthropologist, A. Radcliffe-Brown saw cultural adaptation simply as an extension of biological adaptation, writing that "What must be emphazised is that [biological and cultural adaptation] are only different aspects [by which] the total adaptational system [of humanity] can be looked at for convenience of analysis and comparison" (Radcliffe-Brown 1952: 9). Recent work in developing an evolutionary approach to culture change also sees it as at least partially adaptive; a 2016 review paper titled '*Multilevel Processes and Cultural Adaptations: Examples from Past and Present Small-Scale Societies*' states that "cultural (mal) adaptive dynamics relate to multilevel socio-ecological processes occurring at different scales, where the lower levels combine into new units with new organizations, functions and emergent properties or collective behaviors" (Reyes-Garcia et al. 2016) The history, approaches and results of culture-evolution theory are the topic of many books, and for our purposes I simply wish to point out that in the modern approach, culture is seen simply as an adaptive tool by which humanity, in particular, adjusts itself and (often) its environment for survival and propagation (Smith et al. 2018). This does not mean that all cultural behaviors or features may be explained directly as environmental adaptations; human behavior is so complexly-driven that it can stray far from the material requirements of survival. And, this view does not see all cultural features as perfect adaptations; in *Sick Societies*, anthropologist R. B. Edgerton explores maladaptation, in which cultural behaviors are detrimental to culture practicing them (Edgerton 2010). This should not be surprising as biological adaptation can also be imperfect, with a behavior having both advantages and disadvantages or costs to the organism. But enough cultural behavior is adaptive that the study of this adaptation has been a useful analytical framework for many anthropologists.

7.2.2 Adaptive Phenomena

Adaptation is carried out by living organisms in three main ways. *Behavioral* adaptations include moving out of the sight of a predator, or diving to a certain depth of water for a certain food. Such behavior is largely instinctual in most animal life, sometimes supplemented by learning and/or innovation during the course of life. Behavioral adaptations are of course a significant aspect of human behavior and they differ mainly in that instructions for them are cumulative over time; that is, no other animal life passes on their innovations and lessons learned, to the next generation, in a way remotely as rich and complex as humans. *Structural* adaptations are largely anatomical and are largely the result of evolutionary variation and selection, outlined in Chap. 2; an example is the flower-fitting beak of the hummingbird. Finally, *physiological* adaptations are *bodily processes* that benefit an organism due to their better fitting the organism to the environment; these include highly efficient blood oxygenation in diving mammals. In most animal life these modes of adaptation are entirely genetically transmitted (aside from some learning) from one generation to the next in a reactive way; life forms are unaware of the evolutionary process and are simply born with these features that fit the environment for better or worse. Unique to humanity, it seems clear, is the understanding of the process of evolution and the cognitive capacity to envision a multigenerational future and to proactively prepare for that future by inventing adaptations. While these inventions have largely been technological (an igloo, a stone arrow-tip, a lunar suit) and behavioral (planting seeds, consciously migrating from one resource patch to another, unseen and known only through detailed traditions that describe a landscape), recently we have developed the ability to genetically alter our offspring (and the properties of many other life forms) as well. As with any technology, there are moral implications of genetic alteration by humanity that are beyond the scope of this book; they may be explored in many other works including the good introduction, "*Genetic Engineering: Moral Aspects and Control of Practice*" (Genetic Engineering 2017). Some instances and details of adaptation are presented in Table 7.2, which may serve as a useful heuristic for space settlement planners.

A chief lesson of Table 7.3 is just how rich and potentially capable is the human adaptive suite, which includes both reactive and proactive adaptation, the capacity to adapt very rapidly and very specifically. No other life forms possess such adaptive capacity, and only humanity, that we know of, possess an understanding of the passage of multigenerational time, evolutionary principles and the proactive capacity to specifically shape their elective environment, as explored in Table 7.3, which

Table 7.3 Aspects of niche construction

	Perturbation	Relocation
Inceptive	Organisms initiate a change in their environment by physically modifying their surroundings	Organisms expose themselves to a novel selective environment by moving to or growing into a new place
Counteractive	Organisms counteract a prior change in the environment by physically modifying their surroundings	Organisms respond to a change in the environment by moving to or growing into a new place

Humanity has been very effective in altering its social and material environments to benefit its offspring. That this has been morally unfair (favoring some offspring, e.g. caste members, over others) and is currently becoming unsustainable on a global scale are in a theoretical sense immaterial; the point is that humanity is superior at niche construction. Space settlement is best considered in the table cell indicating Inceptive Relocation. Table adapted by the Author from Table 2.1 of Odling-Smee et al. (2003)

summarizes the phenomenon of *niche construction*. In this field of study, a niche is a range of behavior and/or material adaptations that actively promote the survival of a current generation and shape the predicted selective environment of the future generations to their benefit. Niche construction, then, is shaping the environment to one's benefit, and for the benefit of descendants. Humanity is the most prolific of niche constructors (and has done plenty of niche destruction as well).

Tables 7.1, 7.2 and 7.3 indicate that if any known Earth species has a chance at successful expansion beyond the boundaries of Earth—which, I argue, are more psychological and philosophical than material—it is the human species. If we are to use these adaptive tools, we should know as much as we can about them. I have mentioned this earlier in this book, but here we bring the issue to a point: to use these tools we cannot proceed with the common misconception of evolution as an internally-guided process in which inevitable progress is made and, when applied to humanity, essentially presumes that space settlement will happen naturally, just as a consequence of time and our 'innate drive to explore'. I argue that, rather, while there are long-term trends in evolution, currently it is not clear that space settlement is inevitable to succeed, and it is more likely to succeed if we realize the breadth and power of the adaptive tools at our disposal, and deploy them. This will require cultural motivation.

Adaptation takes place on a number of biological and cultural timescales, as outlined in Fig. 7.1. A chief lesson here is that that to ensure long-term success in space settlement, we must plan for and carry out plans on the ecological timescale, not just the shorter timescales of political attention or even the lifetime of the individual. Space settlement will not be successful until it is successful over multiple generations. When would we deem a population 'safe'? I suggest several centuries, a timescale on which significant biocultural adaptation may occur and may have undergone shocks, to which such adaptations have been resilient.

A timescale so far determined for various medical issues related to human exposure to microgravity is seen in Fig. 7.2. This indicates that whether or not symptoms

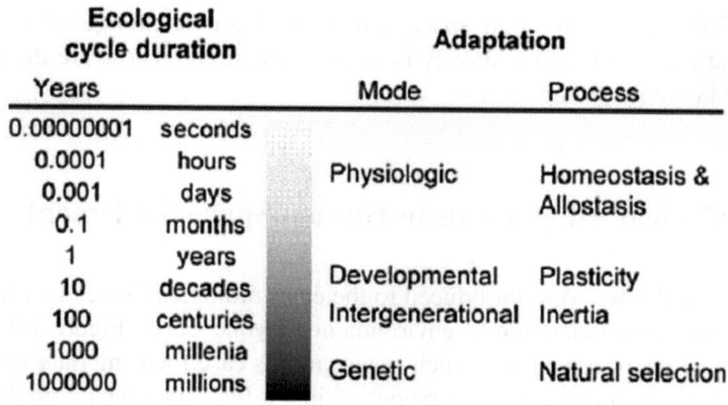

Fig. 7.1 Timescales of biological and cultural adaptation. Diagram by the author

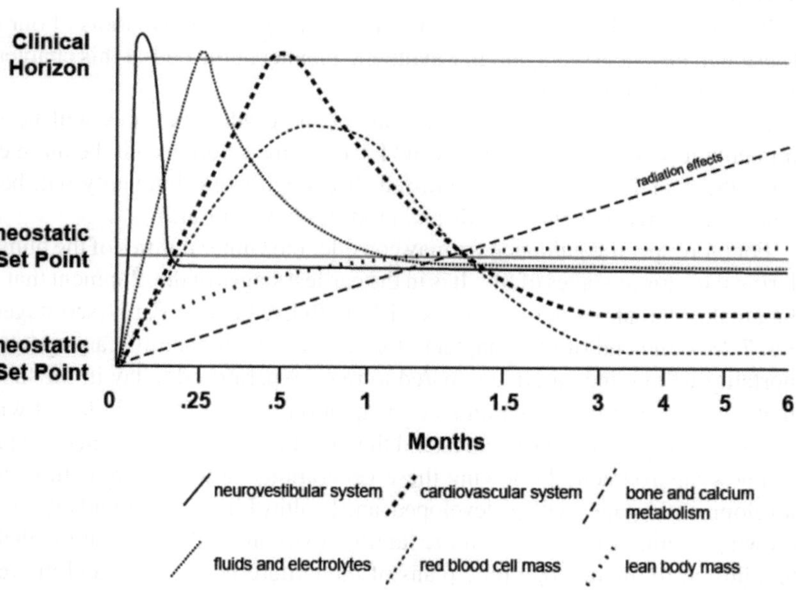

Fig. 7.2 Timescales of human adaptation to microgravity conditions. Diagram adapted by the Author after Wotring (2017). See text for discussion

of disequilibrium resulting from human exposure to microgravity conditions appear (in the diagram, crossing the line of *clinical horizon*) adaptive mechanisms tend to reach an equilibrium or homestatic state after several months (aside from radiation effects, which are cumulative over time and may be partly mitigated by poorly-known DNA repair phenomena). We may apply such an appreciation for the timescale of biological adaptation to beyond-Earth conditions to the question of biocultural space adapation. On what timescales can we expect human biology and culture change not just over a standard 6-month space station tour, but over *multigenerational* time, and in relatively small and relatively isolated populations? These are the questions addressed in the following section.

7.3 Biological Adaptations in Human Space Settlement

In Chaps. 2 and 3 we were introduced to the chief biological issues involved in the prospect of human adaptation to environments beyond Earth. Figure 7.3 indicates the extremely nuanced nature of such issues, in this case focusing only on the periods of embryonic and fetal development of just a few functional complexes. The complexities indicated here suggest that at present a summary portrait of the human biological adaptive system may not be completed, but only fleshed out, as research is itself in a very early phase.

To this end in Table 7.4 we see the chief biological adaptive tools of our species. These will be discussed again in a synthetic manner at the end of this chapter, which focuses on cultural adaptation.

For the moment we may say that all of these adaptive tools will be used by humanity to settle permanently beyond Earth. Some solutions will be more effective than others, and some attempts will fail. If enough work, humanity will be able to adapt to the environments it finds and modifies beyond Earth.

For more specific guidance we may consider the vulnerabilities of the human body during the various stages of life. It is in the earliest stages of development that perhaps the greatest threats exist for humans wishing to live beyond Earth (see stages 1–4 in Fig. 7.23 at the end of this chapter). It seems inevitable that miscarriage and infant mortality will be increased, compared to rates experienced today in the first world, in at least the first few generations of adaptation beyond Earth; in fact it will likely take at least a generation to understand the inevitably unexpected selective pressures of a new environment. Following these generations, however, protections for these developmental stages will be developed, and healthy birthrates should return, perhaps allowing a return to conditions more familiar to us on Earth today, although they will be adjusted to the demographic plans of the culture in question, and the resources available. One likely scenario is that when healthy birth rates recover, a rapid population growth will be promoted to safeguard the population from the dangers of low populations (outlined in Chap. 3). An equally-likely scenario is that despite a high rate of infant mortality and miscarriage, populations will be encouraged to grow rapidly from the beginning of settlement time. As mentioned frequently in this book, such plans are not my goal at this stage. Currently I wish to outline the main problems that will have to be solved, and guidelines to studying these problems for devising adaptive, evolutionarily-informed plans.

Later in the course of life, cultural means are much more involved (stages 4–7, inclusive, of Fig. 7.23) in protection of the life course. These include activities in many realms of cultural activity including enculturation, the finding and sanctioning of appropriate marriage partners and activities related to rearing the next generation,

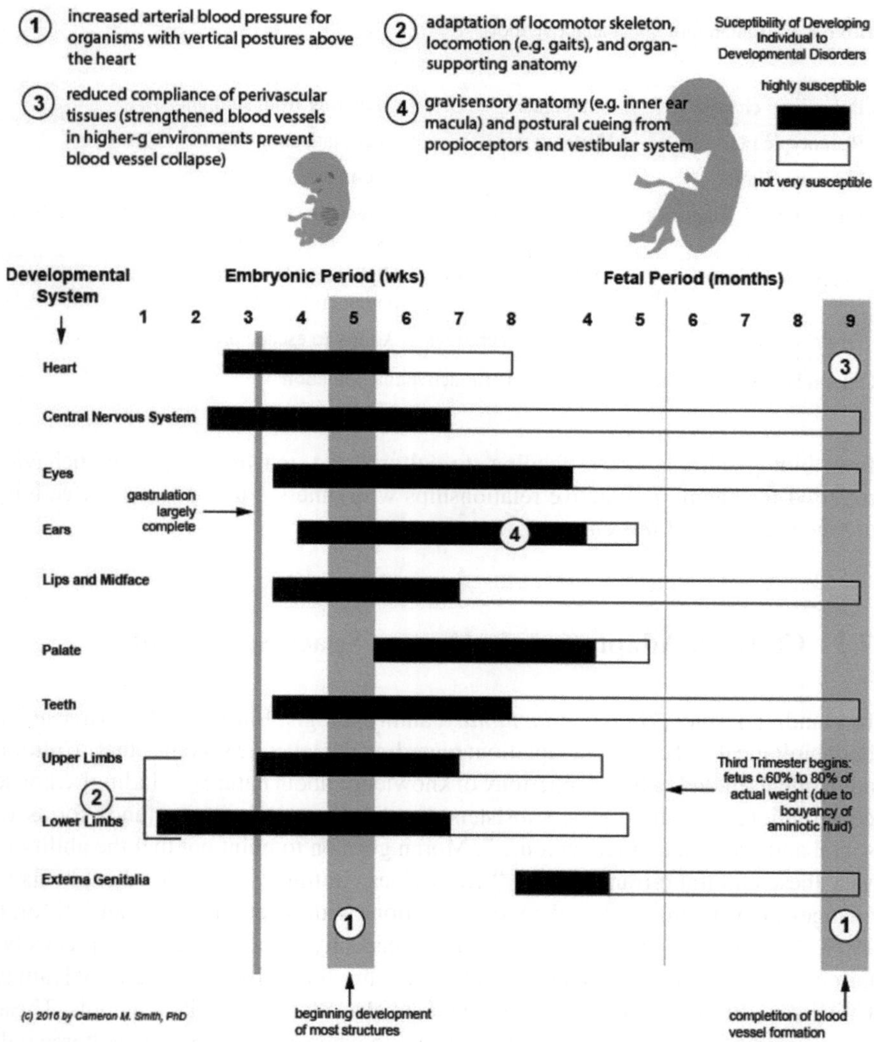

Fig. 7.3 Developmental in human embryo and fetus showing suceptibility to interruption by teratogens and altered gravitation. 1, 2, 3 and 4 indicate periods when the gravitationally- influenced biology of the vertebrate system is affected early, middle and late in development. Diagram by the author based on Carlson (2014) Fig. 8.6 and resources identified in Chaps. 2 and 3

Table 7.4 Human biological adaptive tools

	Timescale	Example
Behavioral change	Immediate	Sheltering from sunlight
Phenotypic plasticity	Hours to days	Acclimatization
Tolerance limits	Seasons	Acclimation
Rapid genetic adaptation	One generation	Epigenetic adaptation and natural selection (particularly on young)
Reproductive rate	Multiple generations	Highly regulable according to conditions; perennial fertility
Dispersal capacity	Multiple generations	Ability to escape predicted catastrophe

Only in humanity are these tools available for active manipulation

from biological, e.g. breast feeding, to cultural, e.g. teaching norms of behavior required for having productive relationships with others. Such domains of activity are the real focus of this chapter.

7.4 Cultural Adaptations in Human Space Settlement

In a landmark study, *Human Adaptability*, anthropologist Emilio F. Moran discussed both biological and cultural adaptation; regarding the latter he specifies that "Cultural adjustments include a broad repertoire of knowledge about nature [including]...house construction, clothing styles, subsistence technology, and ritual...[and] forms of social and economic organization...". Moran goes on to point out that the ability to vary these cultural 'arrangements' "....provides...infinitely variable adjustments to changes in the habitat and to changes in relations with other human groups" (Moran 1979: 95). This really is the key to humanity's capacity for proactive, rapid, precisely-tailored adjustments or adaptation. In Chap. 4 of this book I introduced a set of Human Universals, cultural domains of action that are also adjusted as adaptive tools. These were introduced as a guide to thinking, because when we wonder how cultures will adapt to new environments beyond Earth, it is easy to get lost—should we think about the religion, the economy, the artistic expression...where do we start? I suggested beginning to structure our thinking by thinking first of these Human Universals that are used by all human groups in their adaptation. After a brief introduction to an evolutionary approach to this topic we will survey what is known of the field.

In his evolutionary approach to human culture, W. H. Durham listed three reasons for developing a 'sequential [read 'evolutionary'] transformation theory of cultural change': (1) to give a realistic time dimension to living cultures, (2) to use this dimension to understand "...the historical processes through which people have composed, edited and revised the [symbols] that give meaning and direction to their lives", and (3), to "account for trends in the historical emergence and divergence of ideational systems" (Durham 1991). What specifically could such a theory of cultural evolution explain? Generally, the patterns and variation in human cultural expression which would in part be manifest as the universals and domains of adaptation noted above. Such broad patterns are useful to review briefly, below, as further guides to what is most widely shared in human cultures, as these features or behaviors are likely to represent adaptations or fields or domains of behavior that may be thought of as adaptive tools. One further way to identify such broad patterns is to identify the organization of anthropology—which has the function of studying human culture—itself.

A refinement of the concept of adaptation by cultural universals is appropriate at this stage. In Chap. 5 I introduced Human Cultural Universals as 'adaptive tools'. These, derived by my own work from the work of anthropologist Brown (1991) are listed below again for convenience and for contextualization with such other related topics in this section of this book:

- Language
- Ethics
- Social Roles
- The Supernatural
- Styles of Bodily Decoration
- Family Structure
- Course of Life Stages
- Reproductive Behavior
- Food Preferences
- Aesthetics
- Ultimate Sacred Postulates.

A similar conception of widely-recurring *adaptive cultural domains* was proposed in 1968 by anthropologist Yehudi Cohen. The following list indicates these domains of cultural behavior in order of their degree of shaping by the culture's environment, both physical and cultural. That is, items nearer the top of the list are *least* affected by the environment (though they may well be effected by the environment) and are more likely to be conditioned by such unpredictable variables as culture contact, spontaneous cultural innovation and the varied, individual history of a given culture, aside from its more general and universal dynamics. In contrast, items lower on the list are *more* structured by the environment. In this way the ordering reiterates the conception that *infrastructure conditions superstructure*, with items nearer the lower end of the list being infrastructure (Cohen 1968). We may also note a similarity here to the levels of Maslow's 'Heirarchy of Needs' of the individual, though that concerns the individual rather than the culture.

- Symbols
- Rituals and Etiquette
- Language
- Literature and other Arts
- Cultural Values (moral system)

- Law and Religion
- Economic Organization
- Settlement Pattern
- Technology and Political Organization.

A review of these domains with the Human Cultural Universals reveals significant overlap, again suggesting that these scholars were observing the essential adaptive domain of human behavior. As another guide to the essential features of human culture that recur structurally, but are varied in expression across all known cultures, is the organization of Cultural Anthropology textbooks. Just as all texts on biological evolution will carry chapters on genetic inheritance, genetic variation, phenotypic expression, selection, adaptation and so on, most modern Cultural Anthropology texts convey the findings of several main topics. For example, a review of several modern, widely-used cultural anthropology texts (see Table 1.2, Chap. 1) revealed a similarly-consistent set of themes explored by that discipline, seen below:

- The Culture Concept
- Language
- Kinship/Descent
- Power Relations
- Sex and Gender
- Equality and Inequality
- Religion
- Economy/Subsistence
- Myth, Ritual and Symbol.

Broadly speaking, these are the topics that cultural anthropology is used to explain. They recur here (and through the history of academic anthropology) not because they are just concerns of the present day or even recent, theoretically-biased anthropologists, but because human behavior is not random; rather it is to some variable degrees patterned in ways that address the essential requirements of biologically- and behaviorally-modern humanity. They are here identified because these texts' organization—just as the organization of topics in introductory mathematics or physics texts, for example—reveals the overarching issues explored.

Based on these surveys, in Table 7.5 I indicate the apparent correspondence of the stated Human Cultural Universals and Cohen's Adaptive Domains; I also comment on the nature of the correspondence between these, and finally I place in the third column the recurring anthropological textbook topic that could conceivably encompass the corresponding items in columns 1 and 2.

A review of Table 7.5 indicates that significant, structuring variables of human cultural organization and behavior recur worldwide and are expressed differentially worldwide. Such wide recurrence globally in modern and ancient cultures suggests that these fields of activity do not recur as coincidence. They likely recur as a result of common descent and common use as adaptive tools, addressing the 'heirarchy of needs' of humans based on the variety of life that humans are (sexually-reproducing multicellular large social primates, and so on) in the kinds of environments we inhabit

Table 7.5 Correspondence of human cultural universals with Cohen's adaptive cultural domains and recurring topics in cultural anthropology

Human cultural universal based on Brown (1991)	Corresponding adaptive domains (Cohen 1968)	Note on nature of correspondence (of items in columns 1 and 2)	Corresponding field of study (as used in this book)
Language	Language	Equivalent concerns	Language
Ethics	Law and Religion, Cultural Values (moral system)	Ethics are practical application of moral foundations	Power Relations
Social Roles	Rituals and Etiquette	Social roles are reiterated and reinforced in ritual	Kinship and Descent, Power Relations
The Supernatural	Law and Religion	In many societies, secular law is heavily shaped by religious concepts	Myth, Ritual and Symbol
Styles of Bodily Decoration	Literature and other Arts	The 'arts' in general reflect and contribute to styles of bodily decoration	Myth, Ritual and Symbol
Family Structure	Economic Organization	Family structure is strongly conditioned by economic concerns	Kinship and Descent, Power Relations
Course of Life Stages	Rituals and Etiquette	Life stages are reiterated and reinforced in ritual	Sex and Gender
Sexual Behavior	Rituals and Etiquette	Sexual behavior norms are reiterated and reinforced in ritual	Sex and Gender, Equality and Inequality
Food Preferences	Economic Organization	Food preferences are strongly conditioned by economic concerns	Economy and Subsistence
Aesthetics	Literature and other Arts	–	Myth, Ritual and Symbol
Ultimate Sacred Postulates	Law and Religion	–	Myth, Ritual and Symbol
–	Settlement Pattern	No clear equivalent	Economy and Subsistence
–	Technology and Political Organization	No clear equivalent	Economy and Subsistence

See discussion in the text for important details

(providing an omnivorous diet, in a certain temperature and atmospheric range). These seem to be the domains of action necessary for human cultural existence and also malleable enough for adjustment to the conditions of new environments (both physical and social or cultural). They seem to be the adaptive toolset of human cultures.

7.4.1 A Refined Set of Human Adaptive Universals

For convenience and to provoke thought, below in list form I propose a set of Adaptive Domains of Human Culture derived from the discussion above; this is a hybrid of the Cultural Universals and Adaptive Domains, informed by the organization of the field of Cultural Anthropology. Further, in Table 7.6 I tabulate these items, in order of their direct influencing by the material environment (least at the top and most at the bottom of the list). I also indicate, where appropriate, the magnitude of *options* and *constraints*. Options generally refers to whether the domain in question has been observed to rather mutable or malleable, in which case there are many options, and constraints indicates the nature of constraints that might be most responsible for dampening the field of options just mentioned. Adaptive domains of action that can manifest in many ways and have few constraints to their shape are highly adjustable to new environments, whereas those that exhibit few options and have many and/or strong constraints are less adjustable.

The Adaptive Domains of Human Culture I suggest as a starting point for serious space-settlement planning, based on our survey of human cultures in this book, are listed below in bullet form, and presented with more details in Table 7.6. While it may seem that subtle changes in these domains, to accommodate new conditions beyond Earth, are trivial, in sum they will constitute the evolution of new human cultures.

- Language
- Food Preferences
- Arts and Aesthetics
- Morality, Ethics and Justice
- Religion and Mythos
- Enculturation
- Kinship and Descent
- Social Organization and Power Relations
- Family Structure
- Economic-Political Structure
- Settlement Pattern
- Subsistence Mode.

Are the domains of adaptive behavior, corroborated as adaptive features of human cultures between themselves and as they have conditioned the fields of study of human culture, sufficiently organized in the list proposed above? Sufficient, that is to allow

Table 7.6 Adaptive domains and subdomains of human culture

Adaptive domain	Significant subdomain	Options	Significant constraining variables
Language	*Grammar*	Many	few; largely cultural
	Syntax	Many	few; largely cultural
	Vocabulary	Many	few; largely cultural
	Dialect	Many	few; largely cultural
	Gesture	Many	few; largely cultural
Food Preferences	*cuisine tools*	Many	Many
	cuisine ingredients	broad human palate	import cost if not local
Arts and Aesthetics	*styles of bodily expression*	Many	Few
	music and singing	Many	Few
	static visual arts	Many	Few
	static sculptural arts	Many	Few
	motion arts	Many	Few
Morality, Ethics and Justice	*moral system*	Many	Few
	ethical system	Many	Few
	administration of justice	Many	Few
Religion and Mythos	*Ultimate Sacred Postulates*	Many	Few
Nurturing and Enculturation	*stages of life*	Few	developmental biological and demographic
	system of enculturation	Many	some, related to
Kinship and Descent	*Linearity*	Many	Few
	marriage prohibitions	Few	Demographic
	property transfer & inheritance	Few	Demographic
	sexual activity regulation	Many	Demographic
Social Organization and Power Relations	*social ranking ethos*	Several	Demographic
	power relations	Several	Economic
Family Structure	*household size and composition*	Many	subsistence, political organization
Economic-Political Structure	*essential organization*	Several	resource availability and demand

(continued)

Table 7.6 (continued)

Adaptive domain	Significant subdomain	Options	Significant constraining variables
	technological means of production	Several	resource availability and demand
	organizational mode of production essential political organization	several	resource availability and demand
Settlement Pattern	*essential organization*	Several	resource availability and demand
Subsistence Mode	*essential organization*	Few	resource availability and demand

Note these are necessary for multigenerational success of human cultures; they are as noted adjustable, to varying degrees, for the conditions of habitats beyond Earth. The table is organized with features most shaped by the physical environment at the bottom (lower rows), but this does not mean that features higher cannot be strongly shaped by environment, or that features lower on the list cannot be influenced by input from features higher on the list; if such 'determinism' were evident, human cultures would not have defied easy analysis for the past century of scientific investigation. Note also that adaptive features are normally strongly intertwined, each shaping, to some degree, others, and that such 'entanglements' vary by culture and through time. For these reasons the Adaptive Domains listed, while legitimate, may best be considered as broad domains that may have 'fuzzy' boundaries, but are useful and real adaptive tools of our species. Significant Subdomains (second column) are commonly-varying facets of the adaptive domain noted in the first column

human adaptation to environments beyond Earth? Of course this depends on the environment's detailed properties, and those of the adapting population, but we must begin somewhere. Below I discuss each of these variables specifically with regard to space settlement and where appropriate I suggest fields of study that may help refine understanding of these variables and use of them as adaptive tools. Note that while these general domains of activity were introduced in Chap. 5, in the discussion of Human Cultural Universals, here I focus on adaptive diversity and potential and the ways in which these 'adaptive tools' vary, showing us something of how they change, as an aid to considering their roles in space settlement.

7.4.2 Language

All human communication may be considered under the umbrella term *language*. Languages may be highly distinctive and culture-specific, as that of the Inupiat of the high Alaskan Arctic, however they normally have some overlap with other languages due to a shared ancestry and/or culture contact; the Inupiat Alaskan language has enough overlap with adjacent Canadian Arctic Inuit language that speakers of each can communicate passably. Languages are used to describe a culture's physical and social environments and they are a key adaptive tool; the better a language maps to or represents a given physical and social environment, the better the members of that language's culture may fit their actions to these environments for their various goals. A good example is the use of the term 'wasteland'. For early European Arctic explorers, the Arctic was often referred to as a 'wasteland' or a 'howling waste'; this precluded its being thought of seriously as a landscape that could support large expeditions, so the many British attempts to find a Northwest Passage through the Canadian archipelago were 'heavy', bringing all their food and supplies from Europe, and they did not find a passage. The Norse explorer Amundsen, however, wrote of the Arctic as a landscape rich in resources such as freshwater fish, caribou, even snow that could be built into an emergency shelter or *iglu*. Amundsen dropped the terminology of the Anglo explorer of the times, reconceptualized the Artic as a place that could be survived, and travelled light, living off the land by copying native customs; he was able to find a Northwest passage.

The lesson here is not that we can 'talk our way into space', but that language serves an important function in defining one's reality, even if it is only a perceived reality rather than a true reality.

Language communication in humans is conditioned by three major structuring variables, as seen in Table 7.7. It is unlikely that these will be altered by settlement of new habitats beyond Earth, and they are examples of what will likely not change over time. However, changing the human environment from that of Earth will shape the *outcomes* of the structuring variables in ways too subtle to predict at present.

Table 7.7 Structuring variables of human language

Synchronicity	Time aspect: levels of synchronicity, one-way or two-way communication
Means of expression	Dimensions of expression; primary, secondary, tertiary
Situation	Ability to perceive and understand
	Physical environment and context
	Ability to perceive and understand
	Physical environment and context
	Ability for rational, coordinated interaction
	Type of activity
	Motives for communication
	Relation between communicators
	Pragmatic knowledge
	Addresse(s)
	Topic or subject

The variables *synchronicity*, *means of expression* and *situation* are unlikely to change significantly in human space settlement as they are situated in the context of communicating individuals, and are thus more cognitively- than environmentally-shaped. See text for discussion. From Hård af Segerstad (2003)

Languages may be altered in many ways to adapt to new conditions; adding new words to the vocabulary, for instance, or subtly altering archaic definitions of certain words. Language diversity has been mapped to ecological diversity, indicating that at least some of the content and form of languages has been used to adapt to specific environments; having said this, the rapid movement of humanity worldwide easily ‘blurs’ such a picture, and many historical, unpredictable events can shape a language. Figure 7.4 indicates this visually, showing a broad correspondence of language family to various broad biomes of Africa. The arid Sahran region is characterized by Afroasiatic languages quite distinct from those of the Niger-Congo basin, where environments, including rivers and dense forest, are quite different; however some of this is driven by migration; the Congo basin was largely populated from the North West (in the region of Niger) about 3000 years ago, likely resulting in replacement of Congo-specific language that had evolved there much earlier. The point for our purposes is that we may expect complex language change during space settlement, but that some of it will likely reflect such different new realities as life on the surface (or subsurface) of Mars, or in floating, rotating colonies.

Languages may change rapidly or slowly, depending on many variables, and no universal rate of language change has been identified in research into this field, though

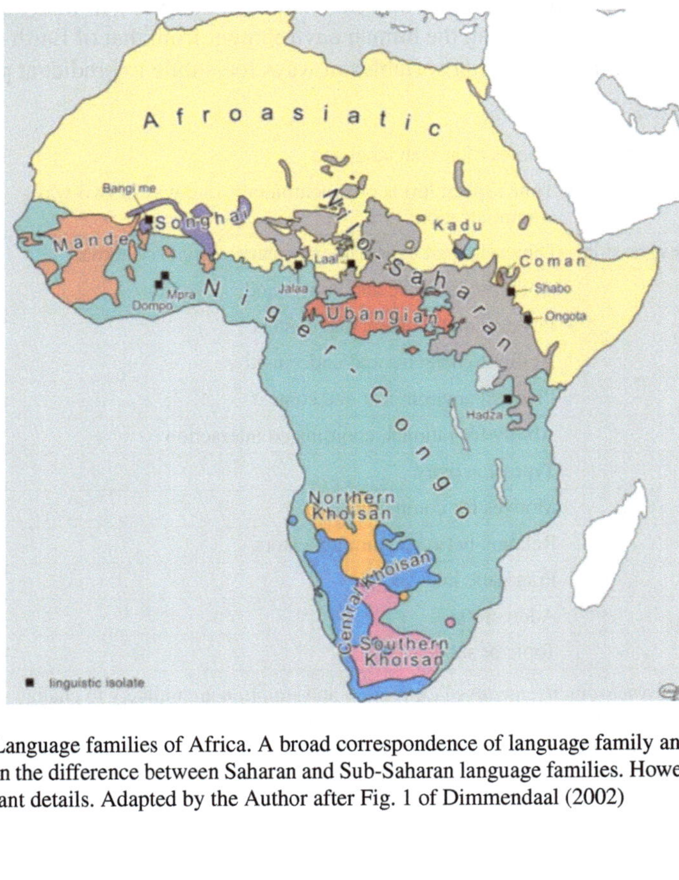

Fig. 7.4 Language families of Africa. A broad correspondence of language family and biome are indicated in the difference between Saharan and Sub-Saharan language families. However, see text for important details. Adapted by the Author after Fig. 1 of Dimmendaal (2002)

theoretical models have shown that language is more likely to change rapidly among small populations, as linguistic innovations are easier to spread to fewer people than more people (Nettle 1999). Note that conscious efforts may be made by the members of a culture to use, or not use, new terms; in modern Quebec there is a federally-organized 'filter' to reduce the introduction into Quebecois words from English or other languages. The opposite also occurs; the introduction into the Oxford English Dictionary of new terms, at a rate of dozens to hundreds per year, promoting linguistic change, at least in expansion of the lexicon. Over 900 new terms were added in 2018, including *endergonic, energeticism, energeticist, energy vampire, epitext, exergonic* and *Facebook*. High social rank or status of a language innovator may increase the likelihood of the innovation being spread and adopted, at least superficially. At the same time, *underground* or *subversive languages* may develop to subvert values of a dominant culture.

Table 7.8 indicates some language change timelines. It seems inconceivable that language would not change substantially when people begin to live beyond Earth. How that change occurs is explored in the following pages.

Languages are organized hierarchically and structurally as seen in Table 7.9, a useful summary of the adaptive sub-tools (structures or components) of language.

Table 7.8 Language change examples

Original language	Derived dialect or language	Time to derived language (years)	Mutually intelligible?	Nature of changes
French	Quebecois	100	Moderate	Pronunciation
Old English (c.1000 AD)	Middle English (c.1600 AD)	600	No	Pronunciation, grammar
Middle English (c.1600 AD)	Modern English (c.1800 AD)	200	No	Pronunciation
Old Norse (c.1000 AD)	Modern Icelandic	800	Moderate	Pronunciation, grammar, spelling

See text for discussion

Table 7.9 Heirarchical and structure organization of human language

Level	Concerns
Pragmatics and Grammar	The uses of language
Semantics and Lexicon	Meanings of words
Syntax	Clause and sentence structure
Morphology	Forms and words
Phonetics	Sounds

The table highlights spoken language, but other forms of communication, including gesture, are implied here. Note the rows are arranged according to the most structuring (at the bottom) to the most situational (at the top). Changes in these levels, subtle or obvious, accumulate over time, resulting in dialects and, ultimately, new languages

Of these, the *lexicon* or *vocabulary* is easiest to predict to change as an adaptation to new environments beyond Earth; new phenomena, such as cloud formation (consider this on Mars, or in a rotating free-floating colony), will demand new words, subltly- or radically-different from the descriptions of these on Earth.

While structural changes (to phonetics, morphology and syntax, for example) are rather slow to change, lexicons can expand rapidly. The most common processes in new word formation are listed below:

- Derivation: adding or prefix or suffix, e.g. democratize
- Back formation: removal of an affix, e.g. 'donate' from donation
- Compounding: combining two extant words, e.g. bailout
- Repurposing: changing the context of a word, e.g. mouse (computer from rodent)
- Conversion: changing the class of a word, e.g. giant meant a large creature until used as an adjective for anything particularly large
- Eponymy: using a placename as a new word, e.g. wellington (boot) from popularizer Arthur Wellesley, Duke of Wellington
- Abbreviation: shortening a word, e.g. van from caravan
- Borrowing: taking a word from another language, e.g. mammoth (to English from Russian)
- Onomatopeia: imitating a sound, e.g. cukoo
- Repetition: repeating a word or sound, e.g. flip-flop
- Nonce word: wide and repeated use of a word invented for simply one use, e.g. quark, fleek
- Error: wide and repeated use of an erroneous communication, e.g. helpmate from biblical transcription error of 'help-meet'
- Portmanteau: trimming a word and adding a new word or word fragment, e.g. sitcom.

As new words are formed by the processes above, they are introduced to the socially-selective environment of communications between individuals, whether face-to-face or via a variety of media. If the innovation is not rejected out of hand (essentially cultural selection against that variation), it may begin to go through the following stages of word spread:

- Stage 1: innovation of a word in one mind, shared among a small group
- Stage 2: new word is adopted by wider community, but is rarely used beyond the subculture of its origin
- Stage 3: new word is adopted by yet wider community, beyond the subculture of origin
- Stage 4: new word is adopted and widely-used by most or all of the culture's subcultures and other.

The complex world of human socialization is often the 'locus' of language change, in the case above, word choice, as it occurs on a moment-to-moment basis of communication between individuals. Linguist Salikoko S. Mufwene highlights this in writing that "...individual speakers are critical unwitting agents of language evolution. This occurs through the day-to-day accommodations which speakers make to one another, the adjustments they make to new communicative needs, and the simple condition of imperfect replication during language transmission. Accommodation emphasizes the significance of idiolect contact within a population of speakers and the central role it plays in language change. While interacting with one other, speakers contribute features to a pool from which they make their selections that can affect the evolutionary trajectory of a language. The features they contribute can be from the same language or dialect, or from different ones. The selections they make are not necessarily constrained by the origins of the features, and each idiolect reorganizes its selections on the model of blending inheritance in biology" (Mufwene 2004: 18).

By these processes new words are formed and spread, slowly or rapidly, tailoring a culture, in most cases, to a better reflection of its physical and social environments. Where innovations improve the correspondence between environment and organism or culture, they may be thought of as adaptations.

In the specific case of space settlement, in Table 7.9 we see significant subdomains of language as an adaptive domain. *Grammar*, referring to the rules of communication (e.g. subject-verb-object sentence construction—vs. subject-object-verb), is foundational to a given language and therefore unlikely to be adjusted significantly in the first few generations of settlement. Grammar is "a device for making communication more efficient and streamlined" (Corballis 2009: 35) and so may be expected to change if that change is adaptive, but normally that change will be rather gradual. Research on grammatical change in human language has shown that it is rather unpredictable, often resulting from historical contingencies: in a 2017 research paper, the authors indicate a slow grammatical change in English over 500 years, as illustrated in Fig. 7.5a. Similarly, *syntax* is a rule system for sentence formation, and therefore rather basic to a given language, but it may be modified somewhat more easily than grammar.

Vocabulary is more easily modified, by addition of new words or terms, and ceasing the use of others. Figure 7.5b shows the evolution of some American-English words between 1825 and 2000. A generalization regarding vocabulary change is that rarer words "are more prone to replacement than common ones" (Newberry et al. 2017); that is, it is easiest for rarely-used words. Words used very commonly are an 'investment' that people are less prone to change than less-commonly-used words which are less 'enmeshed' in a given language and culture.

Dialect (also *pronounciation* and *accent*) is modified over time, in some ways more readily than grammar and syntax. A spatial approach to dialect has resulted in a complex model of dialect evolution; the gist being that:

> Since language change is inherently stochastic, at small spatial scales we can only expect predictions of a statistical nature. At larger scales an element of deterministic predictability emerges. Within our model, all stochasticity arises from the randomization of initial conditions. The effect of this randomness is strongest in the early stages of language evolution, when the typical size of dialect domains is small...[as population increases and populations spread out spatially, dialects] begin to bundle and the continuum resolves into distinct dialect regions. (Burridge 2017: 18)

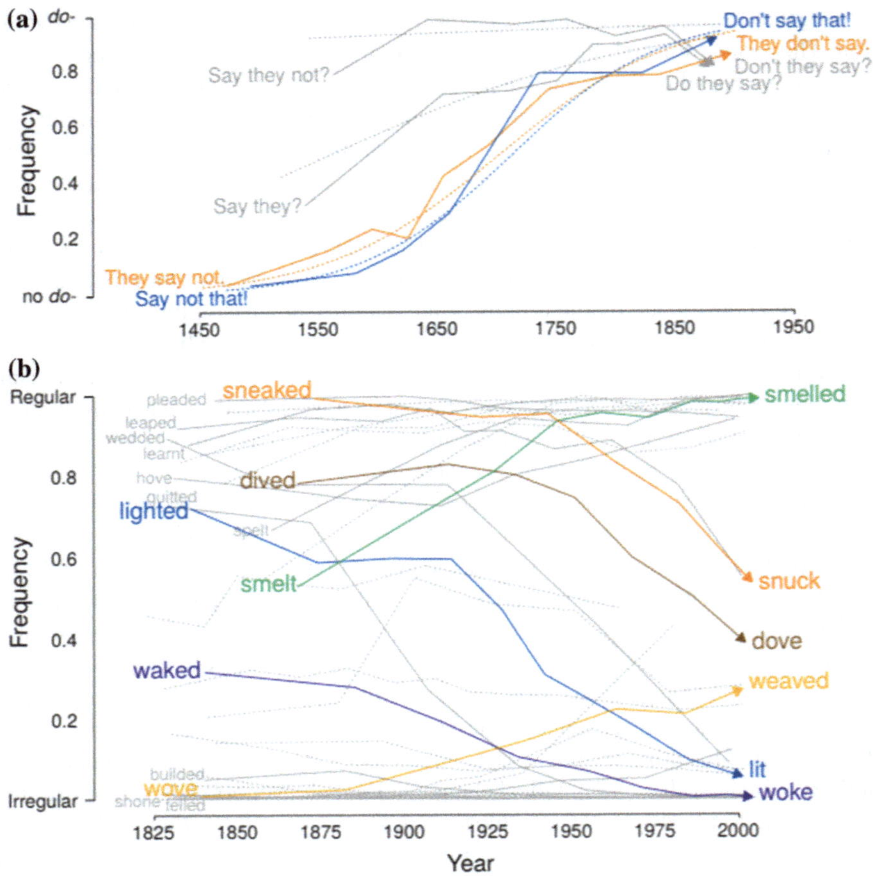

Fig. 7.5 Vocabulary change examples in English, 1425–2000 AD. See text for discussion. Adapted by the Author from Newberry et al. (2017)

A large portion of human communication is nonverbal, so *gesture*, or "visible action as utterance" (Kendon 2004), is important, and it is only one of a variety of nonverbal communication modes, as seen in Fig. 7.6. Certainly we can imagine postures and body language to differ in Mars'.38 g field, the 1/12th g of the moon and whatever gravity field is established in settlement that rotate to provide gravity (Fig. 7.7) .

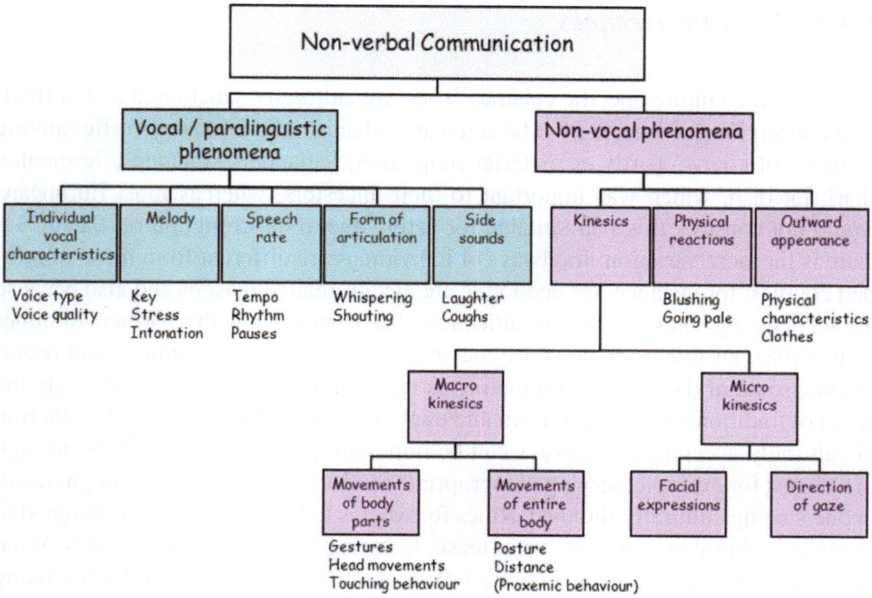

Fig. 7.6 Gesture is one forms of nonverbal communication. See text for discussion. Diagram by the Author

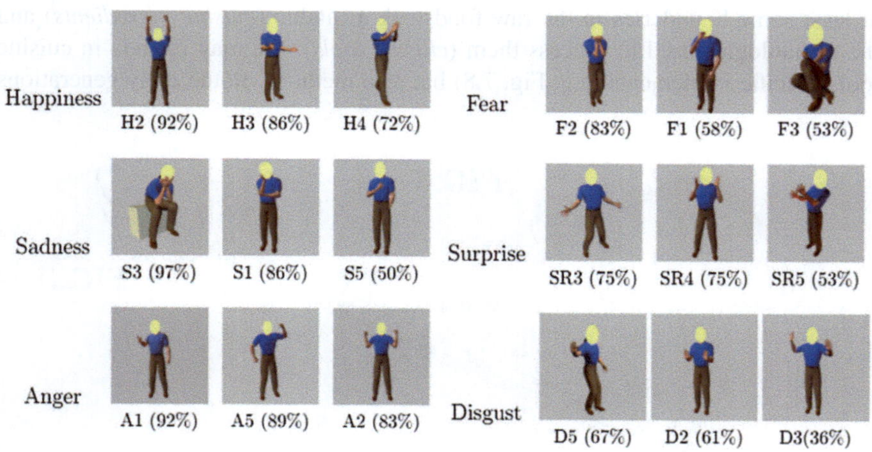

Fig. 7.7 Major human bodily postures. Note these are cross-culturally somewhat variable, but the essential point is that posture will be conditioned by new gravity fields experienced beyond Earth. Adapted by the Author from Shaarani and Romano (2007)

7.4.3 Food Preferences

Foodways are culture-specific cuisines. They are culturally-sanctioned and normalized conceptions of what is fit to be eaten and what is unfit. Foodways differ among cultures, of course, partly as material adaptation; Icelanders venerate a fermented shark fat dish, which was important to their ancestors, whereas most Europeans would not consider this *svid* suitable for eating due to its strong, putrid flavor. But there is far more variation: foodways of individuals are different from those of married couples, for instance, for economic and health-related reasons and also because the act of eating takes on a new significance when carried out with a partner (Kemmer et al. 1998). Different religions, of course, also have food prescriptions and prohibitions, some of them likely originating as material adaptations but some likely the result of traditions with an unknown and unpredictable past historic event. Adhering to culturally-acceptable foodways can be important to a person's progress through life stages; for example, serving inappropriate food at an important feast might result in one's being culturally shunned. Rules for what is to be eaten are often designed to reinforce cultural conceptions, as in feasts featuring certain traditional foods. What is considered fit to be eaten may be influenced by, reiterate, and influence many cultural variables including *socio-economic rank* (haute cuisine—vs. peasant food) or *course of life stage* ('baby foods', foods consumed at life-stage rituals).

While the guides to what is appropriate to eat are tremendously varied, they have at least some boundaries in the raw foodstock available (*cuisine ingredients*) and the technologies used to process them (*cuisine tools*). We may include in cuisine tools specific implements (e.g. Fig. 7.8) but also methods. In the early generations

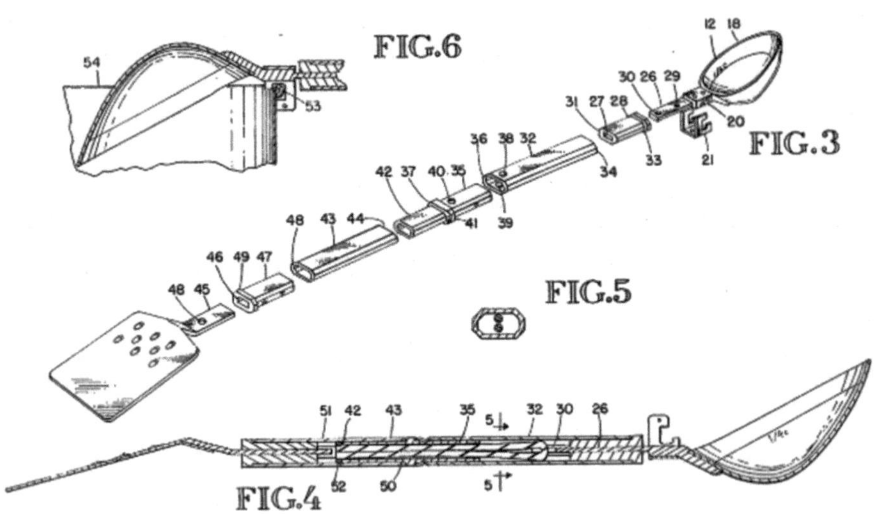

Fig. 7.8 US Patent 4,823,419. This is a novel, multifunctional food-preparation tool. Certainly new such tools will be devised for environmental variables as the .38 g gravity field of Mars, with effects on preparation methods and cuisine overall

of a Mars settlement, for example, it is likely that natural cow's milk butter will be rather rare, considering the cost of such large animals, and that goat's milk butter will be more likely available. Frying ingredients with goat's butter, however, is very different from using cow's milk butter, which has a stronger flavor and a yellow color. As I have mentioned many times, such subtle differences between Earth and off-Earth cultures are not in themselves normally significant, but taken entirely, they constitute overall cultural differences that, in sum, have effects on such important biological aspects of human life as finding mating partners.

Regarding food ingredients, naturally the global variety of foods available on Earth today will not for some generations be available to economically-independent settlers beyond Earth. Their daily menu will differ, often substantially, from that of the people of Earth. If you were to travel to a Mars or Lagrangian settlement, with the expectation that you would eventually go home, your diet would be significantly different from what you consider customary, and like many travelers you might well look forward to getting home. First generation settlers with no expectation of a return to Earth would have to think about things quite differently, and would likely be highly-motivated to experiment substantially with their ingredients and tools to make a new cuisine. Second-generation residents of settlements beyond Earth would be born into a given cuisine, never 'missing' an Earth cuisine they have never experienced, and it would be Earth food that would be unusual and perhaps even upsetting if encountered. We should not underestimate the significance of foodways in shaping cultures and even economies. There is an interesting field to study in the foodways of future space settlers, and food is a popular topic on recreational media, making this a potentially effective topic for engaging the wider public in the prospects of space settlement.

A speculation on cuisine ingredients is seen in Fig. 7.9. This simply illustrates some possibilities, with the many implications sketched out in the paragraphs above. I find these diagrams effective in engaging interest among those otherwise not necessarily interested in humans-in-space activities.

7.4.4 Art and Aesthetics

Culturally-specific conceptions of beauty and order, and their spectra and opposites, and self-expression in general, may be considered in the broad adaptive domain of Art and Aesthetics. Clearly all cultures have such conceptions; in the visual arts, for example, Australian aboriginal styles clearly differ from those of Amazon basin natives, just as Western impressionist paintings differ in their emphases and expressions from East Asian block prints.

Culturally-adaptive functions for art have been argued in anthropology; a recent such argument notes that at least some art is adaptive, serving to communicate ideals and reinforce group cohesion, for the following reasons (Dissanayake 2008):

- Art is found in all cultures
- Art is present among the earliest behaviorally-modern humans of c.100,000 years ago, in the clear use of symbolism

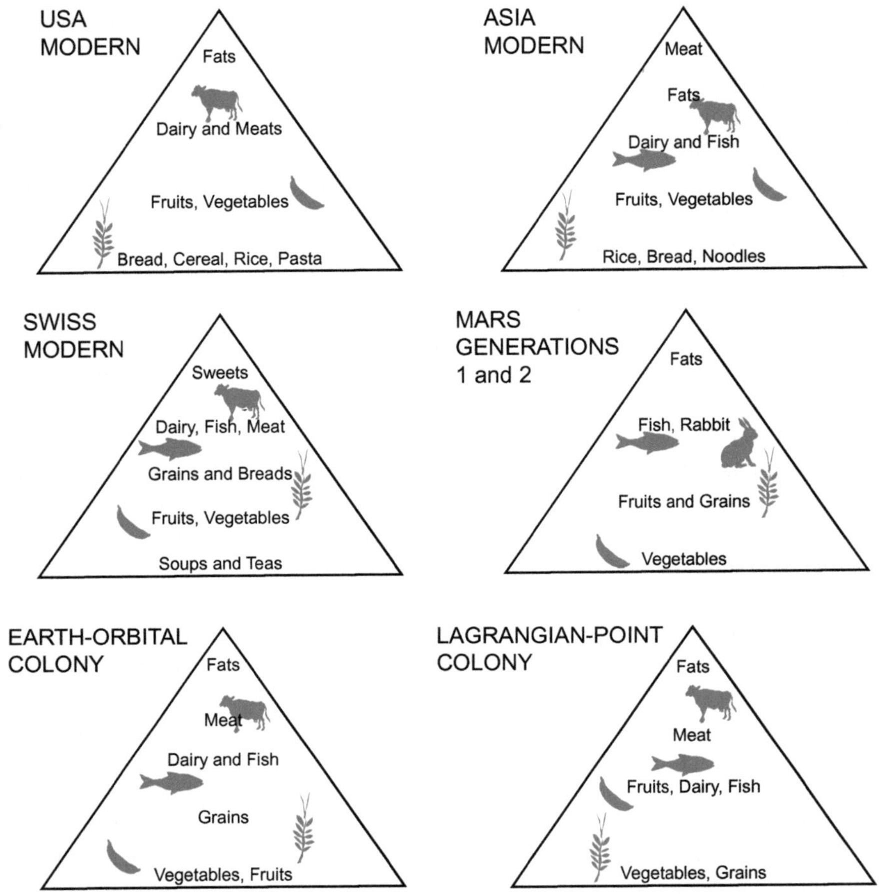

Fig. 7.9 Food pyramids for three modern culture groups and speculative beyond-earth settlement types. What is consumed varies by culture group, and is shaped both by material and cultural concerns, such as raw foodstuff availability, and historical significance (e.g. tradition) of a given food type, respectively. Food pyramids for off-Earth settlements will undoubtedly differ significantly from these, and are worth consideration in space planning. Diagrams from Hastorf (1999) supplemented by the Author

- Artistic rudiments are detectable and easily fostered in the behavior of young children, as when babies and toddlers spontaneously move to music, sing along or alone, make marks, decorate their bodies and possessions, play with words, find pleasure in rhythm and rhyme, or enjoy make believe
- Art is generally an attractant and source of pleasure, like other adaptive behaviors including mating, parenting, resting, or being with familiars in warm and safe surroundings.
- Art occurs under appropriate and adaptive conditions or circumstances—that is, it is often "about" important life concerns, as in ceremonies that mark stages of life or that concern prosperity, safety, and subsistence.

- Art is costly: large amounts of time, physical and psychological effort, thought, and material resources are devoted to the arts as to other biologically-important activities such as sex, finding, preparing, and consuming food, socializing and gaining social acceptance, helping close kin, talking with friends, and acquiring useful information.

Artistic expression is important in leisure time activities (whether creating art or consuming it) and has tremendous potential cultural force. Literary theorist J. Carroll points out this potential: "the arts are indispensable . . . for the organization of shared experience that makes collective cultural life possible." (Carroll 2004: ii). Artistic expression has been used to reiterate and drive transformative culture change, as in the use of cubism in Western painting to communicate a cultural break with more ancient traditions, or a naturalistic expression in portraiture characteristic of the cultural revolutions in New Kingdom Egypt; now consider the potential of literature, music and film to motivate culture change, and we find in this domain a powerful tool of cultural shaping that will undoubtedly be altered by, and likely alter, Beyond-Earth cultures as they shape themselves to new experiences. To reiterate the potential power of the arts as culturally-motivating, recall the long history of censorship and attempts at censorship, and, in the United States, the enshrinement of free expression as a right in the First Amendment, as noted by "Literature and the arts must be protected by the First Amendment. They lead the way toward sensitive and informed appreciation and response to the values out of which the riches of the general welfare are created." (Meiklejohn 1961: 256).

How does artistic expression change through time? Recent work in the field of *creativity* research has resulted in a fruitful recognition that artistic innovation occurs with "The ability to shift between a *focused analytic mode of thought* and a *defocused associative mode of thought* ... known as *contextual focus*" in which "[innovation occurs as an] object of thought [is] held in mind [and is exposed to] more [associations] that have previously gone unnoticed…the ability to shift between a focused analytic mode of thought and a defocused associative mode of thought is known as contextual focus and it plays an important role in creativity" (Gabora and Smith 2018: 4). The relevance for our topic is that creativity is cognitively-grounded (though must work with perceptions, which will be novel in new habitats beyond Earth) and thus should not be directly inhibited by exposure to conditions beyond Earth. In fact, the new perceptions (for example, of the human senses) experienced beyond Earth would inevitably result in new associations, and of these new productive associations, or creative output.

How could the variety of artistic expressions outlined in Fig. 7.10 play beyond Earth? Here I do not mean to make specific predictions, but only demonstrate that we may expect creativity in these artistic fields.

In the fields of *music and singing* we see that these cultural universals have several likely reason for their being widespread. These include facilitating *socialization* of the young (e.g. with instructive songs and music), *caregiving* (soothing debilitated individuals), *sexual selection* (e.g. performances used to attract mates) and *social group cohesion* (e.g. group participation in musical communication of such impor-

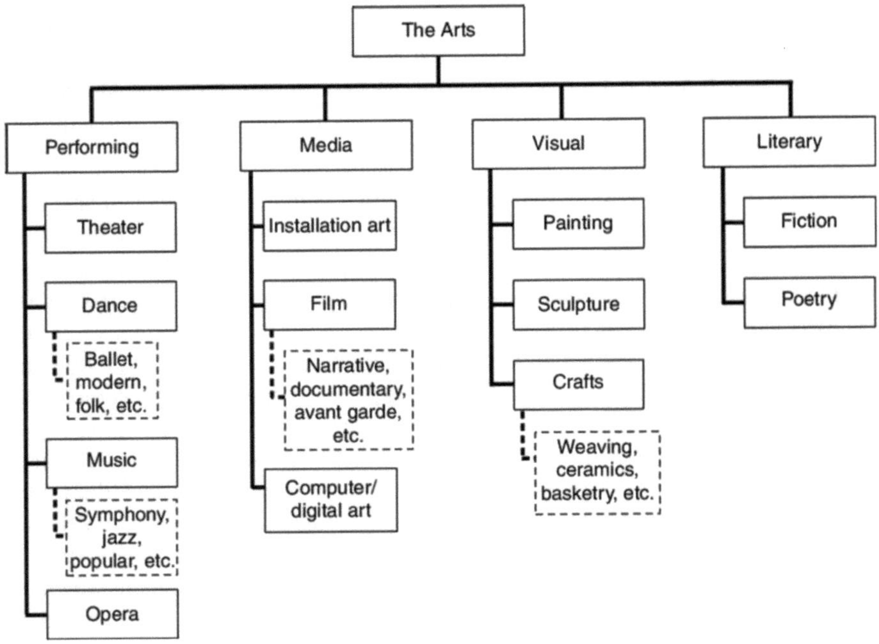

Fig. 7.10 Classification of the arts. The classification is biased towards Western expression, but is generally applicable worldwide. Diagram by the author, adapted from Fig. 2.1 of McCarthy et al. (2001)

tant cultural information as the Ultimate Sacred Postulates (Fitch 2005). How and why music and singing evolve has been studied largely from an historic perspective, identifying the often-unpredictable events at the root of various musical styles. Since music is a Cultural Universal, however, I suspect it to have adaptive value, perhaps as suggested above. Attempts to generalize about the mechanisms of musical style change have not been very fruitful, and again it seems that many such changes are rooted in highly-unpredictable social and even personal events that result in 'spontaneous' innovations which are then widely accepted. Relatively recently musicologist David Temperley proposed a theory of musical style change based on the concept that music serves (at least in part) (Temperley 2004: 314):

> …to convey certain structures to the listener via a surface of notes…For [musical] communication to occur successfully, the structures must be recoverable from the surface [read 'music']. A situation where, given the rules of the style, it was impossible for the listener to infer from the music which structure was intended would not be very satisfactory from the communicative point of view. I will argue that this principle, which I call *communicative pressure*, acts as a significant force on the evolution of musical styles.

Temperley goes on to illustrate the small variations that occur as new styles of communication emerge (example in Fig. 7.11a), describing these as variations on the way of communicating the original message. This is a subtle and complex model of musical change that I am not qualified to evaluate, but again the point is that we

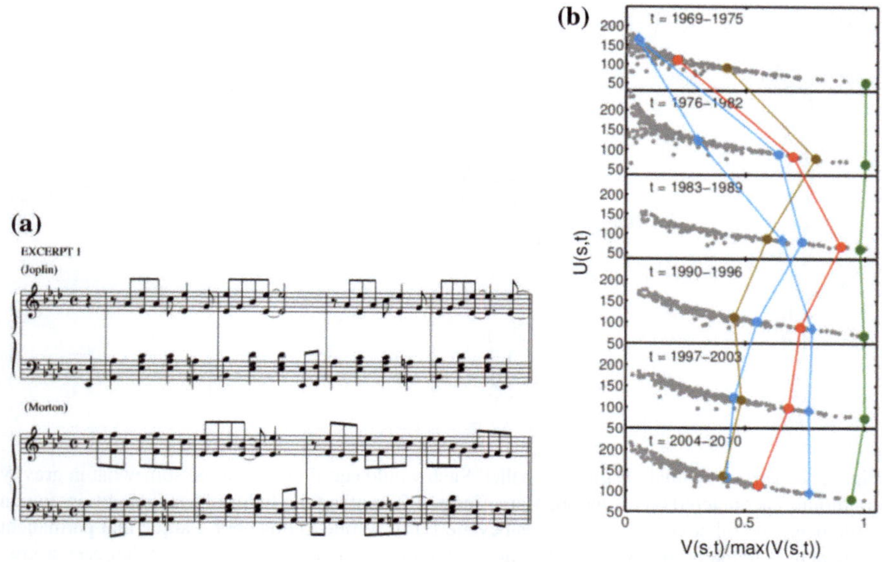

Fig. 7.11 Variations on a musical item. **a** At the top, the notation by Scott Joplin, originator of this music, which has been copied but varied in its expression (at bottom) by the later musician, Jelly Roll Morton. The changes are slight, but noticeable, and result in a slightly new way of communicating the same messages conveyed as the originator. **b** Decreasing musical complexity in various styles, after stylistic 'fixation', with instrumental complexity on the vertical axis and time on the horizontal axis. Diagrams adapted by the Author from **a** Temperley (2004) [L] and **b** Percino et al. (2014)

may expect change, and some of it may be adaptive, and there are ways to think about such change in an evolutionary sense. Music is also seen as conditioned by the need to satisfy contrasting demands for familiarity and novelty; and an analysis from this perspective finds that change and innovation are rapid when a style is being established, but the style becomes 'fixed' or more uniform over time, satisfying the consumer's demand for familiarity (Fig. 7.11b).

On a more specific note we may remember that in *Skylab*'s lowered-pressure (5psi compared to Earth's 14.7psi standard) and 70% oxygen (compared to c.20% on Earth) environment, astronauts noticed that sound propagated unusually (not carrying far) and astronaut Jack Lousma noted also that this atmosphere "made it hard to whistle," (Lindgren 2009: 86) which nobody had predicted. Other unpredictable 'insignificant' variations on humanity's Earth-evolved cultural behavior will undoubtedly emerge as products of new environments.

In the *static visual arts* (including *static sculptural arts*) there will be almost unlimited scope for innovation as humanity extends permanently beyond Earth. Some comments on possibilities just becoming real in the first space race of the 20th Century were outlined by aeronautical engineer-artist Frank J. Malina (1912–1981) who wrote that (Malina 1970: 324):

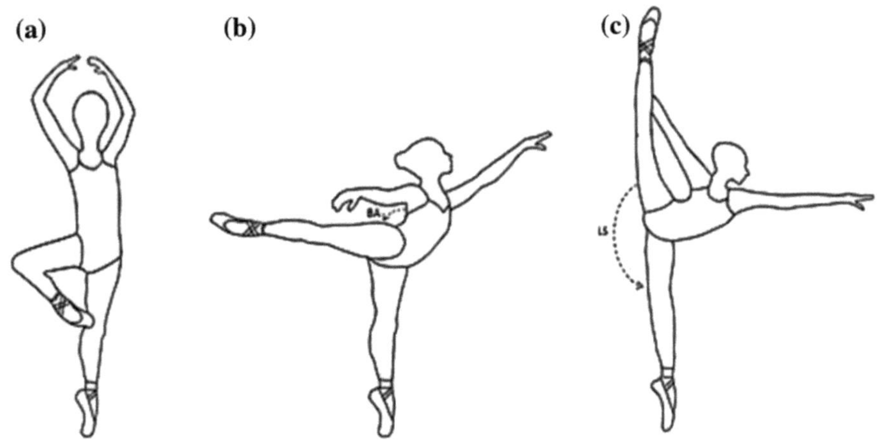

Fig. 7.12 Three essential positions in ballet. Such would certainly be altered somewhat in gravity conditions encountered on the moon, Mars, Ceres, different gravity fields of a spinning Lagrangian settlement, and other conceivable places beyond Earth. While in this book I argue that permanent settlement is unlikely in any zero-g or microgravity environment, it may well be that easy access to microgravity by spacefaring people will result in the use of that unique environment specifically for artistic expression; for example, 'space dance'. Diagram by the Author

> [many artistic] possibilities and permutations [are being exposed for innovation in the space age]… From the point of view of technique, in conditions of zero gravity, painting with a brush will be possible but not pouring of paint: sculpture, whether modelled or constructed, will be difficult; lightweight plastics may be suitable. The peculiar behaviour of liquids might be exploited. On the Moon, an artist will be able to make sculptures of very large size [due to the low gravity field] and the cohesiveness of Moon dust may permit the forming of sculptures of very long life, since there are no problems with either wind or water erosion. Wind-driven kinetic objects will not be possible, except in areas of habitation, however, the intense energy of sunlight under vacuum conditions will drive kinetic sculptures by the pressure of light on surfaces. Paints used on Earth may not withstand space conditions for long enough periods of time; painterly techniques used on Earth should be applicable, although, 'action' painting may be affected.

In the *motion arts*, we can imagine that altered gravity conditions will result in innovations of such artistic forms as ballet: Fig. 7.12 indicates three of the main ballet positions; how would these be translated to or innovated from in different gravity fields?

Personally I believe a renaissance in artistic expression will be among the most valuable and enriching products of permanent space settlement.

7.4.5 *Morality, Ethics and Justice*

Culturally-specific *moral codes* or foundations are deeply-held conceptions of right and wrong that apply to many levels of cultural organization, from the individual to multinational corporations. These moral codes are often founded in the Ultimate Sacred Postulates of a given culture, a concept discussed in Chap. 5. These postulates and their ensuing morality are normally slow to be changed as they have served well enough to carry a culture to where it is at a given moment, and the inconveniences of change. *Ethics* are the practical application of moral codes, and the administration of justice is the practical application of moral conceptions to cases of ethical breaches (Cohen 1986). Communication between individuals requires some degree of trust, and this requires some degree of shared conceptions of morality, ethics and justice. These vary worldwide but all cultures hold at least three ethical behaviors inviolate: one must tell the truth, respect human life, and respect property rights. That is, in all cultures, it is generally agreed that the truth should be told, that human life is of high value and that people deserve to keep that which is theirs. Of course, transgressions of these occur every day, but they are principles that, if eliminated, would cause a breakdown of the trust needed for communication and problem-solving. A practically-oriented, limited (but I think generally applicable) search for moral universals in contemporary cultures has identified the following values expressed rather widely (naturally, exceptions may be found here and there, but I do not subscribe to the postmodern excess that these are only expressions of the researchers' biases) (Kinnier et al. 2000):

Commitment to Something Greater than Self

- to recognize the existence of a Supreme Being, higher principle, transcendent purpose, or meaning to one's existence
- to seek truth
- to seek justice

Self-Respect, but With Humility, and Acceptance of Personal Responsibility

- to care and respect oneself
- to not exalt oneself or overindulge
- to act in accordance with one's conscience and to accept responsibility for one's behavior

Respect and Caring for Others

- to recognize the connectedness of all people
- to serve humankind and to be helpful to individuals
- to not harm others

Caring for Other Living Things, and the Environment

Finally we note that moral systems are learned, aside from the smallest portion of instinctual behavioral guidance. Whatever the details in a given culture, in the development of the individual's moral system of conduct, it has been found that several stages of moral understanding occur in the individual from 'heteronomous morality' shaped by avoidance of personal punishment for transgressions (infancy and early childhood) to mutual, interpersonal and social 'consciences' in which formalized laws are upheld and good social bonds are maintained by adherence to conventionalized behavior (Colby et al. 1983). The lesson for our topic here is that morality is not genetically-transmitted, and is thus cultural and thus highly subject to unpredictable change; however it is universal among cultures and thus will be an aspect of space settlement. The same may be said of the ethical system, which represents the practical application of the moral generalizations underpinning behavior.

The most concrete manifestation of the moral system is the administration of justice, which will certainly require adjustment to non-Earth conditions in all respects. Little has been published on this matter regarding individual colonies or settlements beyond Mars. On a larger scale, a 2016 paper on the legal structures of early Mars settlement is one of many contributions sketching out the future of this field. Its conclusions includes this encapsulation of their proposed legal macrostructure (Bruhns and Misra 2016):

> In response to rising interest in Mars colonization missions from both national space agencies and private enterprise, we propose a pragmatic solution that allows bounded sovereignty claims outside of preserved planetary parks but remains in the spirit of the Outer Space Treaty. Scientists will establish planetary park locations and regulations. Private and governmental parties may land on and occupy limited plots of Mars' surface, which they will govern according to their national laws. They may conduct science, and may even commence commercial resource exploitation missions. Conflicts will be solved diplomatically, or through use of the temporary tribunal composed of representatives of other colonies on Mars. A Mars Secretariat will act as an administrative body to facilitate communication between the colonies.

Currently the scholarly journals *Journal of Space Law* and *Space Policy* are international, peer-reviewed clearinghouses for topics including the administration of justice beyond Earth and they will likely see more activity regarding the legal issues of space settlements in the coming decades.

As we have seen many times so far in the domains of cultural adaptation, there is nothing inherent in the reasonably-expectable conditions of 'humanity beyond Earth' that would preclude space settlers devising practical systems of morality, ethics and justice. The functioning of healthy cultures requires these, and they may be expected to change beyond Earth. Even before leaving Earth, however, it is likely that much consideration will go into this field to help delineate the diffusion of our species throughout the solar system.

7.4.6 Religion and Mythos

Each culture's specific religious conceptions, and historical/mythical tales of origins and, sometimes, destines, are maintained as structuring to the culture by their enshrinement as sacred and thus, largely unquestionable, and thus largely unchangeable. A long anthropological history of the anthropology of religion has shown that the specific forms of religions are so molded by unpredictable, historical processes, that it is difficult to generalize upon them. Here, however, we are not concerned with the specific content of any religious tradition, but rather the functions of religion in a culture, and the ways in which religions change, and whether such change can be adaptive.

Perhaps the widest view of the religious experience has been published by the late anthropologist Roy Rappaport. His well-received *Ritual and Religion in the Making of Humankind* surveys the origin and cultural functions of religion, utilizing an evolutionary, but not eco-deterministic, approach; religions, for Rappaport, may 'map' to environment in some aspects, but are not determined by environment. Rappaport's focus on the function of religion led him to identify that religions arose out of rituals. Religions describe the universe, and how to interact with it; these instructions are formalized in such forms as commandments, and these are communicated and reiterated during rituals in which the self is forgotten, and the group's concerns are foremost, with groups singing, chanting or otherwise acting in union, again with the individual self forgotten for the moment. If Rappaport is correct, and my opinion is that he is at this level, then we may say that a significant function of religion is the communication of the Ultimate Sacred Postulates via communal ritual; because of the investment in the religious or mythic narrative, these are normally rather slow to change. An example of substantial change in the Western tradition was 'Vatican II', a reformation of the Catholic church, in the mid-20th Century AD, which included the novelty of priests delivering religious rituals in languages other than Latin.

It is reasonable to propose that the adaptive function of religion is that it reiterates basic concepts of how to behave appropriately, and thus it serves as a within-group cultural unifier (Rappaport 1999). It has of course been manipulated to drive attack on people *outside* the group using just this unity.

In addition to this group-level function, it has been reasonably proposed that religious traditions have been found to address a number of psychological needs of the individual. A recent cross-cultural investigation found that religious traditions served "distinct psychological processes (cognitive, emotional, moral, and social)" (Saroglou 2011). These needs were served by four main religious prescriptions, as seen in Table 7.10. The details may be endlessly debated, but it is reasonable that these are at least partially adaptive functions of religious traditions. In the table, the four dimensions or prescriptions are *believing* (holding a reassuring faith in Ultimate Sacred Postulates), *bonding* (maintaining social cohesion among close relations), behaving (providing guidelines for behavior) and *belonging* (maintaining social cohesion among large groups).

As with the structural aspects of language, it seems that the high investment in the group-cohering functions of religious and mythical traditions will always keep them relatively resistant to change. Live beyond Earth might lead to changes to the contents of religion, well beyond our scope, but it is likely that religions will continue to be used as guides to essential conceptions of the nature of the universe, and how to behave properly in that universe, as communicated through group-reiterating rituals and Ultimate Sacred Postulates. There is no apparent reason that religious traditions and their rituals and even artifacts could not persist beyond Earth.

Table 7.10 Major dimensions of religious traditions

Prescriptive dimension	Personal relevance	Cultural products	Goals	Transcendence	Common alternate states
Believing	Beliefs	Dogmas	Truth	Intellectual	Literal versus Symbolic
Bonding	Emotions	Rituals	Awe	Mysticism	Negative versus Positive
Behaving	Morality	Norms	Virtue	Moralization	Self- versus Other-Focused
Belonging	Identity	Groups	Totality	Identity	Exclusive versus Inclusive

Adapted by the author from Table 2 of Saroglou (2011)

7.4.7 Nurturing and Enculturation

Enculturation is the process of a person's acquisition of a culture's specific guides to behavior. Since complex human behavior is key to adaptation, as we have seen throughout this book, the process of transmitting these guides to behavior is important to cultural survival; cultural anthropologist Hersovitz highlights this in writing that "enculturation of the individual in the early years of… life is the prime mechanism making for cultural stability" (Herskovitz 1964: 327).

Enculturation, then, provides cultural guides to behavior for humanity whereas instinct provides most guides of behavior to many other life forms. Put somewhat lightheartedly, enculturation converts "the 'barbarian invasion' of the continual stream of newborns" into functional members of a culture (Parson 1991: 143).

As an example, Table 7.11 outlines stages of life, and the rights and responsibilities associated with them, among traditional Hutterites of North America. Note in particular the very specific guides to what is appropriate at various stages of life, which are highly optimized for both group and individual survival. Such specifics are unlikely to be modelled productively by space settlement planners, and it is arguable that such fine 'social engineering' would be inappropriate. Rather, it is members of the space-settling cultures themselves who determine these details (though important details) of the workings of cultures beyond Earth. The point, once again, is that these variables of life are malleable in human cultures, and may be adjusted to accommodate nearly any conceivable conditions.

Anthropologists of *cultural transition* rituals that mark certain stages of life Martha and Morton Fried have proposed that "All of life is a risky business; transition rituals are a way of channeling anxiety," and concluding in *Transitions: Four Rituals in Eight Cultures* (Fried and Fired 1980: 268–270)

Table 7.11 Hutterite life stages, rights and responsibilities

Life stage	Chronological age	Rights/Entitlements	Responsibilities/Expectations
House children	Birth—2 years	Free movement once walking (c.10 months); can be loud; need not do colony work	–
Kindergarten children	3–5 years	Mock fighting; need not do colony work	Quietness; obedience; stillness; memorization & recitation of hymns & prayers; learning roles e.g. females play with house goods, males play 'tractor'; learning English language
School children	6–14	Can always ask for assistance or supervision in colony work; may develop own talents without pressure of school grading system; eldest take first helping at table	Learning social interaction specific to colony harmony; learning absolute obedience; learning Ultimate Sacred Postulates; learning life-stage rights and responsibilities; taking up physical colony work; learning German language; may not speak unnecessarily at meals; may only eat in the children's dining room
Young people	15-Baptism	May eat with adults; is greeted with handshake; receives cloth, tools, utensils for carrying out daily life; boys get tractors; girls begin baking; children may court and date; may visit and work at other colonies; may experience some of the outside world in these 'foolish years'	Sits at lowest-ranked position at the table; undertakes adult colony work; obedience to adults; preparation for baptism.
Baptism (rite more than life stage)	c.19–20 (girls), c.20–26 (boys)	Males receive voting privileges; males and females may attend weddings and funerals in other colonies; formally a part of the *Gemien*	Internalization of community over individualism; submission to church; daily behavior scrutinized for righteousness; must report anything untoward observed in colony
Marriage (rite more than life stage) and Adulthood	c.20–25 for marriage, adulthood to c.45 (female) and 50 (male)	Freedom of 'internal world' of the marriage; freedom to manage social lives	Strict submission to the rules—formal and informal—of the colony; must do colony work well and without complaint; mother, nursing and childrearing; father, economic security largely through agriculture

(continued)

Table 7.11 (continued)

Life stage	Chronological age	Rights/Entitlements	Responsibilities/Expectations
Elder life	>50	Excused from milking, hoeing and other physical colony work; excused from some teaching duties; excused from close control of household economy; eldest capable males become *Brotschneider* (bread-cutter) at the table	Spiritual preparation for death

Tabulation by the Author based on information in Hostetler and Huntington (1967) and Saterlee (1993)

> …no population ever visited by ethnographers, or known in any detail to history, has ever lacked some observance of the four transitions discussed in this book [namely, birth, adolescence, marriage and death]…And that in turn suggests that ritual plays an adaptive role in the process of cultural evolution…[functioning] adaptively at the social level…

It has long been recognized that cultural guides to behavior are not necessarily shackles; that is, they *may* clearly delineate very specific courses of action—like the appropriate way to greet a person of royalty—but they may be rather fuzzy, only prescribing broad outlines to action. For example, a culture's conventions of visual art may only set broad outlines to the artist's behavior. The anthropological turn away from seeking universals and paying more attention to the individual in part was driven by recognizing that people often act in ways that counter such guidelines, however this should have been a supplementary study, rather than a rejection of pattern. In any case, the individual has also long been recognized in anthropology, 'even' the pattern-seeking anthropology often referred to in this book. Writing in 1961, the anthropologist Ralph Linton identified this in the following passage (Linton 1961: 40):

> in spite of the thorough enculturation of the individual, he still retains the capacity to think and to devise new forms of behavior in response to situations for which the patterns of his culture are inadequate. Enculturation merely limits the conceptual tools with which he can operate and the directions in which he will think. The individual is the irreducible variable in every social and cultural situation. The individual is the irreducible variable in every social and cultural situation…the yeast in the cultural brew, and every new element of culture can be traced back ultimately to some individual's mind.

To whatever degree then—and it surely varies among the scope of a culture's myriad guidelines to behavior—cultural information must be transmitted from one generation to the next, and this is carried out by the various processes and mechanisms of enculturation. They may be informal or formal, for instance carried out simply by mild corrections by parents of offspring to reinforce the 'correct' way to behave, or highly formalized, as in training military troops.

Whatever its specific content, enculturation processes beyond Earth will continue to be vital to multigenerational success.

7.4.8 Kinship and Descent

All cultures have ways of recording the relationships between group members. Because all cultures also have some conception of pair-bonding—often formalized in *marriage*, which is usually between a single male and a single female, though variations do exist—in regards to the relationship of males and females generating the next generation, the high investment in rearing offspring, not exclusively by, but significantly by, the bonded pair or group, seems to require keeping track of relations, known formally as *kinship*. How kin are identified specifically varies worldwide (see below) but in its function of managing identity, globally it deals with levels of identity as seen in Fig. 7.13.

While kinship often classifies people according to biological relationships, there are many exceptions. Still, we may identify *consanguinal* kin, who are directly related by blood, such as the parent-to-offspring relationship, and *affinal* kin, related by marriage and other social institutions, such as a brother-in-law. Adopted people and others brought into a family through any of a wide variety of social mechanisms are identified as *fictive* kin. Kinship strongly conditions the transfer of rights, property and other resources from one generation to the next through systems of *inheritance*. The main varieties of such *descent systems* are matrilineal and patrilineal, identifying

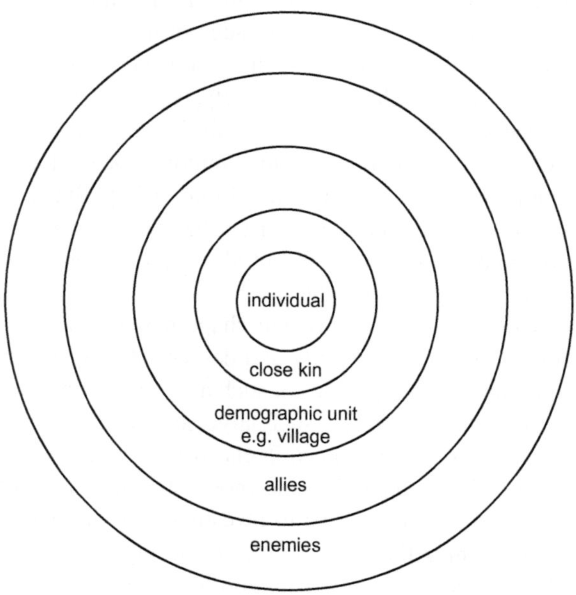

Fig. 7.13 Levels of Human Identity. How these are defined varies per culture, but all cultures make essential identity distinctions between people based on their proximity to the self. Diagram by the author

identity based on the mother's or father's side of the family, respectively; this is the domain of *linearity* and how it would play out beyond Earth is likely unpredictable.

The issue of *marriage prohibitions* based on kinship, in the avoidance of incest, highlights genetic health issues discussed in Chap. 2.

Globally, four main varieties of *kinship-based descent* are identified; others have certainly existed in the past, but are rather difficult to reconstruct from archaeological remains. The type of kinship-based descent is strongly conditioned by the mode of subsistence, as identified in a cross-cultural survey of over 500 non-industrial cultures (Aberle 1961); this identified that hunting-and-gathering (foraging) cultures commonly reckoned descent bilaterally (not strictly according to either mother's or father's lineage) while domesticators, of whatever variation from low-intensity horticulture to high-intensity agriculture, tended to be patrilineally-focused, reckoning descent through the male lineage. For the forseeable future, settlements beyond Earth will have to employ domestication—subsistence on crops and/or animals grown in controlled circumstances—of some variety or other. I do not think, however, that this implies any specifiable system of kinship or descent reckoning as cultures evolve beyond Earth. In the cultures studied above, the modern industrial and economic system was not in effect. Other factors that condition kinship reckoning in the industrialized nations include occupation, socio-economic rank, ethnicity, education and even political affiliation and religion. The arrangements of such a system have altered basic aspects of kinship reckoning and even family structure in ways too complex to understand so far, and thus too complex to model in the future.

For example, the traditional nuclear family of Scandinavia just a century ago has been significantly replaced by non-marriage cohabitation and births 'out of wedlock'. In the United States, we see that "No family change has come to the fore in modern times more dramatically, and with such rapidity, as heterosexual cohabitation outside of marriage. Within three decades in most advanced nations the practice of non-marital cohabitation has shifted from being a widely eschewed and even illegal practice to one which, increasingly, is viewed as a normal part of the life course and a necessary prelude to, or even substitute for, marriage. In America before 1970, for example, cohabitation was uncommon, a deviant and unlawful practice found only among people at the margins of our society. Since 1970 the number of Americans living together outside of marriage has increased more than 1000%, with such couples now making up about 10% of all couples: (Popenoe 2009: 430). Such changes are well-documented in Table 7.12.

Certainly such trends will shape new conceptions of kinship, as defined in its role of specifying identity and descent at least in part to determine conventions of inheritance. The specifics of such new conceptions have not yet been resolved. And, the specific *reasons* for such large-scale changes, on which I do not comment here in any moral sense, are influenced by changes in the culture's values, which are so historically unpredictable (not just in the United States, but globally) that I do not find them productive to investigate for the purpose of making predictions. We see patterning that has occurred regarding descent in the past, and that kinship was likely reshaped as an adaptation to different kinds of subsistence, but this was in non-industrial, non-global cultures.

Table 7.12 Percent changes in family structure likely effecting kinship conceptions

	Changes, Mid-1990s to Mid-2000s		
	% Change in nonmarriage cohabitation	% Births to unmarried women	% Parentage in single-parent households
Australia	48	21	17
Canada	32	−16	6
Denmark	−1	−2	9
France	26	30	49
Germany	37	76	9
Italy	23	90	9
Netherlands	1	113	−7
New Zealand	60	11	4
Norway	7	10	9
Spain	No data	140	79
Sweden	23	3	1
UK	52	23	5
USA	49	19	4

The data are for a selection of Western industrialized nations, mid-1990s to mid-2000s. Positive numbers indicate increase in percentage of the identified variable (columns). The reasons and results of the changes are not under discussion here, only that variables very likely to effect conceptions of kinship are here seen to be in a significant change through time. See text for discussion. Based on Tables 1, 3 and 5 of Popenoe (2009)

Regarding the culturally-adaptive capacity of kinship reckoning for our species, then, it is clearly highly malleable and therefore likely to be reformed as needed to accommodate the novel needs of cultures beyond Earth.

7.4.9 Social Organization and Power Relations

Culturally, *power* may be defined as the ability to constrain the options of others, and while there are many variations on the general theme, human cultures can broadly be described as being *egalitarian* or *ranked*. In egalitarian cultures there is an ethos of equal access to resources by all members of the culture; in ranked cultures such access is not guaranteed. As with kinship systems (see above), in nonindustrial cultures there has been a strong conditioning effect of subsistence mode on power relationships; in small, mobile foraging groups there has typically been a more egalitarian ethos, and strong social sanctions against attempts by any individual to self-aggrandize (Kelly 1996). This has been explained anthropologically in many ways, the most convincing, to me, being as a mechanism to keep individuals free of social and even material entanglements (e.g. debts), which allows them the social and material mobility to move across large foraging territories for often-sparse resources. The commonly-held anthropological explanation for social or economic ranking, concomitantly, is the emphasis on investment in landscapes and other resources made by sedentary, domesticating peoples, where a stronger conception of property rights seems to have evolved in all cases. For whatever reasons, it is clear that intensive farming cultures, from ancient to modern, have always expressed rather strict social and economic ranking, often in the form of *slavery*. Only in the last few centuries in the West, in fact, has a more democratic conception actually materialized, with at least the concepts of equality established as laws, an historic phenomenon we explored in Chap. 6.

A basic classification of social organization as it relates to power is seen in Fig. 7.14; while cultural anthropology has identified many general manifestations of cultural organization, some described in Chap. 5 of this book, I find the work

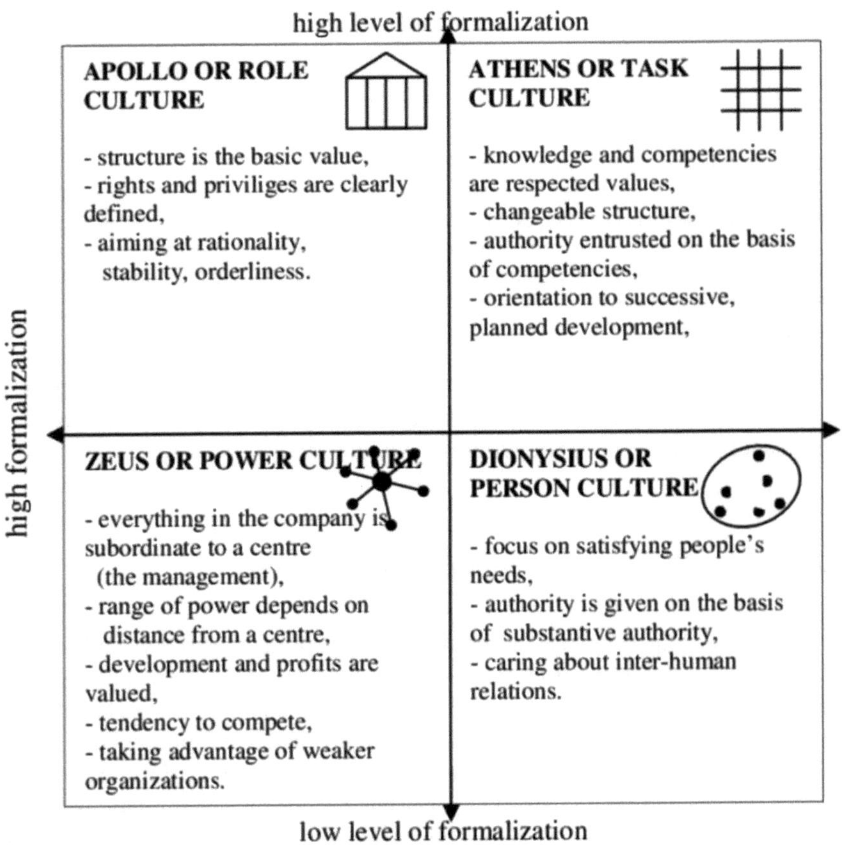

Fig. 7.14 Four socio-cultural organizations relating to power. Adapted by the author from Handy (1993)

of Irish philosopher Charles Handy quite applicable to the general organization of modern, industrialized cultures.

The economic, human rights, political and legal structures of humanity are so diverse as to require legions of information specialists—lawyers and their assistants—today. How these will be arranged for new conditions beyond Earth is, again, of interest, but again, difficult to predict. Some have anticipated the issue and have even established draft constitutions for communities on Mars.

7.4.10 Family Structure

A fundamental unit in all human cultures is the *family*, the closest of individuals in terms of identity to a given individual. A review of human family system evolution, highlighting the system's arrangement as an adaptive tool, indicates that "Humans are [uniquely complex among the primates in] the coevolutionary development of (a) stable and moderately exclusive breeding bonds, (b) bilateral kin recognition and relationships, and (c) reciprocity and kin links among coresident families ….the family environment, including care [from parents, to teach the offspring the methods of survival]…is a primary source and mediator of the [development] of social competencies [for survival]" (Flinn 2011: 18–26).

For at least part of their duration, families normally cohabit single structures, and their total membership is sometimes referred to as the *household*. Functionally, the household is typically organized to promote the survival, health and propagation of the family's lineage. Anthropologically, the functions of the household—whether that of ancient Maya (organized as nuclear families inhabiting small houses) or that of the Chinookans (organized as extended families inhabiting large lodges)—may be generalized as listed below. Note that while states or higher-level social entities may be involved in the activities of families in carrying out these functions, the family remains primary:

- **reproduction**: the family is organized in a way that is healthy for the young and promotes their eventual finding of mating partners
- **consumption**: resources are consumed on scales that reflect the household size rather than the individual
- **transmission**: the family is the primary unit of transmitting rights and property from one generation to the next.

Human family structure is strongly conditioned by resource availability and distribution in space and time. Anthropologically we may generalize that where resources are few and far apart, family units are typically small and culturally members are quite ready for the unit to split (fission) or rejoin (fusion) as the resources of a given season dictate; where resources are more stable, among farming cultures, for example, and where larger populations are promoted to carry out the simultaneous tasks of farming, families tend to be larger and less amenable to fission. For space settlers of the forseeable future, intensive domestication (farming and, possibly, the raising of animals) will be the mode of subsistence, so family sizes and stabilities familiar with the last few thousand years of settled human existence will likely continue. At any rate, as seen in Chap. 6, we may say that cultures of some thousands of farming people, organized in villages composed of moderately-small families (of up to about 10 individuals at a given time) have persisted for many generations in the past, and may well be emulated in some regards by the earliest settlers beyond Earth.

However space settlers organize their family units, it is not realistic that they would persist for long if they did not serve at least the function of protection of the offspring (raising them through various life stages as discussed). Regarding the adaptive variety and potential of human family organization, it seems wide and potent. On Earth, for example, human family structure varies, as mentioned, with *subsistence mode* (smaller families among foragers, larger families among farmers); it may also be shaped by *political* concerns (traditional Polynesian cultures sent any sons after the first son physically away from the household) or other variables including the religious traditions of a given culture. In short we may say that some variety of family structure will continue beyond Earth, though its particular shape, in population and membership over time, will vary. Such expected variation from 'how we have done it on Earth' should not, as we have seen in many cases, be a cause for concern. Some kind of family structure is evident in all social animals, and none is so social as the modern human being. Reshaping family systems will likely be a healthy sign of adaptation.

7.4.11 Economic-Political Structure

A culture's *economy* may be summarized as its means of acquiring energy and other resources, and they ways in which these are distributed among the culture members; these are the study of *production* and *distribution*, respectively. As mentioned in Chap. 5, the main economic varieties among humanity historically and prehistorically have been foraging, horticulture, and farming, with these different ways of acquiring resources strongly conditioning their distribution. As we have seen many times, however, for the forseeable future, space settlers will be farming people, with the many emphases this subsistence requires. Economic structure may be seen then as highly flexible in humanity, malleable for a wide variety of modes of subsistence and other concerns. Economic systems may be resistant to change due to high investment initially.

In the parlance of industrialized Western cultures, we may suggest that the earliest settlers beyond Earth will have at least two economies, a non-market economy of local production and consumption—for example household-controlled farming and fabrication of household goods with such technologies as additive manufacture (e.g. 3d printing)—and a market economy of special products produced on the individual, household and/or village/community scale for export. This would establish rather new economic realities, and we may be certain, based on the highly fluid nature of human ability to acquire and distribute resources and energy, that they will be tailored as needed for conditions encountered beyond Earth.

7.4.12 Settlement Pattern

The ways that a culture's members are distributed in a given geography, across both space and time, is referred to as the *settlement pattern*, which is strongly conditioned by the subsistence mode (discussed below), economy, proximity to resources, territorial issues of neighboring groups and many other variables.

Foragers and herding folk have been quite mobile, but for any forseeable future, the subsistence mode will be farming, which means, largely, residential sedentism. This conditions architecture and many other variables, as we have seen.

One aspect of the settlement pattern we may well envision in early colonies beyond Earth will be a transition from smaller to larger settlements. As I have mentioned, this would make sense if population growth is insulated or protected by ample subsistence surpluses; a larger population is always more secure than a smaller population. However the details are arranged, it is likely that the settlements of early Mars or Lagrangian sites will grow from small hamlets (of less than 100 people) to larger villages (into the hundreds and perhaps low thousands of people) and then city-like sites. This transition from smaller to larger populations will likely be accompanied by another transition, sometimes termed the *demographic transition* (DeJong 1972: 39–40) which involves higher populations due to increased infant survival (less infant mortality) as a new settlement region is adapted to and its particular hazards identified and mitigated. As mentioned in Chaps. 2 and 3, I expect that the early postfertilization stages of development are most likely to be vulnerable to new selective pressures beyond Earth, and this will result in higher infant mortality than we experience today in the developed world. However, this will ease as the problems are identified and mitigated. Through time, then, the settlement structure beyond Earth will certainly change from the originating experimental colony character due to (a) increasing food surpluses making it safer to increase the population, (b) increasing populations due to reduced infant mortality rate resulting from better understanding of the specific environment and (c) increasing specialization of activity by colony members, involving an agricultural character shift from the domestic, internal economy to an increasingly-larger domain of trade and interaction. Such well-known shifts can only be introduced here, and are worth further investigation as serious space settlement planners increasingly take (I hope) a multigenerational approach.

Whatever the settlement size and distribution, note that human settlements range from individual houses to megacities. All have links with others for trade, to facilitate marriage connections outside the family, and so on. While definitions vary globally, the classification listed below should help to structure our thinking about settlement patterns beyond Earth. There are few universally-accepted definitions of settlement types, so I discuss them in general terms below.

The *independent dwelling* is a relatively rare variety of Earth settlement, and as mentioned is something of a misnomer because, aside from the unusual cases of castaways, human dwellings are always part of a group of dwellings, even if they are spatially widely-separated. Examples of independent dwellings include the long-houses or lodges of native peoples of the Pacific Northwest. On the Lower Columbia River, for example, Chinookan peoples in the precontact era sometimes lived in these wooden structures, often >20 m in length and >10 m in width, occupied by multiple families. The largest of these are sometimes considered 'one-house villages'. But even these large structures' populations were linked by trade and marriage to others, including multiple-lodge villages with populations in the many hundreds or a thousand. Other independent dwellings include traditional Norse (Scandinavia and Iceland) turf houses that housed a nuclear or extended family, but normally less than 10 persons. These 'crofts' could be operated as surprisingly-independent socio-economic units, but even they had contacts with neighbors for social and economic reasons. Some early Central-European 'LBK' houses (ranging from 7500 years ago to about 5000 years ago) occupied by nuclear or extended families but normally less than about 20 persons. All of these (NW Coast, Norse, LBK) were sedentary cultures, the Norse and LBK people subsisting by farming, whereas the NW Coast people were foragers, sometimes sedentary but sometimes moving seasonally from one resource patch to another. More independent dwellings are known among highly mobile foraging people, but these have little significance for our consideration of early space settlers, who will undoubtedly be sedentary, closely attached to their farms. Perhaps a useful analogue for an early Mars settlement is the traditional extended-family German/Bavarian *hof* or 'estate'. The term 'hof' has several meanings but here refers to the whole of a family's physical estate, largely being the house (dwelling) and its associated buildings and farmland.

Hamlets are small, multi-family settlements of farming peoples, often in populations less than a hundred, composed of somewhat independent households. The have a particularly domestic economy and outlook but do maintain links with other settlements for social and some economic reasons. Early space settlements might well emulate the hamlet structure for extended nuclear families, such that these subpopulations can support themselves and even be quarantined if necessary. Still, hamlet members might work to contribute to a communal surplus of staples, via farming, in order to support a planned larger population.

Villages, as we have seen in Chap. 6, typically number less than a few thousand inhabitants, but are more unified in identity than hamlets. The many similarities between rather independent farming villages of the past, and early human settlements beyond Earth, have been examined in Chap. 6.

Towns are yet larger population centers than villages, and often include many specialist workers who are disengaged from food production; these include artisans and traders whose existence implies yet more contact with neighboring groups and a more complex economy than that of the independent village. Figure 7.15 illustrates the town of Davos, Switzerland.

The term *city* refers to yet larger populations, perhaps into the tens of thousands; defining characteristics include their economic interdependence rather than indepen-

Fig. 7.15 The town of Davos, Switzerland. Early settlements beyond Earth will likely have the character of towns and larger villages for some generations, before growth to city-like settlements. Public domain image from CC BY-SA 3.0, https://commons.wikimedia.org/w/index.php?curid=234149

dence. This is seen in the fact that cities normally house more people than can be supplied with staples (food and water) from even the immediate environment, such that these are imported with technologies such as pipelines and long-distance trade networks. City-dwellers are often not engaged in food-production and include the artisans and traders of towns, but include also information-managers such as accountants, lawyers, religious specialists and civic leadership hierarchies. The development of complex human aggregations are difficult to predict, but multigenerational city planning would be a worthwhile study among space settlement planners.

Beyond the scale of the city are the *conurbation* (merged city settlements) and the *megacity*, but it will be some generations before they are a special topic of space settlement planning.

Table 7.13 summarizes the properties of some of the settlements discussed, and some other population structures sometimes considered as early space settlement analogues by planners. Note that many are economically or otherwise unsuitable as early space settlement analogues, such as aircraft carriers. While these are rather independent, they are constantly supplied by tenders, are populated by crews who have a distinct conception of returning to a 'real' home at some time in the future, carry out no significant agriculture, and have a military rather than family community

Table 7.13 Selected human settlements with possible analogues for mars or moon settlement

Settlement class	Settlement type	Average population	Chronology
Independent Dwelling	LBK (early Farming) House, Central Europe	Nuclear & extended family: <30	c.7000–5000 BP
	Pre-Roman roundhouse e.g. Iberia, Britain	Nuclear & possibly small extended family: <20	c.4000 BP-Roman period
	Norse/Icelandic croft	Nuclear & possibly small extended family: c.5–10	c.670–1800 AD
	European Hof	Nuclear & extended family: c.10 to <30	Centuries around 1800 AD
Independent Village	Independent Neolithic Village	100 to c.1000	Globally c.10000–5000 BP
	Greek *polis* European Monastery	Many hundreds to low 000s A dozen to many dozens (monks)	Aegean area, 2400 BP Early centuries AD—1500s AD
	Tibetan Monastery	A dozen to many dozens (monks)	300 BC–present
	Kibbutz	Low hundreds of (kibbutzim)	1910–present
	Holden village	Tens to hundreds of individuals (may be families)	20th C AD–present
Somewhat Independent Special-Function Settlement *• = does not normally produce its own foodstuffs, though may be supplied with surpluses*			
	European *oppida*: regional trade and manufacture center for local tribal/village populations	Dozens to hundreds of artisans, elites, servants, some families; may be self-supporting for staples	c.200 BC to Romanization (c.50s BC)
	Roman *villa*:	Extended family or, offseason, workers only; strong economic function though may be self-supporting for staples	c.50 BC–500 AD
	Hacienda/Ranch: colonial economic manufactory	Tens to hundreds of individuals (may be families); economic, not community function	c.1500 AD to present

(continued)

Table 7.13 (continued)

Settlement class	Settlement type	Average population	Chronology
	McMurdo station, Antarctica: research station	<1500 individuals; employees & researchers rather than families	1956–present
	Mining village: extractive facility	<1000 individuals; employees & managers rather than families	
	Nuclear submarine: military functions	<300; crew, not families	1950s AD–present
	Nuclear aircraft carrier: military functions	<5000; crew, not families	1950s AD–present
	Remote marine oil rig: extraction function	<300; crew, not families	1950s AD–present

See text for discussion. Table by the Author

structure. At present it is somewhat independent, small towns and villages that serve as the best analogues for what we can expect when families begin to live beyond Earth (Fig. 7.16).

7.4.13 Subsistence Mode

All reasonable plans for space settlement presume a subsistence mode of some form of domestication of plants and animals. The main difference we may expect from that of Earth farming is that early settlers will have a distinctively low-intensity variety of farming, often termed *hortiuculture*, as their focus will be on providing supplies for the immediate population (beginning with that of the houseold) and, later, a larger population via stocked surpluses. From this low-intensity mode of farming, as populations increase, intensity and mechanization will increase, with a shift from *horticulture* to *agriculture*. Agriculture will serve larger populations, place more stocks in surplus, and may even be engaged in a long-distance foodstuffs trade.

The success of human settlements beyond Earth for the forseeable future will depend on the quality of the farming arrangements; they are currently being modelled and empirically tested by space agencies worldwide, as mentioned variously in this book.

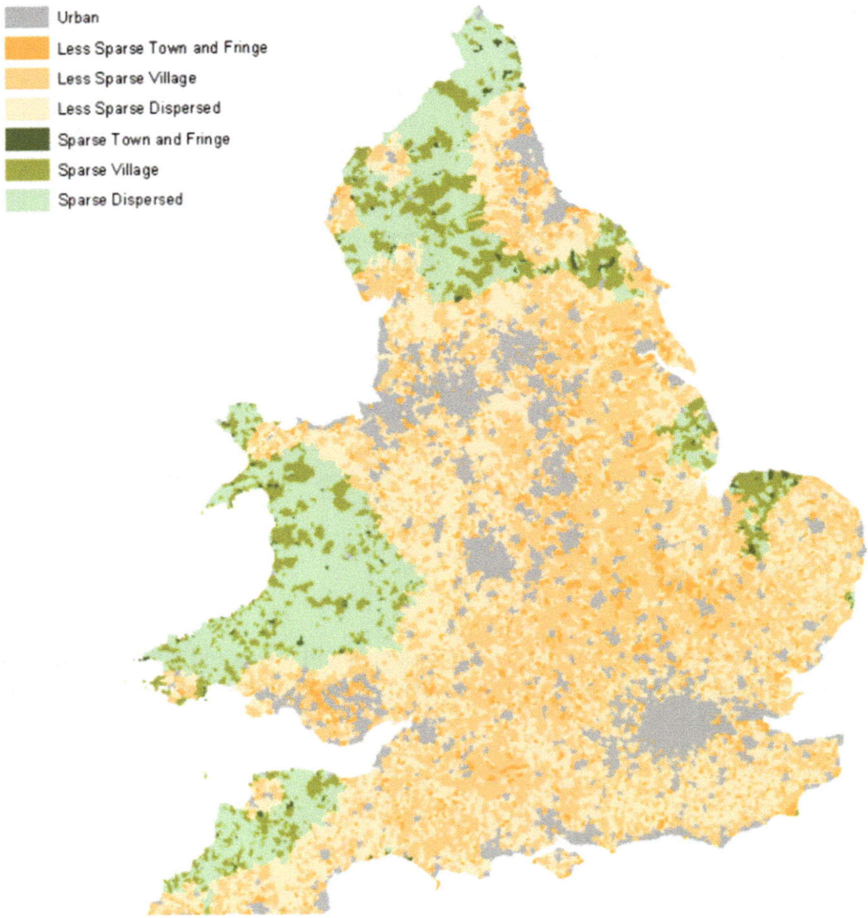

Fig. 7.16 Settlement classification and distribution in modern England. The distribution and size of residences, and larger aggregations of people, are strongly conditioned by resource distribution and strongly condition many aspects of human culture and behavior. See text for discussion

7.5 Plausible Human Cultural Universal Adaptations in Early-Generation Settlements Beyond Earth

In Sect. 4.3, **Mechanisms and Rates of Culture Change: The Process of Ethnogenesis**, I described the root of the cultural-adaptive process, innovation, that can, after a period of turning innovations to adaptations, lead to *ethnogenesis*, the formation of new cultures. In this section I apply what we learned of such processes to the specific cultural and physical environments we might reasonably expect in early space settlement on Mars and a Lagrangian settlement. I follow my own suggestion, explained in Sect. 1.6.1 *Prescription and Prediction in Space Settlement Studies*, that we not predict or prescribe too much or in too much detail. I preface these informed speculations with a note on what is meant by biocultural adaptation.

7.5.1 The Simple Meaning of Biocultural Adaptation

In recent years the terms *resilience* (the ability to return to equilibrium after disruption), *innovation* (the new, productive association of ideas) and *adaptation* (as used throughout this book) have become widely-used in business and commerce. I think this is partly a fad and partly a genuine attempt to move towards biological models for design and processes, as popularized by the inventive solutions resulting from the field of biomimicry, discussed elsewhere in this book. Here I simply wish to point out that adaptation as it will be used to allow humanity to live beyond Earth is simply adjustment of our biology and culture to accommodate new conditions. What is adaptation for 'life in space'? Simply, change; adjustment of our species for that 'life in space'. As we and nature adjust our biology, and as we adjust our priorities, technologies, customs, languages, beliefs and so on, for new conditions beyond Earth, that itself will constitute adaptation. So it is not an exotic thing, but it is an approach to space settlement that I think is important to follow. It highlights intelligent reaction to the unexpected rather than trying to force Earthborn concepts. Particularly for the first generations of space settlers there will have to be an essential ethos of the acceptance of change, of dynamism. Not everything will work, and as in any innovative endeavor, we will learn sometimes by our mistakes.

7.5.2 Options and Constraints in Adaptation to Life Beyond Earth

Any prospective behavior is conditioned in part by both options and constraints. If I wish to carry out behavior A, there may be options to do so in ways A^1, A^2 and A^3. Constraints, however, preclude ways A^4, A^5 and A^6, so while they exist in theory, they are not really options. In the parlance of human adaptation for space settlement, the options and constraints to our adaptive capacities may be identified by identifying our adaptive tools; that is, the biological and cultural domains that are sufficiently mutable to accommodate human biology and culture in conditions encountered beyond Earth. This book has surveyed our biology and behavior, seeking these adaptive tools, and has identified important elements of our biological and cultural adaptive tools. Some informed speculation on how these cultural dynamics might play out in early space settlements is presented in Table 7.14.

Table 7.14 Adaptive Options in Adaptive Behavioral Domains on Mars and a Stanford Torus. Generally speaking we see that in most adaptive domains there are many viable options for the forseeable period of early space settlement, in the ways human activity may be arranged for suitability conditions beyond Earth

Behavioral adaptive domain	Significant subdomain	Early Mars settlement adaptive options	Early stanford torus adaptive options
Language	*Grammar*	Many	many
	Syntax	Many	many
	Vocabulary	Many	many
	Dialect	Many	many
	Gesture	Many	many
Food Preferences	*cuisine tools*	Many	many
	cuisine ingredients	few; must be locally grown	moderate: most local, some may be imported
Arts and Aesthetics	*styles of bodily expression*	Many	many
	music and singing	Many	many
	static visual arts	Many	many
	static sculptural arts	Many	many
	motion arts	Many	many
Morality, Ethics and Justice	*moral system*	Few; must function for small but dynamic and isolated colony	Moderate; must function for moderately-sized but dynamic colony
	ethical system	as above	as above
	administration of justice	as above	as above
Religion and Mythos	*Ultimate Sacred Postulates*	highly contingent on founding beliefs	highly contingent on founding beliefs
Nurturing and Enculturation	*stages of life*		
		Few; must carefully monitor and promote controlled reproductive health	Few; must carefully monitor and promote controlled reproductive health
	system of enculturation	Many options though with increased responsibilities and perhaps life-stage awareness	Many options though with increased responsibilities and perhaps life-stage awareness
Kinship and Descent	*lineal priority*	Many	Many
	marriage prohibitions	Must reflect reproductive health concerns	Must reflect reproductive health concerns

(continued)

Table 7.14 (continued)

Behavioral adaptive domain	Significant subdomain	Early Mars settlement adaptive options	Early stanford torus adaptive options
	property transfer & inheritance	Moderate; must reflect property ethos of small, dynamic, isolated community	Moderate; must reflect property ethos of moderately-sized, dynamic, somewhat isolated culture
	reproductive activity regulation	Must reflect reproductive health concerns	Must reflect reproductive health concerns
Social Organization and Power Relations	*social ranking ethos*	Several	Several
	power relations	Several	Several
Family Structure	*household size and composition*	Earliest will be small but promoting large families/households as agricultural surplus grows	Depends on demographic/populating schedule
Economic-Political Structure	*essential organization*	Several	Several
	technological means of production	Much local/household production for own use e.g. additive manufacture, export fabrication unlikely for some time except for local trade	Much local/household production for own use e.g. additive manufacture, but also export fabrication likely
	organizational mode of production	Several; dependent on local resources rather than imports	Several; dependent on local resources and imports
	essential political organization	Several; must work for small, isolated village/community population that is dynamic and will promote larger population as agricultural surpluses grow	Several; must work for village-to-small city population that is somewhat isolated, and is likely to favor a stable population size
Settlement Pattern	*essential organization*	Residential sedentism	Residential sedentism
Subsistence Mode	*essential organization*	High-tech, low-volume horticulture transitioning to high-volume agriculture over some time	High-tech, high-volume agriculture

See text for discussion

7.6 Applying Adaptive Tools for Designing Genuinely Adaptive Space Settlement

To best plan space settlement in a way informed by the lessons of the processes and products of evolution, I propose the approaches introduced in Table 7.15. These approaches are philosophically rather well-delineated and have emerged in the last two decades or so as ways of research and development that are evolutionarily-informed, have produced concrete results in a variety of engineering and other fields, and I feel are applicable to a variety of stages or human projects in the study, planning and implementation of human space settlement.

I begin with proposing that human needs, biological and cultural, are identified with the tools of biology and anthropology.

Table 7.15 Tools for adaptive space settlement planning

Stage	Field of study	Philosophy of study	Tools and resources
Identify human biological and cultural needs	Human Biology and Anthropology	Global biological studies of human biology and cross-cultural studies to identify human adaptive universals	NIH and associated health databases; field study; eHRAF
Design adaptive processes for settlement beyond Earth	Biomimicry	Emulation of naturally-evolved processes and materials	Encyclopedia of Life, AskNature, other biological databases
Design adaptive tactics/physical items for settlement beyond Earth	Evolutionary Computing (and Biomimicry)	Utilize adaptive capacity of evolutionary principles, expressed in computing routines, to arrive at design optima	None to date
Model off-Earth habitats for testing	Multiagent Simulation	High-fidelity simulation of off-Earth habitat components in operation; statistical analysis of system properties over time	SIMOC, OLTARIS, Theseus
Physical Modelling	Analogue Studies	Learning (expected and unexpected) from Earth-based simulations using prototype hardware	Analogue hardware testing, e.g. hardware items such as space environment suits, or entire communities such as BIOSPHERE

See text for discussion

7.6.1 Anthropology and Space Settlement Planning

The adjustment of cultural domains of behavior as adaptation was established in Chap. 4 and refined above in Sect. 7.4. While there we have glanced briefly at how some cultural variables might change in living beyond Earth, the focus of this book has not been on such speculation, rather in outlining the best ways to design an evolutionarily-based approach to space settlement. An important tool for this effort will be anthropological analysis and understanding of humanity's cultural adaptations to date and range of potential adaptations, and suitability of such adaptations for adjusting to the conditions of space settlements.

Based on global cross-cultural research initiated by G. P. Murdock in the 1940s, Yale University's Human Area Research Files (HRAF) encode a vast number of human domains of activity, revealing both patterns and variation. These files are available in database format, or may be searched online with customizable search tools. These records remain in use, and are being refined rather than set aside as result of an earlier form of anthropology, and their wide use following their digitization has led to a resurgence in cross-cultural research, and even a journal the *Journal of Cross-Cultural Anthropological Research.* Murdock and his intellectual descendants had, of course, to organize their survey of human behavior and did so by categorization and labeling of human activities; though the definitions and boundaries of such variables of human action are always worth re-examination, the bulk have been found to be real enough for productive comparison across cultures. Such comparison is decidedly not an attempt to rank cultures according to one another or some arbitrary standard, but to identify the patterns and variation in human cultural behavior. In this effort, observations of human cultural behavior were are currently classified into just over 70 cultural variables that are either universal or nearly-so, and will serve as useful guides to the aspects of culture that may be adjusted as adaptations to conditions in habitats beyond Earth. These may be thought of as fine-grained refinements of the Cultural Human Universals introduced earlier in this book. They are far too numerous to examine in this book alone, and require the expertise of many minds to consider in terms of space settlement. These variables include such familiar domains of behavior as COMMUNICATION, which has subheadings indicating commonly-recurring manifestations (of communication) as *Gestures and Signs*, *Non-Electronic Transmission of Messages*, *Dissemination of News and Information*, *Press*, *Mail*, *Telephone and Telegraph*, *Radio and Television*, *Public Opinion*, *Proxemics* and *Internet Communications*, and other, less-familiar variables such as INTERPERSONAL RELATIONS, the subheadings of which are *Social Relationships and Groups, Friendships, Cliques, Visiting and Hospitality, Sodalities, Ettiquette, Ethics, Ingroup Antagonisms* and *Brawls, Riots and Banditry*. I would caution researchers from expecting highly-specific results from the study of these records, but rather more general patterns, and thus also from over-critique of these terms and their definitions. Again it is my feeling that we are not best used to predict specifically what will occur in human space settlement, but to better plan and carry it out according to what we know of the range and nature of human adaptations on Earth to date.

Certainly no such listing will satisfy every anthropologist or person interested in space settlement, but it remains a useful guide, as general or specific as one wishes it to be considered, for recurring ways that humanity has organized its activities across the globe and through time. In this recurrence of phenomena is their essential reality as adaptive tools for humanity. Manipulating the manifestation of the universals such as COMMUNICATION—which must be done in some way—into such varieties as *Mail, Gestures and Signs* and *Internet Communications* has of course implications for the overall characteristics of a given culture. The full list of 606 Human Cultural Universals (in my terminology) indexed in the eHRAF is presented in Table 7.16, below, as an introduction to the student to the many domains of behavior that may be addressed in space planning. Critically, the list allows the ordering of thought so that the prospect of space settlement is not overwhelming, but may be seen at least in part as simply an adjustment of these variables of human behavior to new conditions beyond Earth. If the student or researcher wishes to reclassify some of the presented terms, that is entirely possible, just as is the insertion of new classes of behavior if needed. I would caution however against doing this lightly; the listed universal domains and their manifestations are real phenomena observed in many cultures through space and time and have an objective reality, I feel, beyond the reach of the postmodern critique that 'all knowledge is socially fabricated'. That critique, a topic for another discussion entirely, is rather a nonstarter for such an applied effort as an anthropology of space settlement. With these provisos in place I invite the reader to examine Table 7.16 both critically—with an eye to modifying terms and so on according to one's own expertise and field of knowledge—and uncritically, with an eye to identifying fertile areas of applying one's own expertise and field of knowledge.

7.6.2 Biomimicry and Space Settlement Planning

How do we emulate natural adaptation for our project of permanent space settlement? The field of *biomimicry* gives us a guide. The Biomimicry Institute outlines a six-step process for applying lessons from Earth evolution to a given human design problem. These are graphically communicated in Fig. 7.17 and are summarized in text below.

Step 1 is to "*articulate the impact you want your design to have in the world and the criteria and constraints that will determine success*". This is the delineation of a specific goal, what I term the *human project*. Definition of the goal focuses research and prevents unwanted straying. A space-settlement goal may be to *develop a closed ecosystem habitat*. Defining parameters would include the number of people the habitat is meant to support; constraints that would be involved in determining success of the design might be variable states e.g. carbon dioxide level in the breathing atmosphere, ambient temperature, and so on.

Table 7.16 Extended list of human material cultural universals

140 Human Biology	520 **Recreation**
141 Anthropometry	521 Conversation
142 Physical Descriptions	522 Humor
143 Genetics	523 Hobbies
144 Population Identification	524 Games
145 Ontogenetic Data	525 Gambling
146 Nutrition	526 Athletic Sports
147 Physiological Data	527 Rest Days and Holidays
150 Behavior Processes and Personality	528 Vacations
151 Sensation and Perception	529 Recreational Facilities
152 Drivers and Emotions	530 **Arts**
153 Modification of Behavior	531 Decorative Art
154 Adjustment Processes	532 Representative Art
155 Personality Development	533 Music
156 Social Personality	534 Musical Instruments
157 Personality Traits	535 Dance
158 Personality Disorders	536 Drama
159 Life History Materials	537 Oratory
160 Demography	538 Literature
161 Population	539 Literary Texts
162 Composition of Population	5310 Verbal Arts
163 Birth Statistics	5311 Visual Arts
164 Morbidity	540 **Commercialized Entertainment**
165 Mortality	541 Spectacles
166 Internal Migration	542 Commercialized Sports
167 External Migration	543 Exhibitions
168 Population Policy	544 Public Lectures
170 History and Culture Change	545 Musical and Theatrical Productions
171 Cultural Revitalization and Ethnogenesis	546 Motion Picture Industry
172 Prehistory	547 Night Clubs and Cabarets
173 Traditional History	548 Illegal Entertainment
174 Historical Reconstruction	549 Art and Recreational Supplies Industries
175 Recorded History	550 **Individuation and Mobility**
176 Innovation	551 Personal Names
177 Acculturation and Culture Contact	552 Names of Animals and Things
178 Sociocultural Trends	553 Naming

(continued)

Table 7.16 (continued)

179 Economic Planning and Development	554 Status, Role, and Prestige
180 Total Culture	555 Talent Mobility
181 Ethos	556 Accumulation of Wealth
182 Functional and Adaptational Interpretations	557 Manipulative Mobility
183 Norms	558 Downward Mobility
184 Cultural Participation	560 **Social Stratification**
185 Cultural Goals	561 Age stratification
186 Cultural Identity and Ethnocentrism	562 Gender Status
190 Language	563 Ethnic Stratification
191 Speech	564 Castes
192 Vocabulary	565 Classes
193 Grammar	566 Serfdom and Peonage
194 Phonology	567 Slavery
195 Sociolinguistics	570 **Interpersonal Relations**
196 Semantics	571 Social Relationships and Groups
197 Linguistic Identification	572 Friendships
198 Special Languages	573 Cliques
200 Communication	574 Visiting and Hospitality
201 Gestures and Signs	575 Sodalities
202 Non-electronic Transmission of Messages	576 Etiquette
203 Dissemination of News and Information	577 Ethics
204 Press	578 Ingroup Antagonisms
205 Mail	579 Brawls, Riots and Banditry
206 Telephone and Telegraph	580 **Marriage**
207 Radio and Television	581 Basis of Marriage
208 Public Opinion	582 Regulation of Marriage
209 Proxemics	583 Mode of Marriage
201 Internet Communications	584 Arranging a Marriage
210 **Records**	585 Nuptials
211 Mnemonic Devices	586 Termination of Marriage
212 Writing	587 Secondary Marriages
213 Printing	588 Special Unions and Marriages
214 Publishing	589 Celibacy
215 Photography	590 **Family**
216 Audiovisual Records and Equipment	591 Residence
217 Archives	592 Household

(continued)

Table 7.16 (continued)

218 Writing and Printing Supplies	593 Family Relationships
220 **Food Quest** (Subsistence)	594 Nuclear Family
221 Annual Cycle	595 Polygamy
222 Collecting	596 Extended Families
223 Fowling	597 Adoption
224 Hunting and Trapping	600 **Kinship**
225 Marine Hunting	601 Kinship Terminology
226 Fishing	602 Kin Relationships
227 Fishing Gear	603 Grandparents and Grandchildren
228 Marine Industries	604 Avuncular and Nepotic Relatives
230 **Animal Husbandry**	605 Cousins
231 Domesticated Animals	606 Parents-in-Law and Children-in-Law
232 Applied Animal Science	607 Siblings-in-Law
233 Pastoral Activities	608 Artificial Kin Relationships
234 Dairying	609 Behavior toward Nonrelatives
235 Poultry Raising	610 **Kin Groups**
236 Wool Production	611 Rule of Descent
237 Animal By-Products	612 Kindreds and Ramages
240 **Agriculture**	613 Lineages
241 Tillage	614 Sibs
242 Agricultural Science	615 Phratries
243 Cereal Agriculture	616 Moieties
244 Vegetable Production	617 Bilinear Kin Groups
245 Arboriculture	618 Clans
246 Forage Crops	619 Tribe and Nation
247 Floriculture	620 **Community**
248 Textile Agriculture	621 Community Structure
249 Special Crops	622 Community Heads
250 Food Processing	623 Councils
251 Preservation and Storage of Food	624 Local Officials
252 Food Preparation	625 Police
253 Meat Packing Industry	626 Social Control
254 Refrigeration Industry	627 Informal Ingroup Justice
255 Canning Industry	628 Inter-community Relations
256 Cereal Industry	629 Inter-ethnic Relations
257 Confectionery Industries	630 **Territorial Organization**

(continued)

Table 7.16 (continued)

258 Miscellaneous Food Processing and Packing	631 Territorial Hierarchy
260 Food Consumption	632 towns
261 Gratification and Control of Hunger	633 Cities
262 Diet	634 Districts
263 Condiments	635 Provinces
264 Eating	636 Dependencies
265 Food Service Industries	640 **State**
266 Cannibalism	641 Citizenship
270 Drink, Drugs, and Indulgence	642 Constitution
271 Water and Thirst	643 Chief Executive
272 Nonalcoholic Beverages	644 Executive Household
273 Alcoholic Beverages	645 Cabinet
274 Beverage Industries	646 Parliament
275 Drinking Establishments	647 Administrative Agencies
276 Recreational and Non-therapeutic Drugs	648 International Relations
277 Tobacco Industry	650 **Government Activities**
278 Pharmaceuticals	651 Taxation and Public Income
280 **Leather Textiles, and Fabrics**	652 Public Finance
281 Work in Skins	653 Public Works
282 Leather Industry	654 Research and Development
283 Cordage	655 Government Enterprises
284 Knots and Lashings	656 Government Regulation
285 Mats and Basketry	657 Public Welfare
286 Woven Fabrics	658 Public Education
287 Nonwoven Fabrics	659 Miscellaneous Government Activities
288 Textile Industries	660 **Political Behavior**
289 Paper Industry	661 Exploitation
290 **Clothing**	662 Political Intrigue
291 Normal Garb	663 Public Service
292 Special Garments	664 Pressure Politics
293 Paraphernalia	665 Political Parties
294 Clothing Manufacture	666 Elections
295 Special Clothing Industries	667 Political Machines
296 Garment Care	668 Political Movements
300 **Adornment**	669 Revolution
301 Ornament	670 **Law**

(continued)

Table 7.16 (continued)

302 Toilet	671 Legal Norms
303 Manufacture of Toilet Accessories	672 Liability
304 Mutilation	673 Wrongs
305 Beauty Specialists	674 Crime
306 Jewelry Manufacture	675 Contracts
310 **Exploitative Activities**	676 Agency
311 Land Use	677 Organized Crime
312 Water Supply	680 **Offenses and Sanctions**
313 Lumbering	681 Sanctions
314 Forest Products	682 Offenses Against Life
315 Oil and Gas Wells	683 Offenses Against the Person
316 Mining and Quarrying	684 Sex and Marital Offenses
317 Special Deposits	685 Property Offenses
318 Environmental Quality	686 Nonfulfillment of Obligations
320 **Processing of Basic Materials**	687 Offenses Against the State
321 Bone, Horn, and Shell Technology	688 Religious Offenses
322 Woodworking	689 Social Offenses
323 Ceramic Technology	690 **Justice**
324 Lithic Industries	691 Litigation
325 Metallurgy	692 Judicial Authority
326 Smiths and Their Crafts	693 Legal and Judicial Personnel
327 Iron and Steel Industry	694 Initiation of Judicial Proceedings
328 Nonferrous Metal Industries	695 Trial Procedure
330 **Building and Construction**	696 Execution of Justice
331 Construction	697 Prisons and Jails
332 Earth Moving	698 Special Courts
333 Masonry	700 **Armed Forces**
334 Structural Steel Work	701 Military Organization
335 Carpentry	702 Recruitment and Training
336 Plumbing	703 Discipline and Morale
337 Electrical Installation	704 Ground Combat Forces
338 Miscellaneous Building Trades	705 Supply and Commissariat
339 Building Supplies Industries	706 Navy
340 **Structures**	707 Air Force
341 Architecture	708 Auxiliary Corps
342 Dwellings	710 **Military Technology**
343 Outbuildings	711 Military Engineering

(continued)

Table 7.16 (continued)

344 Public Structures	712 Military Installations
345 Recreational Structures	713 Ordinance
346 Religious and Educational Structures	714 Uniform and Accouterment
347 Business Structures	715 Military Vehicles
348 Industrial Structures	716 Naval Vessels
349 Miscellaneous Structures	717 Military Aircraft
350 **Equipment and Maintenance of Buildings**	718 Special Military Equipment
351 Grounds	719 Munitions Industries
352 Furniture	720 **War**
353 Building Interiors and Arrangement	721 Instigation of War
354 Heating and Lighting Equipment	722 Wartime Adjustments
355 Miscellaneous Building Equipment	723 Strategy
356 Housekeeping	724 Logistics
357 Domestic Service	725 Tactics
358 Maintenance of Non-domestic Buildings	726 Warfare
360 **Settlements**	727 Aftermath of Combat
361 Settlement Patterns	728 Peacemaking
362 Housing	729 War Veterans
363 Streets and Traffic	730 **Social Problems**
364 Refuse Disposal and Sanitary Facilities	731 Disasters
365 Public Utilities	732 Disabilities
366 Commercial Facilities	733 Alcoholism and Drug Addiction
367 Parks	734 Invalidism
368 Miscellaneous Facilities	735 Poverty
369 Urban and Rural Life	736 Dependency
370 **Energy and Power**	737 Old Age Dependency
371 Power Development	738 Delinquency
372 Fire	740 **Health and Welfare**
373 Light	741 Philanthropic Foundations
374 Heat	742 Medical Research
375 Thermal Power	743 Hospitals and Clinics
376 Water Power	744 Public Health and Sanitation
377 Electric Power	745 Social Insurance
378 Atomic Energy	746 Public Assistance
379 Miscellaneous Power Production	747 Private Welfare Agencies
380 **Chemical Industries**	748 Social Work

(continued)

Table 7.16 (continued)

381 Chemical Engineering	750 **Sickness**
382 Petroleum and Coal Products Industries	751 Preventive Medicine
383 Rubber Industry	752 Bodily Injuries
384 Synthetics Industry	753 Theory of Disease
385 Industrial Chemicals	754 Sorcery
386 Paint and Dye Manufacture	755 Magical and Mental Therapy
387 Fertilizer Industry	756 Shamans and Psychotherapists
388 Soap and Allied Products	757 Medical Therapy
389 Manufacture of Explosives	758 Medical Care
390 **Capital Goods Industries**	759 Medical Personnel
391 Hardware Manufacture	760 **Death**
392 Machine Industries	761 Life and Death
393 Electrical Supplies Industry	762 Suicide
394 Making of Heating, Lighting Equipment	763 Dying
395 Making of Optical and Photo Equipment	764 Burial Practices and Funerals
396 Shipbuilding	765 Mourning
397 Railway Equipment Industry	766 Special Burial Practices and Funerals
398 Manufacture of Vehicles	767 Mortuary Specialists
399 Aircraft Industry	768 Social Readjustments to Death
400 **Machines**	769 Cult of the Dead
401 Mechanics	770 **Religious Beliefs**
402 Industrial Machinery	771 General Character of Religion
403 Electrical Machines and Appliances	772 Cosmology
404 Household Machines and Appliances	773 Mythology
405 Measuring and Recording Devices	773 Animism
406 Weight-moving Machinery	775 Eschatology
407 Agricultural Machinery	776 Spirits and Gods
408 Computer Technology	777 Luck and Chance
410 **Tools and Appliances**	778 Sacred Objects and Places
411 Weapons	779 Theological Systems
412 General Tools	780 **Religious Practices**
413 Special Tools	781 Religious Experience
414 Miscellaneous Hardware	782 Prayers and Sacrifices
415 Utensils	783 Purification and Atonement
416 Appliances	784 Avoidance and Taboo
417 Apparatus	785 Asceticism

(continued)

Table 7.16 (continued)

420 **Property**	786 Ecstatic Religious Practices
421 Property System	787 Revelation and Divination
422 Property in Movables	788 Ritual
423 Real Property	789 Magic
424 Incorporeal Property	790 **Ecclesiastical Organization**
425 Acquisition and Relinquishment of Property	791 Magicians and Diviners
426 Borrowing and Lending	792 Prophets and Ascetics
427 Renting and Leasing	793 Priesthood
428 Inheritance	794 Congregations
429 Administration	795 Sects
430 **Exchange**	796 Organized Ceremonial
431 Gift Giving	797 Missions
432 Buying and Selling	798 Religious Intolerance
433 Production and Supply	800 **Numbers and Measures**
434 Income and Demand	801 Numerology
435 Price and Value	802 Numeration
436 Medium of Exchange	803 Mathematics
437 Exchange Transactions	804 Weights and Measures
438 Domestic Trade	805 Ordering of Time
439 Foreign Trade	810 **Sciences and Humanities**
440 **Marketing**	811 Logic
441 Mercantile Business	812 Philosophy
442 Wholesale Marketing	813 Scientific Method
443 Retail Marketing	814 Humanistic Studies
444 Retail Businesses	815 Pure Science
445 Service Industries	816 Applied Science
446 Sales Promotion	820 **Ideas About Nature and People**
447 Advertising	821 Ethnometeorology
450 **Finance**	822 Ethnophysics
451 Accounting	823 Ethnogeography
452 Credit	824 Ethnobotany
453 Banking	825 Ethnozoology
454 Saving and Investment	826 Ethnoanatomy
455 Speculation	827 Ethnophysiology
456 Insurance	828 Ethnopsychology
457 Foreign Exchange	829 Ethnosociology

(continued)

Table 7.16 (continued)

458 Business Cycles	830 **Sex**
460 **Labor**	831 Sexuality
461 Labor and Leisure	832 Sexual Stimulation
462 Division of Labor by Gender	833 Sexual Intercourse
463 Occupational Specialization	834 General Sex Restrictions
464 Labor supply and Employment	835 Kinship Regulation of Sex
465 Wages and Salaries	836 Premarital Sex Relations
466 Labor Relations	837 Extramarital Sex Relations
467 Labor Organization	838 Homosexuality
468 Collective Bargaining	839 Miscellaneous Sex Behavior
470 **Business and Industrial Organization**	840 **Reproduction**
471 Ownership and Control of Capital	841 Menstruation
472 Individual Enterprise	842 Conception
473 Corporate Organization	843 Pregnancy
474 Cooperative Organization	844 Childbirth
475 State Enterprise	845 Difficult and Unusual births
476 Mutual Aid	846 Postnatal Care
477 Competition	847 Abortion and Infanticide
480 **Travel and Transportation**	848 Illegitimacy
481 Locomotion	850 **Infancy and Childhood**
482 Burden Carrying	851 Social Placement
483 Weight Moving	852 Ceremonial During Infancy and Childhood
484 Travel	853 Infant Feeding
485 Travel Services	854 Infant Care
486 Regulation of Travel	855 Child Care
487 Routes	856 Development and Maturation
488 Warehousing	857 Childhood activities
489 Transportation	858 Status of Children
490 **Land Transport**	860 **Socialization**
491 Highways and Bridges	861 Techniques of Inculcation
492 Animal Transportation	862 Weaning and Food Training
493 Vehicles	863 Cleanliness Training
494 Highway Transportation	864 Sex Training
495 Auxiliary Highway Services	865 Aggression Training
496 Railways	866 Independence Training
497 Rail Transport	867 Transmission of Cultural Norms

(continued)

Table 7.16 (continued)

498 Terminal Facilities	868 Transmission of Skills
499 Highway and Railway Construction	869 Transmission of Beliefs
500 **Water, Air and Space Transportation**	870 **Education**
501 Boats	871 Educational System
502 Navigation	872 Elementary Education
503 Waterways Improvements	873 Liberal Arts Education
504 Port Facilities	874 Vocational Education
505 Water Transport	875 Teachers
506 Aircraft	876 Educational Theory and Methods
507 Aviation	877 Students
508 Airport Facilities	880 **Adolescence, Adulthood and Old Age**
509 Air Transport	881 Puberty and Initiation
510 **Living Standards and Routines**	882 Status of Adolescents
511 Standard of Living	883 Adolescent Activities
512 Daily Routine	884 Majority
513 Sleeping	885 Adulthood
514 Elimination	886 Senescence
515 Personal Hygiene	887 Activities of the Aged
516 Postures	888 Status and Treatment of the Aged
517 Leisure Time Activities	890 **Gender Roles and Issues**

Note important discussion in the text

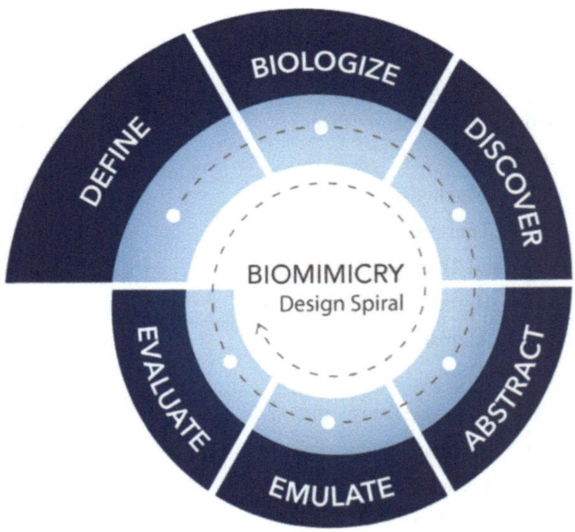

Fig. 7.17 The biomimicry design process. Note that as in science, where induction and deduction continually inform each other, in this process what is learned in any given process step might well require refinement of a 'prior' step. Note also that the author suggests two additional steps at the end of the process, as discussed in the text, specific to human project of space settlement. Image courtesy of the Biomimicry Institute

Step 2 is to "*analyze the essential functions and context your design solution must address [and] reframe them in biological terms, so that you can 'ask nature' for advice*". This step involves 'biologizing the project—whether this is appropriate or not should be determined in this step, for example there may be no significant evolved natural analogue to consider for a particular human project. If it is appropriate to seek guidance from evolution, here the researcher would specify the desired outcome, e.g. greater lift or aerodynamic efficiency of a wing. To 'reframe" this in biological terms is to begin considering where in nature such criteria have already been evolved, in this case in the wings of efficiently-flying birds. A space-settlement goal example might be to identify a strong, transparent tissue that could be grown, replacing the manufacture of polycarbonates.

Steps 3 and 4 are to "*look for natural models (organisms and ecosystems) that need to address the same functions and context as your design solution [and] identify the biological strategies that support their survival and success* [and learn the] *mechanisms that make the biological strategies successful*'" This step involves a search through the literature describing the variety of living systems on Earth, plant, animal and otherwise, globally and through time. Since the body of literature is large, productive search criteria should be specified. Once appropriate natural analogues have been identified (which can only be achieved through a broad familiarity with the literature and species of interest), the researcher would identify how the structures or processes analogous to the human project actually work. A space settlement example may be the search for a method of keeping surfaces clear of dust, for example the surfaces of solar panels on Mars; and a biological analogue might be found in the 'self cleaning' properties of some plant leaves.

Steps 5 and 6 are to identify the "*patterns and relationships among the strategies* [identified in steps 3 and 4 to inform the design concept so that it achieves the criteria delineated in Step 1]." Here the researcher would formally apply the lessons of the evolved system to the human project by emulating the evolved structure or process, *on a spectrum from less-to-more-directly-applied*, in the design of the human project.

An important step, late in the process, is to ensure viability of the concept in its environment, particularly over some time and considering the properties and resources of that environment. In the early stages of space settlement many resources will be rare and continually recycled or regenerated, and design and strategies will have to account for these realities. Each biomimicry-designed aspect of space settlement will of course have to be so evaluated. As designs are finalized, they may be digitally modelled and introduced into computer simulations of off-Earth habitats, such as the aforementioned SIMOC.

I would add two final steps to the biomimicry design process as it applies to space settlement planning. First would be to clearly identify the built adaptation and its value in what we might term the *currency* of adaptation. Adaptation, recall, as a noun is a thing or process that allows life where, without that thing or process, life would be impossible. At this stage of design I would recommend clarifying *how the designed adaptation allows life in the given environment*, and *how well it does so*, suggesting recommendations for improvement. Second would be to build a generic digital model of the adaptation for integration with multiagent digital simulations, as discussed below. This would foster some uniformity in space settlement design efforts, which would increase productive cooperation between researchers.

This process may be applied to the search for useful analogues, evolved in nature, for the things or processes humanity wishes to construct, which I refer to *human projects*. A good example is adhesive materials. The *human project* was to design a dry adhesive material; the search of evolutionary analogues identified the gecko's foot pads as superior dry adhesive properties; and the application of the principle copied the Velcro-like, fine, hooked surfaces of the gecko's foot pads in a new, dry-adhesive material.

This biomimicry approach may be used on any scale, from strategic goal definition to architecture in human settlements to genetic therapies and nanomaterials that may all assist in space settlement. It will be up to the planners of real space settlement to apply this method. In this book can only outline some useful applications and lessons of adaptive biology to various scales of action. Table 7.17 summarizes some such levels or scales of action, some examples of biomimicry application, and suggestions for the domains of the evolutionary literature to search for useful analogues to the desired adaptive processes and products of evolution.

In considering the exciting possibilities of biomimicry suggested above, it should be remembered that analogues in the biological world must be shown to be *appropriate to the human project*. That is, if an evolved analogue process or material is only superficially analogous, and cannot materially assist with the human project, then it is should be recognized as such and the search moved on. I would not suggest entirely deleting the results of the search however, as the search result may be a useful heuristic or thinking device to organize efforts, even if it does not produce materially-useful results for the human project. I suggest that the results of an evolutionary analogue study be divided, in fact, into at least two broad categories; *heuristic* and *practical* results. Heuristic results can shape further thought and stimulate creativity. Practical results allow more direct development of strategy, tactics and lower-level phenomena such as materials development. How far biomimicry will be taken in planning space settlement is yet to be seen. Growing starships, for example, has been proposed many times, harnessing the power of biological self-assembly for the assembly of different materials.

A further caution is noted here for students of this approach; not finding useful analogues for human projects does not necessarily mean that no analogue exists. About four to 30 million species exist on Earth (May 1990), and modern evolutionary biology is only some decades old. Many species remain undescribed and possibly unknown, and vast Earth ecosystems, such as those of the oceans and the polar realms, are only just now being explored with evolutionary perspectives. Much remains to be learned of the workings and results of Earth's evolutionary processes. Still, lest we be paralyzed by an ever-expanding or ever-narrowing search, we must at some time conclude that for the moment, the productive analogues are such and such (or do not exist), and carry on with our projects. We should always keep current with what is being learned, but waiting until we have found the perfect analogue could easily lead to an endless research or study period. As seen in Chap. 5, however, this research stage of the emigration process must conclude before the actual movement

Table 7.17 Scales of planned action and potential biomimicry resources for planning permanent human space settlement

Sample goal and/or Scale of space settlement planning (the *human goal*)	Analogous biological scale	Domain of evolutionary and/or Adaptive biology to search for suitable analogues	Example of appropriate and productive analogue
Multigenerational Success/Long-Term (*strategic*) Goals	Global patterns of adaptation across genera and *between* species. For example the *segmented* body plan of arthropods (c.1 million species today). The domain or level of *adaptive patterns*	Comparative studies highlighting adaptive successes over long time spans, e.g. millions of years and despite variation in selective agents in space and time	Anatomical segmentation allows redundancy of components, allowing survival despite injury or breakdown of individual components. Architectural segmentation may be an appropriate emulation
Intergenerational Success/Short-Term (*tactical*) Goals	Repeated patterns of adaptation *within* species, even though there are genus-level differences. For example skin coloration within *Homo sapiens*. The domain or level of *adaptive variation*	Comparative studies highlighting adaptive successes over shorter time spans, e.g. thousands of years and to more specific selective agents in space and time	North American bison vary significantly in hump size and chap length (among other traits) indicating adaptability of the basic bison body plan. Adaptive clothing or architecture, whose configuration can be altered according to immediate conditions, may be an appropriate emulation
Materials or Processes for Extraterrestrial Conditions	Characteristics, physiological or behavioral traits of a given life form at any scale, from molecular (genes) to behavior	There is a vast body of literature on Earth life anatomy, physiology and behavior; however much of this is descriptive rather than explanatory and contemporary field studies, including long-term studies, continue to improve our understanding of the natural world	Gliding bird wings are particularly aerodynamically efficient and have been emulated to increase aircraft wing efficiency

The field of biomimicry has no formal methods for identifying appropriate and productive analogues between evolved life forms and human projects, and it remains up to the individual researcher to estimate these in any given study. Some analogues will be found to be more appropriate metaphorically, or as heuristic thinking devices, than practically, while others will be of more practical use

beyond Earth's boundaries may be carried out. To extend this caution, students should note that even observed patterning might not be well understood, highlighting the critical difference between scientific *description* and *explanation*. For example, it is well-documented (*described*) that in the past 50 million years, bird species have diversified; they might have canalized (lost diversity), but in nearly all bird genera there has been diversification. This observed pattern, however, has not been very well *explained* and so a lesson for eager students of biomimicry might not be evident. Again I must caution that what has been observed of the evolutionary process on Earth by naturalists to date may be erroneous in its interpretation, and that much remains to be learned. Having said all this, I feel confident that biomimicry will grow as a field, rather than wither, and that it will be a significant and enduring tool for planning and carrying out successful space settlement.

In addition to biomimicry, some researchers highlight the use of living systems themselves—including bioengineered species—to promote the unique, multigenerational objectives of space settlement. In a recent paper, for example, titled '*Getting there and staying there: supporting and enabling persistent human life on Mars using synthetic natural rubber, self-healing materials, and biological batteries*' the authors write (author's emphasis in bold) (Acharya et al. 2002):

> Planetary exploration requires a balance between preemptive planning and financial feasibility…the advantages of biological solutions, namely the self-sustaining abilities of low-mass organisms, to make planetary exploration more self-sufficient and economical [are investigated]. **Prioritizing repair over replacement**, we are developing self-healing materials embedded with *Bacillus subtilis*. For longer-lasting energy, we are designing a "biobactery" using linearly oriented *Escherichia coli* to generate power. For renewable materials, we are engineering bacteria to synthesize and degrade rubber. Individually, these projects offer sustainable alternatives for repair, power, and materials. But when combined, these consolidated insights can provide us with the power to get to Mars and resources to sustain us while we're there.

In addition to bioengineering nonhuman life, bioengineering our own species has been mentioned in Chaps. 2 and 3 of this book.

7.6.3 Evolutionary Computing

Evolutionary computing uses the principles of biological evolution to generate products or processes according to desired performance criteria, rather than design from the top down. Recall that the biological process is essentially one of *replication, variation* and *selection*. More technically it may be defined as "…a type of machine learning that employs a stochastic search of a solutions space, genetic operators, a fitness function, and multiple generations of evolved programs to resolve a user-defined task, such as the classification of data" (Staats 2016: iii).

For example, to design an aircraft with increased lift at certain speeds, the designer begins with a computer model of a generic wing and sets in motion a program that replicates that wing, with variations, and then selects among the variants based on

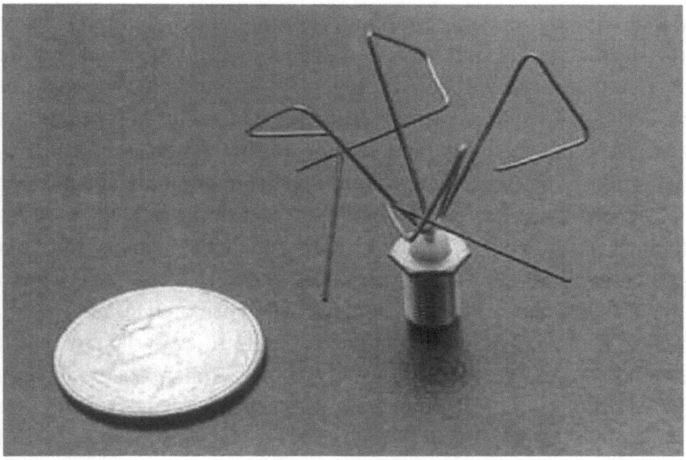

Fig. 7.18 X-Band antenna designed with evolutionary programming. From Lohn et al. (2005)

their performance. The method is widely used and has at least a 20-year history. On well-known example is that of an X-band antenna for a NASA spacecraft (Fig. 7.18), the authors summarizing the project as follows: "We present an evolved X-band antenna design and flight prototype currently on schedule to be deployed on NASA's Space Technology 5 (ST5) spacecraft. Current methods of designing and optimizing antennas by hand are time and labor intensive, limit complexity, and require significant expertise and experience. Evolutionary design techniques can overcome these limitations by searching the design space and automatically finding effective solutions that would ordinarily not be found. The ST5 antenna was evolved to meet a challenging set of mission requirements, most notably the combination of wide beamwidth for a circularly-polarized wave and wide bandwidth" (Lohn et al. 2005). Evolutionary computing has also been applied in artificial intelligence (Sipper et al. 2017) and medicine (Andres-Ena and Sipper 2017).

Evolutionary computing is not specially qualified for adaptive space settlement planning simply because it uses evolutionary and adaptive means. It is simply an approach that happens to use adaptive principles. I mention it only to introduce the field to the reader. Other computing methods remain useful; evolutionary computing is simply another tool available. Many useful approaches and reviews are available in the *International Journal of Applied Evolutionary Computation*.

7.6.4 Multiagent Simulation

The SIMOC model was introduced in Chap. 5, as a potentially powerful tool for assisting in planning space settlement. The key to multiagent simulation is that it can model the interactions of any number of system variables at each computing cycle, and that with modern computing, cycles may be accelerated from real time to provide long-term datasets suitable for statistical investigation and characterization. Like any model, the results will only be as good as the modelled components and their interactions, so much work in such projects goes into accurate computer modelling of these variables and their interactions; the formalization and standardization of units, for example, is particular important, as is the mathematical expression of their interactions per computing cycle. An exiting aspect of the SIMOC model developed by Kai Staats as part of Arizona State Univeristy's Interplanetary Initiative is that it will be publically released, with a web interface, so that many people can run the program in parallel, thus harnessing the power of distributed computing much in the way that the SETI at Home project handles vast quantities of data by distributing the computing load among many home-owned computers that process the data on local PC's before returning the results to a centralized data repository.

At present the SIMOC tool is designed to model, and then analyze the results of, the physical characteristics of an off-Earth colony; in particular the habitat's agricultural, life-support, recyclable and consumable variables are modelled, as is the health of each 'colonist' placed in the system. The project's development is at the stage of release (within a week of this writing), and summarized below:

- **Phase I**: In Phase I development, emphasis was placed on the location of the simulated habitat, in low-Earth orbit, on the Moon, in free space, or on Mars. Attention was given to specific locations, such as a valley, mountain top, or polar cap as each would inherit a particular in situ resource utilization (ISRU) parameter. In each setting the habitat itself maintained a relatively stable internal configuration of human inhabitants, structures, and ECLSS functions based upon what was to have been a real-world baseline.
- **Phase II**: Phase II provides the user with control over the physico-chemical ECLSS and bioregenerative systems design. A tour of Biosphere 2, attendance of the International Conference on Environmental Systems (ICES), and a deeper literature review determined that habitat design alone provides ample complexity for an engaged learning environment. Further discussion with experts at NASA and Paragon confirmed that the most critical issues in long-term, isolated human habitation are life support systems.
- **Phase III**: Phase III welcomes higher quality agent data, an improved Agent Configuration Editor, and updated web dashboard. It then takes us into the integration of environmental sensors and the SIMOC user interface, such that in place of the data generated by the agent model, we have instead a live data feed.
- **Phase IV**: In Phase IV the user is able to expand the habitat to support an increase in the human population. This might include the addition of greenhouse, crew habitat, and solar panel array. To determine the rate of expansion, we track resources by kilogram and determine what construction materials shipped from Earth vs were manufactured locally, via ISRU. Each expansion task is restricted by the cost of energy and time.
- **Phase V**: In Phase V we introduce a degree of realism with events unfolding out of the user's control. This represents aging of the systems and stochastic (entropic)

breakdown such as an airlock leak, a solid waste processor failure, or a meteorite strike resulting in catastrophic failure of the greenhouse and all crops therein. The user-player is then making decisions about how to allocate resources to save lives, or to abandon ship and return to Earth before everyone starves.

SIMOC is currently configured to model, as mentioned, physical rather than social dynamics, although the designers have expressed an interest in the interactions and emergence of social phenomena. Such social phenomena have been addressed in the field of multiagent social simulations. Capably defined as "…the intersection of three scientific fields, namely, agent-based computing, the social sciences, and computer simulation," (Davidsson 2002). Agent-Based Social Simulation (Fig. 7.19) has significant promise, but I fear at present the field is somewhat theoretically lacking, specifically lacking in a unified theory of evolutionary culture change. I believe such a theoretical body is possible, as discussed elsewhere (Gabora and Smith 2018), and anticipate it to be of use for future social system modelling. Currently researchers may monitor progress and results of the field in journals including the *Journal of Artificial Societies and Social Simulation* [*an interdisciplinary journal for the exploration and understanding of social processes by means of computer simulation* (JASS 2019)] and *Cliodynamics*, defined as "a transdisciplinary area of research integrating historical macrosociology, cultural and social evolution, economic history/cliometrics, mathematical modeling of long-term social processes, and the construction and analysis of historical databases" (CLIODYNAMICS 2019).

Very recently sociologist W.S. Bainbridge has published *Computer Simulations of Space Societies* but as of this writing I have not had an opportunity to review this work. Its abstract reads "At the intersection of astronautics, computer science, and social science, this book introduces the challenges and insights associated with computer simulation of human society in outer space, and of the dynamics of terrestrial enthusiasm for space exploration. Never before have so many dynamic representations of space-related social systems existed, some deeply analyzing the logical implications of social-scientific theories, and others open for experience by the gen-

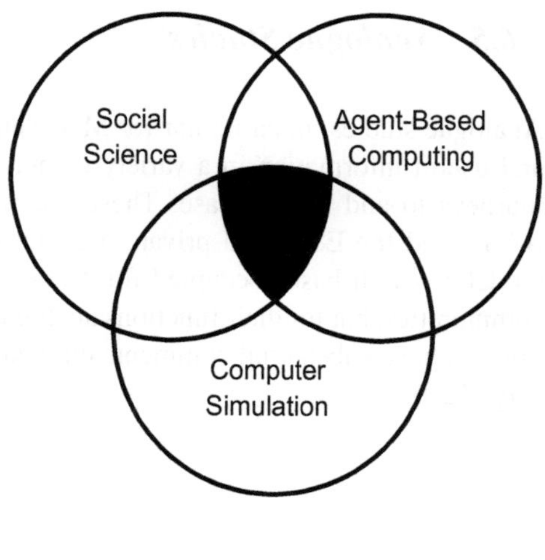

Fig. 7.19 Visual summary of agent-based social simulation. The overlap of social sciences, agent-based computing and computer simulation is the domain of Agent-based social simulation. From Davidsson (2002)

eral public as computer-generated virtual worlds. Fascinating software ranges from multi-agent artificial intelligence models of civilization, to space-oriented massively multiplayer online games, to educational programs suitable for schools or even for the world's space exploration agencies. At the present time, when actual forays by humans into space are scarce, computer simulations of space societies are an excellent way to prepare for a renaissance of exploration beyond the bounds of Earth" (Bainbridge 2018).

The simulation of social systems has at least one central goal in understanding patterning and variation in human group behavior. While this goal may be of use in planning space settlement, I would for the moment forego using the results of such simulations for much substantial planning. Rather I would suggest planning guidance from what anthropology has identified as patterns and variation in human action and adaptation, some of which is sketched out in this book. This data, based on observation of real rather than simulated cultures, should take precedence until it is demonstrated that artificially-modelled communities behave in ways significantly similar to real human communities. It may well be that this is possible, and I do not discount the possibility, but I would use caution in this field at its current state of development. The chief difficulty here is that anthropology has been so culturally fragmented in its own development and short history that, while it has identified interesting patterns and variations globally among human cultures, it has also taken many theoretical diversions and generated much misleading and sometimes erroneous information on these phenomena. Not only that but more critically, cultural anthropology currently does not have a unified, general model of human culture change—which I think can and should be an evolutionary model. Consequently much computer simulation of human cultures is heavily freighted with the various personal biases of individual researchers rather than generally- or widely-agreed-upon terms and processes regarding human behaviors and interactions. Until such is available, I would hold social simulations of cultural phenomena beyond Earth in a category of *highly speculative* rather than letting it influence research or planning.

7.6.5 Analogue Studies

Analogue studies, in particular for Mars settlement plans, have returned interesting and useful information in a variety of fields, for example, scheduling exploratory journeys to and from a base. These are carried out by space agencies including NASA and the ESA, and private organizations including the Mars Society. They model research bases occupied for relatively short periods, of course, rather than communities, but for their functions analogue studies are of great interest. At present I do not have substantial comments on what part they will play in space settlement (Fig. 7.20).

Fig. 7.20 Analogue study site, Utah, December 2018. Photo courtesy of 2018 MDRS 200 Crewmember Makiah Eustice

7.7 Designing a Genuinely Adaptive Framework for Human Space Settlement

In prior publications I have advocated the establishment of an *evolutionary paradigm for space settlement*, describing this as a framework or scaffold, a cognitive guide to the field of space settlement planning. This has been something of a struggle, although I think it has been good to promote, because I have had to be rather vague about 'an adaptive framework for space settlement'; I have also not known how to best describe such a framework and suggest how it be implemented. In the research for this book, however, I have sharpened focus on what such a framework would look like as discussed below. This framework is currently being developed as a discrete model that will be published in the research literature, analogous to various 'roadmaps' for space exploration released by federal agencies.

I begin with two examples of conceptual frameworks used to guide creative efforts in complex projects, to assist us in formulating an adaptive framework for human space settlement.

In aviation operations the SHELL conceptual framework is used to plan and diagnose flying systems and operations. This conceptual model was introduced F.H. Hawkins to unify and organize what he considered an important but disparate and poorly-defined field of aviation-related activity referred to as *human factors*. In an introductory chapter to his influential book *Human Factors in Flight*, Hawkins introduces SHELL as "a conceptual model of human factors" involving "the relationship of human factors and the aviation environment" (Hawkins 1993: 22). This conceptual model is illustrated in Fig. 7.21a, which shows the acronym SHELL read out as four letters, starting with S indicating the software or flying procedures, H the Hardware of

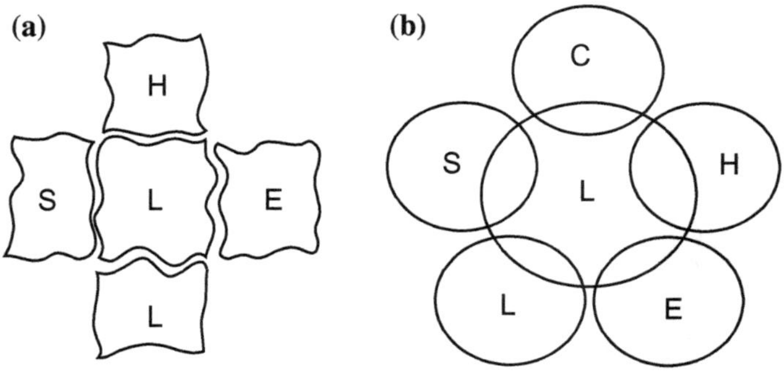

Fig. 7.21 SHELL and SCHELL conceptual models. See text for discussion. Adapted by the Author from diagrams in Hawkins (1993) (**a**) and Brown (2016) (**b**)

the flying system at the top of the diagram, the Environment (E) in which the system flies to the right, and the lower L indicating Liveware or personnel associated with flying, and at the center, the Liveware of people actually flying the system. The edges of the components interlock as irregular puzzle pieces, a graphical device Hawkins uses to indicate that the boundaries of these system components may be fuzzy but must be designed to interface well if the system is to work well. Figure 7.21b shows the SHELL concept expressed in a different visual manner; here there is also an elaboration, introducing C, for 'culture', which recognizes that the culture of people in the system is important to consider, e.g. to facilitate coordination between the FAA and France's equivalent civil aviation authority, the DGAC (Brown 2016). In this diagram the domains S, C, H, E, L and L are depicted as overlapping with the L at the center, but I think this is inadequate, as is the first diagram (7.21a) a, in that of course there can be integration between, for example, C and H (in SCHELL) or E and S (in SHELL) that are not depicted. Rather than picking nits here, I am grappling with how we visualize and then use such models as guides to decision-making in complex projects such as space settlement.

In the end the SCHELL and SHELL concepts are comprehensible and useful ways to guide behavior. For example, they remind practitioners to always consider the interface of the element they are working on with other elements of the system; the likely prevents waste or duplicated efforts. The model has also been used to diagnose system failures, like the loss of Air France flight 447, analyzed by one author using SCHHELL as a framework for understanding the series of system failures (Kneightley 2004) (Table 7.18).

For a second example we may look at an explicitly adaptively-aspiring model of management. In this case, the goal was to devise a more proactive, adaptive landscape management approach. The approach is seen in Fig. 7.22, where it is contrasted with an older, more reactive landscape management model. The conventional and adaptive models are very similar, 'adaptation' simply being application of learning from the evaluations to modify the original plans and goals. This is easier said than done as resources are often committed at the early stages in plan/policy formation in ways that

Table 7.18 SCHELL components

Model component	Definition	Example
Software	Procedures involved in the flying system	Charts, checklists
Culture	Values, norms and practices of society/organization involved in the flight system	United States' FAA, France's DGAC
Hardware	Tangible, physical objects involved in the flight system	Radio, engine
Environment	Ambient conditions in which the flight system is operated	Weather, aircraft cockpit layout
Liveware (crew)	Individuals directly operating the flying system	Pilot, other aircraft operators
Liveware (other personnel)	Individuals aiding operation of the flying system	Ground crew, radar operators

See text for discussion. Adapted by the Author from Kneighley (2004)

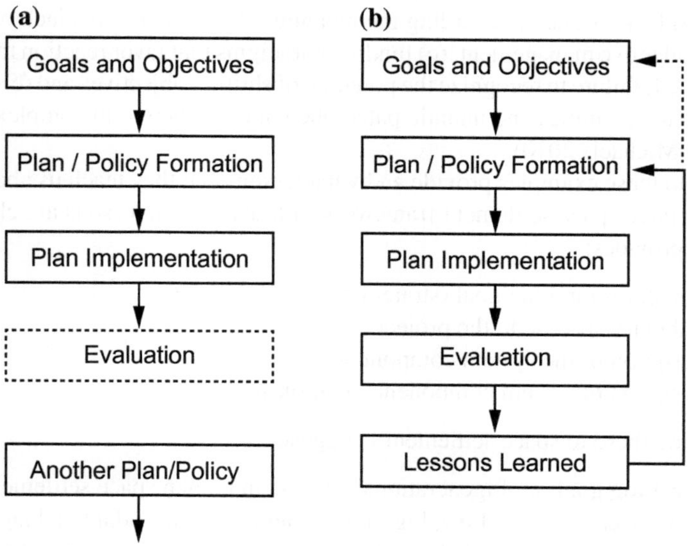

Fig. 7.22 Conventional and adaptive planning models. See text for discussion. A is a conventional management model and B is an adaptive management model. Adapted by the Author from a diagram in Kato and Ahern (2008)

preclude them being redirected by lessons learned by experience—but the structure and logic of the system here make it clearly more likely to adjust to new conditions, that is, 'smarter' or 'more adaptive'. The authors of this approach name its four main practices:

- conceiving management actions as experiments
- conducting several plans/experiments at once for fast learning
- monitoring being the key
- 'learning by doing'.

For the US National Research Council, adaptive planning is defined as "flexible decision making that can be adjusted in the face of uncertainties as outcomes from management actions and other events become better understood. Careful monitoring of these outcomes both advances scientific understanding and helps adjust policies or operations as part of an iterative learning process…Adaptive management is most visibly associated with big picture applications with a high degree of complexity" (Williams et al. 2009: v)

Finally, it has been noted that challenges to adaptive management implementation include (a) lack of clarity regarding the meaning of adaptation, (b) lack of success stories in adaptive management, (c) funding paradigms that favor reaction rather than proaction, (d) failure to recognize the potential of shifting objectives and (e) failure to accept the uncertainties and unanticipated phenomena inherent in complex systems (Tyre and Michaels 2010).

What do these examples provide and what lessons can they teach us about building an adaptive space settlement framework? I think several lessons are clear; each approach provides:

1. a rational, overarching goal (strategy level)
2. a vocabulary specific to the project
3. a description of the system components
4. a description of system components relationships.

Applying these to space settlement I suggest:

1. **overarching goal**: multigenerational success in human space settlement
2. **project-specific vocabulary**: highlight evolutionary and adaptive language
3. **system component description**: delineate components as adaptive tools
4. **description of system components relationships**: establish relationships and interactions of components to be genuinely adaptive, meaning that they are able to change according to certain conditions and in a known and useful way.

To implement the strategic goals, the system must be accepted. Authors working to have one adaptive management model accepted suggest the following ways to gain adaptive management model acceptance (Kato and Ahern 2008: 549–551):

- **reduce fear that the experimentation and unpredictability inherent in proactive management is a detriment**; remind planners that no experiment is a failure so long as one learns from it

- **try to moderate short-term thinking reactions to adaptive management models** by highlighting the participant's inclusion in a larger, longer project
- **highlight shared goals of members of different areas of the system** rather than suggesting competition between them
- **highlight shared goals of members of different areas of the system** rather than suggesting competition between them.

These methods may be implemented in shaping the adaptive framework for space settlement. A draft of such is currently underway, a project I expect to take 1–3 years, requiring collaboration with my colleagues in various space sciences. Important to implementing such a program is the establishment of a timeline for accomplishment of various milestones. Such a timeline is dependent on many factors including technology readiness levels of various components necessary for the project, also currently under investigation for this project.

7.7.1 The Overarching Philosophies of Adaptive Human Space Settlement

The following general philosophies derive from the many adaptive and evolutionary phenomena discussed in this book. They are structural concepts that will condition the shape of the adaptive framework for human space settlement. Each requires (a) operational definition, (b) placement with definite relationships to other concepts and (c) integration into a whole conceptual framework. They are listed here for the benefit of others seeking the overarching lessons of adaptive evolution as they pertain to human space settlement.

- Human settlement of space will be an exercise in cultural and biological adaptation
- To make this most likely to succeed, we should apply the lessons of evolutionary adaptation
- Expect change during space settlement; evolution and adaptation are characterized by change
- Expect imperfections on the path to adaptation; every adaptation has pros and cons
- Some change will occur as we expect, other change will be a surprise
- Do not expect to accurately predict many changes on the timescale of the second to third generations of space settlers; rather, focus on supplying the first settlers with the best lessons of evolution and anthropology to make for success
- It is in the first and second generations of space settlement that the greatest selective pressures—biological and cultural—will bear upon the settling population
- The earliest settlers will have to sacrifice, so they will have to *believe in the project*
- Innovations are often resisted as they, by definition, violate the established order which benefits many or powerful individuals
- Innovations are most commonly accepted when they meet a perceived need or want

- Expect some facets of the human experience to change rapidly, while others change slowly
- We need not design the details of human settlements beyond Earth; these will be shaped in-place by the emigrating populations
- We can usefully supply the lessons of adaptation to help with the project of human space settlement
- The field of human space settlement offers a field of unbounded creative potential for the human mind
- Settling beyond Earth will be worth it for the arts alone
- The settlement of space is not meant in this book to be an escape from problems on Earth or the establishment of a Utopian society; it is meant as a security measure for the advances of the Enlightenment and a stimulant to human creativity in all fields
- One of the most significant results of human space settlement will be the inevitable advances in philosophy and science; just as they flourished in the 17th Century with the revelations of the microscope, telescope and discovery of two previously-unknown continents
- Another significant result of the human settlement of space will be that it will constitute a significant reorganization of the cultural and genetic composition of humanity (among settlement populations), in the long run manifest as the ninth evolutionary transition of Earth life
- The human settlement of space will be about the perpetuation of Earth life
- The boundaries of 'sky', 'space' and such exotic locales as the surface of Mars or Pluto are much more social and philosophical than they are material, just as the boundaries of human 'races' are more social than biological; one way to begin dispelling these limiting concepts is to routinize space travel and access via commercialization and privatization of space access, rather than leaving it exclusively the domain of federal governments and militaries; this development, now a decade old, is well underway, and promising
- Powerful symbols involved with meeting humanity's individual and collective needs can be deployed to reveal the significance and sharpen the appeal of space settlement.

Lessons of Evolutionary Adaptation

- The most long-lived species have been numerous, widespread and diverse
- The fastest and most subtly-tailored responses to environmental stimuli will be most effective in adaptation; humanity possesses these AND the capacity for proactive—rather than just reactive—adaptation
- Most evolutionary products—developmental schedules, phenotypes, physiologies, cultural traditions including domestic architecture, to name a few—are by necessity adaptations to prior or current selective pressures and thus are vulnerable to shock when the selective pressures change; the lesson is that to best employ humanity's rapid capacity for specific adaption to new selective pressures, we should identify the differences between the prior or present selective environments and those to be encountered and adapted to, and then be proactive in tailoring our

adaptations to these identified features and expecting changes in certain realms of behavior. Thus, cultural rejection of social or technological or even biological proactive adaptations will be minimized

- As a rule of thumb, we should make the revolutionary transition to being multi-planetary incrementally rather than abruptly; wherever possible, recommend rather incremental adjustments to the conditions of new environments
- We should limit our prescriptions and predictions to (a) the earliest, most adaptively-stressed generations of space settlers, (b) to the features of human life most vital to survival and propagation and (c) to the aspects of such features of life that are most likely to differ from those of 'Earth Normal' conditions so as to limit the 'system shock' of exposure to new selective pressures
- Evolution requires Replication, Variation and Selection of self-replicators (genes) and other replicable, high-fidelity information (memes); wherever possible promote, and do not impede, any of these stages of the evolutionary process responsible, ultimately, for successful adaptation.
- Wherever possible, broaden the human *reaction norm* or *adaptive range* culturally and biologically; this is done by actively innovating and by protecting naturally-arising variation at least until it has been studied and found to be actively deleterious—but even contemporaneously-destructive or neutral innovations might become useful in the future and their properties should be archived
- Maslow's Heirarchy of Needs is a useful guide to what must be satisfied to guarantee a tolerable individual existence; the Human Universals explored in Chaps. 4 and 5 are similarly significant, on the cultural group level, and can help organize thought and planning.

7.8 Conclusion and Recommendations

In this book I have examined and described humanity's adaptive tools. It is up to policy-makers to internalize the evolutionary and adaptive significance of space settlement if an adaptive framework for it is to be genuinely built and used to guide decisions.

It is important to remember three facts of the space settlement project. First, no matter how similar to Earth people try to make future off-Earth environments, it is not realistic to think they will replicate all Earth conditions; and even slight differences will have effects on human biology and culture. Second, humanity's adaptive tools will be deployed proactively in anticipation of new conditions, but space settlers will undoubtedly encounter unexpected conditions, for which flexibility and adaptability will be the key to survival. Finally, while humanity is very good at adaptation, we do not always get it right, and we must be ready to recognize this and change, once again, while attempting to adapt beyond Earth. All of these issues should be kept in mind as, in this chapter, we examine more specifics of the great project of human adaptation beyond Earth.

The goal is permanent, multigenerational space settlement by viable human populations materially and biologically independent of Earth. This will require biocultural adaptation. The strategy for accomplishing it should be informed by all we know of evolutionary adaptation. The specific tactics for carrying it out should be conditioned by an evolutionarily-informed strategy and, wherever appropriate, emulation of biologically-evolved adaptive features and processes. Figure 7.23 reminds us and summarizes for us these principles of multigenerational success via support of the biocultural human being through all stages of the life course. If we are able to support and preserve these, beyond Earth, we will have accomplished a responsible, moral good in safeguarding the future for our species, many other species of Earth life, and human culture.

7.9 Summary Points

- While early-generation settlements expectable of the surface of the Moon and Mars will have some similarities with some modern and ancient human settlements on Earth, they will be largely unique in the human experience in ways that may be studied, anticipated and mitigated for multigenerational success
- A revised list of Cultural Universals may be used to productively study human multigenerational cultural successes and failures on Earth and their potential for purposive adaptation to conditions reasonably expected of early settlements beyond Earth
- Success in human space settlement will be evident when space settlers are capable of being or are entirely independent of Earth; independence referring to biology, culture and political economy
- Successful adaptation is, by definition, change, and divergence (or at least adaptive change) of space-settling people from those of Earth should be expected and embraced as signs of healthy adaptation.

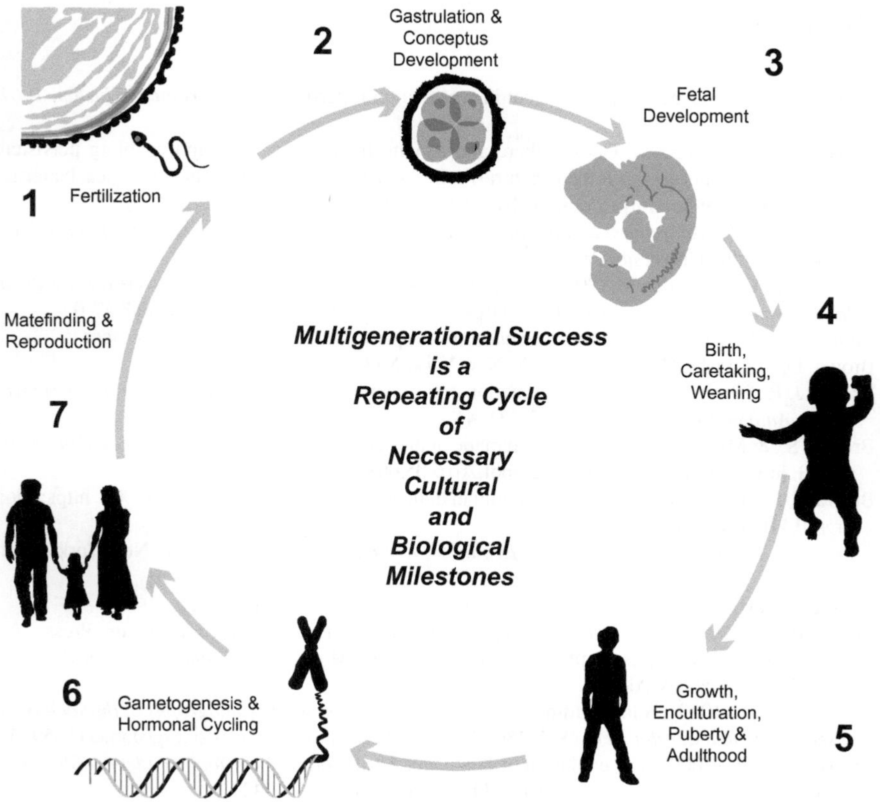

Significant Influences of:

	Biology	Culture
1	sperm & egg viability	-
2	developmental scheduling & oogenesis	-
3	systems development	-
4	early maturation	learning interactions, social & natural history intelligence
5	later maturation & hormonal scheduling	expanding space & time awareness, linguistic & technical intelligence
6	spermatogenesis	-
7	reproductive biology	matefinding & establishment of stable lifestyle

Fig. 7.23 Stages of the biocultural human life cycle. Each must be ensured for healthy, multigenerational success beyond Earth

References

Aberle, D. F. (1961). Matrilineal descent in cross-cultural perspective. *Matrilineal Kinship, 1961,* 655–727.

Acharya, N., et al. (2002). Getting there and staying there: supporting and enabling persistent human life on Mars using synthetic natural rubber, self-healing materials, and biological batteries. *BioRxiv* article at doi: https://doi.org/10.1101/345496.

Alland, A. (1970). *Adaptation and cultural evolution: An approach to medical anthropology*. New York: Columbia University Press.

Andres-Ena, C., & Sipper, M. (2017). Evolutionary computing in medicine: An overview. *Artificial Intelligence in Medicine, 19*(1), 1–23. https://doi.org/10.1016/S0933-3657(99)00047-0.

Bainbridge, W. S. (2018). *Computer simulations of space societies*. New York: Springer.

Brown, D. E. (1991). *Human universals*. New York: McGraw-Hill.

Brown, J. P. (2016). The effect of automation on human factors in aviation. *Journal of Instrumentation, Automation and Systems, 3*(2), 31–46.

Bruhns, S., & Misra, J. H. (2016). A pragmatic approach to sovergnity on mars. *Space Poilicy, 38,* 57–63. https://doi.org/10.1016/j.spacepol.2016.05.008.

Burridge, J. (2017). Spatial evolution of human dialects. *Physical Review X, 7,* 031008. https://doi.org/10.1103/PhysRevX.7.031008.

Carroll, J. (2004). *Literary darwinism: Evolution, human nature and literature*. New York: Routledge.

CLIODYNAMICS. (2019). https://escholarship.org/uc/irows_cliodynamics.

Cohen, R. L. (ed). (1986). *Justice: a view from the social sciences*. New York: Plenum Press.

Cohen, Y. (1968). Culture as adaptation. In Y. Cohen, (Ed.) *Man in adaptation: The cultural present* (pp. 40–60). Chicago: Aldine.

Colby, A., et al. (1983). A longitudinal study of moral judgement. In *Monographs of the society for research in child development* (Vol. 48(1/2), pp. 1–124). https://www.jstor.org/stable/1165935.

Corballis, M. R. (2009). The evolution of language. *Annals of the New York Academy of Sciences 1156*, 19–43 (2009). https://doi.org/10.1111/j.1749-6632.2009.04423.x.

Cosmides, L., & Tooby, J. (1992). Cognitive adaptations for social exchange. In J. H., Barkow, L., Cosmides, & Tooby, J. (Eds.), *The adapted mind: Evolutionary psychology and the generation of culture* (pp. 163–228). New York and Oxford: Oxford University Press.

Davidsson, P. (2002). Agent based social simulation: A computer science view. *Journal of Artificial Societies and Social Simulation 5*(1). See http://jasss.soc.surrey.ac.uk/5/1/7.html.

DeJong, G. F. (1972). Patterns of human fertility and morality. In G. A. Harrison, & A. J. Boyce (Eds.), 1972. *The structure of human populations* (pp. 32–56). Oxford: Clarendon Press.

Dimmendaal, G. J. (2002). Language ecology and linguistic diversity on the african continent. *Language and Linguistics Compass, 2*(5), 840–858. https://doi.org/10.1111/j.1749-818x.2008.00085.x.

Dissanayake, E. (2008). The arts after darwin: Does art have an origin and adaptive function? In K. Sijlmans, & vanDamme, W. (Eds.), *2008. World art studies: Exploring concepts and approaches* (pp. 241–263). Valiz, Amsterdam.

Durham, W. H. (1991). *Coevolution: Genes, culture and human diversity*. Stanford: Stanford University Press.

Edgerton, R. B. (2010). *Sick societies: Challenging the myth of primitive harmony*. New York: Simon and Schuster.

Eriksen, T. H., & Nielsen, F. S. (2013). *A history of anthropology*. New York: Palgrave McMillan.

Fitch, W. T. (2005). The evolution of music in comparative perspective. *Annals of the New York Academy of Sciences, 1060,* 1–21. https://doi.org/10.1196/annals.1360.004.

Flinn, M. V. (2011). Evolutionary anthropology of the human family. In C. Salmon, & T. K. Shackleford (Eds.), *2011. Oxford library of psychology. The oxford handbook of evolutionary family psychology* (pp. 12–32). Oxford: Oxford University Press.

Fried, M. N., & Fried, M. H. (1980). *Transitions: Four rituals in eight cultures*. New York: W.W. Norton & Co.

Gabora, L., & Smith, C. M. (2018). Two cognitive transitions underlying the capacity for cultural evolution. *Journal of Anthropological Sciences, 96,* 1–26.

Engineering, Genetic. (2017). Moral aspects and control of practice. *Journal of Assited Reproductive Gentics, 14*(6), 297–316.

Handy, C. (1993). *Understanding organizations*. Oxfords: Oxford University Press.

Hård af Segerstad, Y. (2003). *Use and adaptation of written language to the conditions of computer-mediated communication* (Unpublished PhD thesis). University of Gothenberg.

Hastorf, C. (1999). *Food and identity*. Find full reference.

Hawkins, F. H. (1993). *Human factors in flight*. London: Taylor and Francis.

Herskovitz, M. (1964). *Cultural anthropology*. New York: Alfred Knopf.

Hostetler, J. A., & Huntington, G. E. (1967). *Hutterites in North America*. New York: Holt, Rinehart, Winston.

JASS. (2019). http://jasss.soc.surrey.ac.uk/JASSS.html.

Kato, S., & Ahern, J. (2008). 'Learning by doing': Adaptive planning as a strategy to address uncertainty in planning. *Journal of Environmental Planning and Management, 51*(4), 543–559.

Keightley, A. (2004). *Human factors study guide*. Palmerston North: Massey University.

Kelly, R. L. (1996). *The foraging spectrum*. San Diego: Kluwer.

Kemmer, D., Anderson, A. S., & Marshall, D. W. (1998). Living together and eating together: Changes in food choice and eating habits during the transition from single to married/cohabiting. *The Sociological Review, 46*(1), 48–72. https://doi.org/10.1111/1467-954X.00089.

Kendon, A. (2004). *Gesture: Visible action as utterance*. Cambridge: Cambridge University Press.

Kinnier, R. T., Kernes, J. L., & Datheribes, T. M. (2000). A short list of universal moral values. *Counseling and Values, 45,* 4–16.

Lankester, E. R. (1915). *Diversions of a naturalist*. New York: MacMillan.

Lindgren, K., et al. (2009). *The skylab medical operations project: Recommendations to improve crew health and performance for future exploration missions*. Online at http://citeseerx.ist.psu.edu/viewdoc/download?doi=10.1.1.548.21&rep=rep1&type=pdf.

Linton, R. (1961). *The tree of culture*. New York: Alfred Knopf.

Lohn, J. D., Hornby, G. S., & Linden, D. S. (2005). An evolved antenna for deployment on Nasa's space technology 5 mission. In U.M. O'Reilly, T. Yu, R. Riolo, & Worzel, B. (Eds.) 2005. *Genetic programming theory and practice II. Genetic Programming* (Vol. 8). Boston, MA: Springer.

Malina, F. J. (1970). On the visual fine arts in the space age. *Leonardo, 3,* 323–325.

May, R. M. (1990). How many species? *Philosophical Transactions of the Royal Society of London. Series B: Biological Sciences, 330,* 293–304.

McCarthy, K. F., Brooks, A., Lowell, J., & Zakaras, L. (2001). *The performing ARts in a new era*. Santa Monica, California: RAND.

Meiklejohn, A. (1961). The first amendment is an absolute. *Supreme Court Review, 245,* 257.

Moran, E. F. (1979). *Human adaptability: An introduction to ecological anthropology*. North Scituate, Massachusetts: Duxbury Press.

Mufwene, S. S. (2004). *The ecology of language evolution*. Cambridge: Cambridge University Press.

Nettle, D. (1999). Is the rate of language change constant? *Lingua, 108,* 119–136.

Newberry, M. G., Ahern, C. A., Clark, R., & Plotkin, J. B. (2017). Detecting evolutionary forces in language change. *Nature, 551,* 223–226. https://doi.org/10.1038/nature24455.

Odling-Smee, F. J., Laland, K. N., & Feldman, M. W. (2003). *Niche construction: The neglected process of evolution*. Monographs in Biology #3, New Jersey: Princeton University Press.

Parsons, T. 1991[1951]. *The Social System*. London, Routledge.

Percino, G., Klimek, P., & Thurner, S. (2014). Instrumentational complexity of music genres and why simplicity sells. *PLOS One*: https://doi.org/10.1371/journal.pone.0115255.g005.

Popenoe, D. (2009). Cohabitation, marriage and child wellbeing: A cross-national perspective. *Social Science and Public Policy, 46,* 429–436. https://doi.org/10.1007/s12115-009-9242-5.

Radcliffe-Brown, A. R. N. (1952). *Structure and function in primitive society*. Glencoe: The Free Press.

Rappaport, R. (1999). *Ritual and Religion in the Making of Humanity*. Cambridge: Cambridge University Press.

Reyes-Garcia, V., et al. (2016). Multilevel processes and cultural adaptations: Examples from past and present small-scale societies. *Ecology and Society 21*(4), 2. http://dx.doi.org/10.5751/ES-08561-210402.

Saroglou, V. (2011). Believing, bonding, behaving and belonging. *Journal of Cross-Cultural Psychology, 48*(2), 1320–1340.

Satterlee, J. (1993). The hutterites: A study in cultural diversity. Bulletins of The South Dakota Stte University Agricultural Experiment Station, Paper 721. Online at http://openprairie.sdstate.edu/agexperimentsta_bulletins/721.

Shaarani, A. S., & Romano, D. (2007). Perception of Emotions from static postures. In *Proceedings of Affective Computing and Intelligent Interaction, Second International Conference*, ACII 2007, Lisbon, Portugal, September 12–14 (2007). https://link.springer.com/chapter/10.1007%2F978-3-540-74889-2_87.

Sipper, M., Olson, R. S., & Moore, J. H. (2017). Evolutionary computation: The next major transition of artificial intelligence? *BioData mining 10*(26). https://doi.org/10.1186/s13040-017-0147-3.

Smith, C. M., Gabora, L., & Gardner-O'Kearney, W. (2018). The extended evolutionary synthesis facilitates evolutionary models of culture change. *Cliodynamics*.

Staats, K. (2016). Genetic programming applied to RFI mitigation in radio astronomy. Master of science thesis, Department of Applied Mathematics, University of Cape Town. Online at https://open.uct.ac.za/handle/11427/23703.

Temperley, D. (2004). Communicative pressure and the evolution of musical styles. *Music Perception: An Interdisciplinary Journal 21*(3), 313–333.

Tyre, A., & Michaels, S. (2010). Confronting socially generated uncertainty in adaptive management. *Journal of Environmental Management, 92,* 1365–1370. https://doi.org/10.1016/j.jenvman.2010.10.014.

Williams, B. K., Szaro, R., & Shapiro, C. D. (2009). *Adaptive management: The U.S. department of the interior technical guide*. Washington, D.C: US Government Printing Office. Online at https://www2.usgs.gov/sdc/doc/DOI-%20Adaptive%20ManagementTechGuide.pdf.

Wotring, V. E. (2017). Fertility Challenges on Spaceflight Missions. Paper presented at the 2017 College of Reproductive Biology Symposium. See https://www.aab.org/aab/2017_CRB_Symposium1.asp.

Appendix
Common Critiques of the Idea of Human Space Settlement

Modern critiques proposing that humanity should not attempt to settle space (e.g. Billings 2018, 'Colonizing other planets is a bad idea') frequently use a small set of emotional or putatively moral arguments against the idea; I have addressed each of these ideas directly and at length in my 2012 semi-popular-science book *Emigrating Beyond Earth: Human Adaptation and Space Colonization* and articles published in *Spaceflight* and *Scientific American Mind*. On the one hand, I do not often engage in this debate as I find its most commonly-cited arguments to be poorly-formulated from the start. On the other, it is important, when one has the luxury of some possibility of cultural influence, to make the genuine case for space settlement. On the third hand, it does not seem to me that a culture that needs to be convinced to do something as difficult and dangerous as attempting space settlement will ever achieve that goal, it will have to be carried out by people willing to take large personal risks for a larger-than-the-individual goal of safeguarding the better of humanity. Such people are not likely to be influenced by critics who suggest that for whatever reason, humans should not attempt to settle space. Such critics will of course not do so and will remain on Earth.

To these ends in this appendix I provide a direct critique of a recent article that uses all the most common arguments that humanity should not attempt to settle space.

"Should a species like ours spread itself to other planetary environments that can be exploited for human gain?" This question uses the 'trigger' word *exploited* in an inflammatory way, even if inadvertently. Exploitation simply means the use of something, and all species, plant and animal, use resources in maintaining the processes of living systems. Also, 'human gain' is here suggested as something distasteful or otherwise negative, while I argue, throughout this book and other writings, that space settlement should be considered a responsible act of species- and civilization-insurance. If humanity and civilization are not considered worthy of preservation, we have a fundamental difference of opinion.

C. M. Smith, *Principles of Space Anthropology*, Space and Society,
https://doi.org/10.1007/978-3-030-25021-8

"Should humans seek to exploit and/or colonize space? If so, how should this be done?" These are vast questions, some of which I address in this book and my other writings advocating for space settlement. In particular in this book and *Emigrating Beyond Earth* I have made significant efforts to cast human space settlement in an evolutionary context and as a net moral good.

"Is it possible that humankind is evolving into 'a terrible new species,' one more destructive than ever before?" It is entirely possible that this is so, and entirely possible that it is not so. The context of the question suggests that the author feels it is the case, but that is an assumption not verified. Any human culture and any technology may be deployed for good or ill.

Would it be ethical to enable people with enough money to buy a ticket to leave our troubled Earth behind? The cost of space access is currently high, but it is being reduced; the expense of voyaging to Mars, or building, for example, Lagrangian-point colonies, will be quite high for some time. The act of departing the troubled Earth is here cast as irresponsible, but may conversely be seen as responsible, as I argue above, as there is no guarantee that Earth civilization will not collapse to the great loss of humanity's many moral goods. The distribution of wealth on Earth will always be an Earthly problem that should not be exacerbated by small numbers of people voyaging beyond Earth.

Would it be ethical for government(s) to subsidize such an enterprise? This is possibly a fair critique and a good argument for private space settlement efforts. I agree that governments should operate first to protect and ensure a safe and fulfilling life for their constituents.

What are the ethics of giving the rich yet another advantage over the poor? If the advantage implied here is the advantage of having a safer and more tolerable life in some habitat beyond Earth, it is not clear how that would occur at the expense of the poor if space settlement were privately funded.

Does this American image of the Wild West extended to off-Earth locations justify cowboy colonization of other planets? In my discussions with serious space-settlement planners I do not encounter a cowboy mentality. Rather I find people interested in self-sustaining settlements that would be among the 'greenest' of all humans settlements because resources would be precious and all processes would be eminently designed for efficiency. I have only encountered the negative visions of the 'Wild West' approach in dystopian fictions, including many depicted by Hollywood.

Does belief in American exceptionalism provide sufficient justification for off-Earth colonization? The conception of settlement beyond Earth originated in the historical period with the Russian Tsiolkovski, and many space agencies—including the Japanese, Canadian, Chinese, Indian, Russian, British, American and European—all express an interest in humanity living beyond Earth. The conception that this idea is entirely American is insular and ignores these many other serious considerations of humanity-beyond-Earth.

Is this near three-centuries-old version of manifest destiny propelling the present generation to colonize other planets? (and) If so, does manifest destiny provide sufficient justification? The conception of space settlement as 'Manifest

Destiny' is rooted in outdated concepts of the 1950s, when space settlement was expressed largely as the ultimate conquest of Nature. I detect no such conception in modern and serious space-settlement planners. The modern conception considers space settlement as a project that would better our understanding of ourselves Nature, through living in exposure to that wider Nature.

Is this near three-centuries-old version of progress propelling the present generation to colonize other planets? (and) If so, does progress provide sufficient justification? As in the prior question, this sentiment of an ancient conception of inevitable progress is outdated and I do not detect it in serious space-settlement plans or discussions.

In other words, is the project or policy going to provide a return on the national investment, if we define 'return' to be the economically sustainable human habitation of space? As in my previous book on this topic, *Emigrating Beyond Earth*, I do not argue that space settlement must provide an economic return on a federal investment, although any investment made at the federal level would be a contribution to the larger project of species- and civilization-insurance.

How much do we actually know about how people's religious beliefs intersect with their thinking about space colonization? I think we know little about this, but am unclear on how 'peoples' (generally) religious beliefs would be affected by space settlement. People of many religious faiths have expressed an interest in space settlement and I see no reason that they would not themselves participate in space settlement. I would no more accept a given religion's opposition to space settlement as a reason to curtail it than I would accept a given religion's opposition to prevent other action by private individuals who have different religious views.

That said, the results of this survey, for those who place weight on such things, do not provide any credible evidence that the U.S. citizenry is in favor of the human colonization of space. The author then should be satisfied that space settlement will not receive federal funding or be attempted. I agree that if the citizenry are not interested in a given project, they should not be compelled by the federal government to take it up. At the same time, individuals pursuing their own course of action should be allowed to carry it out if it does not impinge on the rights of other citizens. Private space settlement, I argue, is in this position.

While Elon Musk may claim that his plan to take people to Mars will be privately financed, it must be noted that Musk built his space business on direct and indirect government subsidies and that government contracts now constitute more than $1 billion of his company's annual revenue. It is disingenuous to imply that the federal government is promoting space settlement by supporting Musk's company with technology development and space-services contracts. The federal government is focused in these contracts on lowering the cost of space access, with an almost exclusive focus on access to Low Earth Orbit. Certainly the federal government is not actively promoting space settlement, and NASA has no office of space settlement, with their mission explicitly described as exploration.

Index

C. M. Smith, *Principles of Space Anthropology*, Space and Society,
https://doi.org/10.1007/978-3-030-25021-8

MIX
Papier aus verantwortungsvollen Quellen
Paper from responsible sources
FSC® C105338

If you have any concerns about our products,
you can contact us on
ProductSafety@springernature.com

In case Publisher is established outside the EU,
the EU authorized representative is:
Springer Nature Customer Service Center GmbH
Europaplatz 3, 69115 Heidelberg, Germany

Printed by Libri Plureos GmbH
in Hamburg, Germany